ANDEAN HYDROLOGY

ANDEAN HYDROLOGY

Editors

Diego A. Rivera

Laboratory of Comparative Policy
in Water Resources Management
Departamento de Recursos Hídricos
Facultad de Ingeniería Agrícola
Universidad de Concepción
Vicente Méndez, Chillán
Chile

Alex Godoy-Faundez

Centro de Investigación en Sustentabilidad
y Gestión Estratégica de Recursos
Facultad de Ingeniería
Universidad del Desarrollo
Santiago, Chile

Mario Lillo-Saavedra

Departamento de Recursos Hídricos
Facultad de Ingeniería Agrícola
Universidad de Concepción
Vicente Méndez, Chillán
Chile

CRC Press
Taylor & Francis Group
Boca Raton London New York

CRC Press is an imprint of the
Taylor & Francis Group, an **informa** business

A SCIENCE PUBLISHERS BOOK

Cover illustrations reproduced by kind courtesy of Dr. Mario Lillo-Saavedra (one of the editors of the book)

CRC Press
Taylor & Francis Group
6000 Broken Sound Parkway NW, Suite 300
Boca Raton, FL 33487-2742

First issued in paperback 2020

© 2018 by Taylor & Francis Group, LLC
CRC Press is an imprint of Taylor & Francis Group, an Informa business

No claim to original U.S. Government works

ISBN-13: 978-1-4987-8840-3 (hbk)
ISBN-13: 978-0-367-78150-7 (pbk)

**Visit the Taylor & Francis Web site at
http://www.taylorandfrancis.com**

**and the CRC Press Web site at
http://www.crcpress.com**

Preface

No other mountain range offers such a diversity of landscapes as the Andes. Stretching for over 4,300 miles, the Andes dissect the South American continent, resulting in a wide variety of climates and ecosystems. However, the Andes are not only important because of the diverse habitats that coexist from north to south at different altitude levels, microclimates and ecological conditions, but also as home to millions of people who share water resources that are crucial to sustainable development.

Water availability in the Andes is mainly driven by spatiotemporal patterns in precipitation. Over the past twenty years, droughts have dramatically increased in number and intensity in many regions, leading to water scarcity situations. Furthermore, floods have also been more severe due to prolonged or intense precipitation. In fact, intense rainfall events—those where a near-record amount of rain falls in just a few hours or days—have become more frequent. As climate continues to change, there is a serious concern for the impact that this will certainly have on the water resources in the Andes, and the implications of this for people, ecosystems and economic activities.

Climate change also affects the hydrological behavior of the river basins in the Andes. Water systems and water resources use generally involve distribution mechanisms that can be based on the open market, direct allocation, prioritization, or other methods. Due to their diverse ecosystems, climates and geomorphological characteristics, the Andean hydrology is extremely complex. This results in difficulties in water management, particularly due to certain socio-environmental issues, such as water scarcity, water contamination and mismanagement of water basins. Therefore, understanding the relationships within hydrological systems and society is crucial to allow the development of adaptations strategies.

This book aims to deepen our understanding of the connections between water resources in the Andes, and social, economic and environmental issues. Based on scientific research and policy assessment, recommendations for decision-making are provided. Challenges regarding a fast increase of water demand are also discussed using a multidisciplinary approach, focusing on the water cycle and water-energy-food nexus.

Water is considered as a basic human right, but the rapid economic and demographic growth rates in South America are putting strong pressure on water demand in the Andes. In this regard, the way in which actors compete for the access and use of water is still an object of study. Industrial activities within a basin convert water resources into products and services represented by GDP, effectively putting a price on water availability and increasing competition.

The authors of the chapters of this book describe how researchers can help local and regional stakeholders with data collection, processing, storage and dissemination of water resources information; highlight the importance of the collaboration between the scientific community and policy makers; and suggest strategies and measures to make collaboration more effective and sustainable.

Water demand can be understood as how water resources connect society with the natural system. This is the so-called socio-environmental link, in which socio-hydrology and water resources combine to act as sources of extraction and receptors of pollution.

The relationship between the mining industry and local communities in Chile, Peru and Colombia is also analyzed, including both legal and regulatory framework. In terms of conflict resolutions between stakeholders, it is essential to understand socio-demographic characteristics. In fact, there are several indigenous peoples in the Andean Region whose traditions and belief systems often mean that they regard ancestral lands and the natural resources they contain with deep respect. However, as this view is not shared by many of the others actors involved, solutions need to be developed in order to allow for social development and economic growth in harmony with local hydrology.

Models such as soil, water and assessment tools (SWAT) have been successfully used for the analysis of watersheds to develop payments for ecosystem services, estimate economic values and generate simulation scenarios for improved management. Nevertheless, unless research centers, NGOs and local governments work together, water resource conflicts will continue, affecting economic activities and communities.

Finally, approaches and methodologies used to generate data for hydrological monitoring and analysis in the Andean regions are also described.

We would like to thank authors for their contributions to better understand how the Andes sustain our life. Also, we would like to thank the Water Resources Research Center for Agriculture and Mining (CONICYT/FONDAP-15130015) and FONDECYT Regular 1160656 for their support. We are also very grateful to Professor Roto Quezada for his continuous support.

September 2017   **Diego A. Rivera**
Alex Godoy-Faundez
Mario Lillo-Saavedra

Contents

Competing Uses and Access to Hydrological Resources in Upstream Peasant Communities of the Cañete River Watershed, Perú

Maria Claudia Tristán Febres,[1,*] *Genowefa Blundo-Canto,*[1]
Gisella S. Cruz-Garcia,[2] *Marcela Quintero*[2] and
Piedad Pareja Cabrejos[3]

"Water is simultaneously a source of conflict and collaboration, of oppression and productive potential."

(Boelens, 2008b)

Introduction

Globally, the pressure on natural resources is increasing, particularly on water. The United Nations' Sustainable Development Goal 6 focuses on ensuring access to water and sanitation for all. Although water is vital for the survival of all living beings, the control and management of water resources entails power relations, conflicts and negotiations between stakeholders

[1] International Center for Tropical Agriculture (CIAT), Av. La Molina 1895, La Molina, Apartado 1558, Lima 12, Perú.
Email: g.blundo@cgiar.org
[2] International Center for Tropical Agriculture (CIAT), Km 17, Recta Cali-Palmira, Valle del Cauca, Colombia.
Emails: g.s.cruz@cgiar.org; m.quintero@cgiar.org
[3] University of Hohenheim/International Center for Tropical Agriculture (CIAT), Schwerzstrasse 1A, 70599, Stuttgart, Germany.
Email: PiedadA_ParejaCabrejos@uni-hohenheim.de
* Corresponding author: mctristan@pucp.pe

across different scales. Water is a natural, social and cultural resource as it meets multiple needs and embodies multiple meanings (Budds, 2010). Therefore the variety of water needs or uses can be translated into rights through the establishment of access and property rights to hydrological resources. While rights can be formal or informal, the shared recognition among users of the resource provides them with legitimacy.

Today, natural environments undergo major changes and transformation processes which result in environmental and social pressures on ecosystem services. Lack of property rights or effective national laws and international treaties can contribute to the depletion of common-pool resources. The absence of rules to limit access and define rights can favor *de facto* free riding by actors who are not concerned with negative externalities caused by their depletion (Ostrom et al., 1999). For instance, unregulated access and subsidies have led to over-exploitation of two-thirds of global fish stocks and have damaged coastal ecosystems (Sukhdev, 2009). The interdependence between social and environmental dimensions is undeniable (Millennium Ecosystem Assessment, 2003; Barrett et al., 2011; Raworth, 2012).

As per needs, practices and demands for water for stakeholders change, the rules of use and access also change. This results in the creation of a landscape of competing or superimposed claims to natural resources, particularly in relation to water. Within such a landscape, the stance of different players diverges strongly (International Water Management Institute, 2007). Such divergences should be solved through negotiation and understanding of the issues faced by different stakeholders and of the potential and opportunities for change (Giller et al., 2008).

This chapter seeks to explore the hydro-social territory (Damonte Valencia, 2015; Boelens et al., 2016) of the upper basin of the Cañete River in Peru, by documenting competing uses and access to water resources in eight peasant communities.[1] In order to better understand the hydro-social territory and its associated conflicts and opportunities, we must answer the following research questions: (1) what are the kind of water resources and what are they used for? (2) who is involved and promotes these uses? (3) what are the current conflicts related to the access and use of water resources? and (4) what are the demands of the different stakeholders? The analysis focuses on the social implications of this competitive hydro-social territory given the presence of both collective and private uses. Power relationships and competing uses and access translate into a number of micro-level conflicts, latent or explicit, between various stakeholders.

The chapter gives a theoretical overview of competing uses and access to water resources, followed by a contextualization of the case study.

---

[1] The term 'peasant' is used to define rural communities such as those from Mesoamerica and the Andes that have well defined corporate and collective structures, with hereditary membership and recognized territorial rights. For further information, see Edelmen (2013).

Subsequently, it presents the results of the analysis that answers our research questions. Finally, it presents a general discussion and conclusion on the research and development implications of the study.

An Overview of the Competing Uses and Access to Water

Water has historically been a resource in dispute. Today, there is a global water crisis and we need to develop policies for water security (Bigas, 2012). The reasons for conflict are diverse: from local conflicts over inequality in access to water and demands of good quality water supply, to water scarcity as a source of armed conflict (Gleick, 1993; Hendrix and Glaser, 2007). While water is seldom the single or major cause of conflict, it can exacerbate existing tensions (Wolf et al., 2005). Conflicts are expected to increase with population growth and increased pressure on natural resources, especially in the context of climate change. For instance, increasing global temperatures appear to be altering global hydrological patterns by changing seasonal trends (Bigas, 2012). Wolf et al. (2005) identified three key elements that can be used in any dispute over water: quantity, quality and timing. Quantity refers to claims over water as a scarce resource, when divergent uses by different stakeholders come into dispute. Quality issues can result in conflict when the resource is polluted and does not satisfy human and ecosystem needs. The timing dimension refers to the synchronization between water uses and the natural water flow. Changes in water flows affect socioeconomic activities that depend on water. When the flow or the timing of use is modified, it can be a source of conflict. Additionally, the needs and demands for water are seldom free from political meaning. Globally, this translates into treaties and agreements. At the international level, they aim to solve transnational conflicts through negotiation and cooperation between states. At the local and regional level, politics permeate negotiations and dialogue between various stakeholders that seek to position their demands. Water availability, water demand, institutional frameworks and the social dimension reconfigure conflicts over water (Rivera et al., 2016).

In the case study presented, the precautionary principle has led international agencies and the Peruvian government to create a Reward Mechanism for Hydrological Ecosystem Services (RMHES). The precautionary principle means that the RMHES rewards the maintenance of the quantity and quality of an available hydrological service. This implies that the main objective is to conserve the provision of hydrological services in order to avoid deterioration in these services, especially when the exact quantification of their benefits, and how their provision is affected by human activity, is uncertain.

The RMHES is a type of Payment for Environmental Services (PES), an incentive mechanism where ecosystem service providers are compensated by the users of those services, conditional on the maintenance or improvement of this provision (Rodríguez de Francisco and Budds, 2015; Wunder, 2015). In a watershed, providers are located in the upper basin, where the hydrological service is sourced, while users (who are the rewarders) are located in the lower basin, where they make use of the hydrological service.

Commonly, PES mechanisms focus on the interaction between downstream and upstream stakeholders, but socio-political conditions, contexts and conflicts can also exist within upstream communities that are providing the service. Overlooking the socio-political complexities within upper catchments might lead to neglect of social equity issues. Among the risks of not taking into account such issues in the implementation of PES lies a potential marginalization of vulnerable groups, capturing of benefits by elites, increased economic inequality, reduced resource access, and increased tenure insecurity (Pascual et al., 2014).

The social dimension, including social organization and collective action, is therefore an essential and constitutive part of an hydro-social territory (Boelens, 2008b). According to Damonte Valencia (2015) and Boelens et al. (2016) an hydro-social territory: (1) integrates both physical and social spaces of a watershed; (2) is defined by the use of water and its social implications; (3) is built upon hydro-social cycles; and (4) includes political and administrative spaces. These territories are different from landscapes as physical borders are defined in the context of power struggles over water.

The analysis presented highlights the implications of competing uses and access to water, acknowledging that, while different stakeholders are defined in relation to their use of hydrological resources, latent and explicit disputes around these uses within and between communities should be examined and understood. Understanding the problems related to water resources within the upper basin of a key watershed in Peru allows us to understand the complexity of a heterogeneous space, where claims on water resources are not only vertical (i.e., upstream-downstream) but also horizontal (i.e., upstream-upstream).

To date, few studies in the Andean region have focused on the political and social aspects of competing uses and access to water within a horizontal dimension. Research on social and economic issues related to water services and their conflicts in the Andes often focus on confrontation and dispute between peasant communities[2] and mining companies (Burneo de la Rocha and Chaparro Ortiz de Zevallos, 2010; Alvarado Merino, 2011; Sosa and Zwarteveen, 2012; Vela-Almeida et al., 2016); land and water appropriation

---

[2] In Peru, peasant communities represent forms of collective organization in the Andes, which are also characterized by their identification within a territory. Peasant families own the local resources. Peasant communities function as a social regulatory body for disputes and internal conflicts, and also defend the territory against external hazards (Diez, 2006).

by agricultural export companies and the implications of water transfer from the Andes to the coast (Oré, 2011; Damonte Valencia, 2015; Vera Delgado, 2015); and peasant communities who claim their control of water resources, especially for irrigation (Boelens, 2008a; Hoogesteger, 2012).

In Peru, there are no clear data on the annual number of conflicts related to water resources. The closest data are the monthly log of the Ombudsman (Defensor del Pueblo) on socio-environmental conflicts. Panfichi and Coronel (2011) conducted a review of these records between 2006 and 2010, identifying 115 conflicts over water resources. They indicate that water conflicts were grouped under the socio-environmental conflict section, therefore detailed statistics on conflicts specifically related to water were unavailable. However, the following trends emerged from the 115 conflicts that were analyzed:

- the main cause of conflict was water pollution, which was already happening or was feared to increase;
- they involved more than one claimant, where several stakeholders claimed irregularities or opposition in relation to the use of water resources;
- the defendant was usually an extractive or energy producing company;
- the conflicts tended to change and take other forms, rather than be solved.

The Ombudsman (Defensoria del Pueblo, 2015) identified 539 social conflicts between 2011 and 2014, of which 153 (28.36%) were linked to water resources. In 70% of cases, the water resource was the central issue in demand; while in the remaining 30%, it was only part of a set of demands of social groups (Defensoria del Pueblo, 2015). Based on this information, the Ombudsman identified three types of stakeholders involved in social conflict: social groups, private companies and the State. The peasant communities are the main social group that demands water rights (i.e., 24% of reported cases). The main demands are the protection of water resources, the remediation of water resources and proper water management for economic activities, among others.

The hydro-social territory of the Cañete River watershed

The basin is in the department of Lima, extending to the provinces of Cañete (lower catchment), Huarochirí (medium catchment) and Yauyos (upper catchment). The source of the Cañete River is the Ticllacocha Lagoon, which is at an altitude of 4429 m.a.s.l., and it flows 235.67 km before it reaches the Pacific Ocean. There are two seasons, dry (from April to November) and rainy (from December to March). Access to the area is limited by difficult road conditions, especially during the rainy season.

According to Quintero et al. (2013) there is an inverse relationship between population density and water precipitation or supply. The coastal area, in the lower part of the basin, is inhabited by 85.6% of the population of the watershed, but receives less than 20 mm of the annual rainfall. The upper area is inhabited by only 4.6% of the total basin population and has an annual rainfall of between 736 and 1169 mm. Water yield is therefore higher in the upper basin as shown in Fig. 1.1. In addition, permanent surface water sources such as glaciers, lakes, springs, secondary rivers and streams, are located in the upper part of the basin. According to the last census carried out by the National Institute of Statistics and Informatics in 2007, the upstream province of Yauyos is one of the poorest of the Lima Department with 61.9% of its population classified as poor.

Natural resource uses in the watershed are multi-purpose and multi-scale. The hydro-social territory of the watershed is therefore complex. The lower part of the basin has the largest demand for water for multiple uses, such as: agriculture (the main water user), domestic use, mining, energy, fisheries, tourism and recreation (Quintero et al., 2013).

The physical space covers the river and its tributaries, lakes and springs, in addition to existent water infrastructure (such as hydroelectric plants,

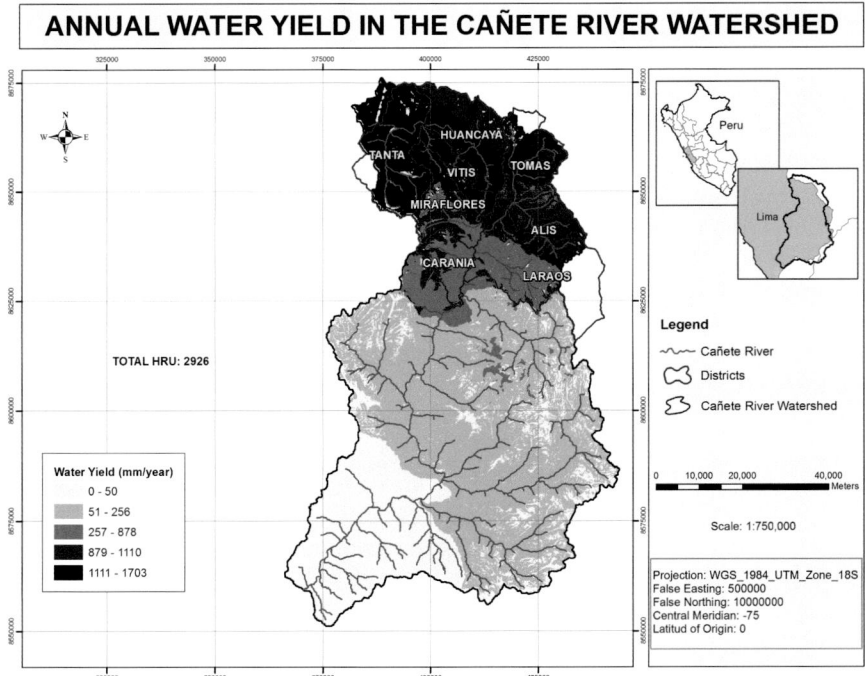

Fig. 1.1 Annual water yield in the Cañete River watershed.

dams and reservoirs). The social space includes: management and access to water by different stakeholders (e.g., crop, livestock and fish farmers, mining companies, hydroelectric companies and the tourism industry). Finally, the political and administrative space is defined by the rules and regulations that govern natural resource use in the territory, the presence of a protected area in the upper part of the watershed, and the establishment of a Reward Mechanisms for Hydrological Ecosystem Services (RMHES).

The eight peasant communities analyzed in this study are located in the upper part of the watershed, between 3,100 and 4,300 m.a.s.l. and their history goes back to pre-Hispanic times. Today, after several transformation processes, the physical territory of these peasant communities is defined by the political limits of eight districts: Alis, Carania, Huancaya, Laraos, Miraflores, Tanta, Tomas and Vitis. These communities mainly practice subsistence agriculture and extensive livestock farming in natural pastures. Agriculture is mostly rain-fed and major crops include maize, Andean roots and tubers and grains. Natural pastures are also rain-fed, while cultivated pastures are irrigated. Two systems of land tenure coexist: communal lands, which can be under collective or family use, where decisions over the whole area are collective but each family takes the decision over their assigned plots and private lands, which are fully under the control of the family who owns them.

These peasant communities share their territory with formal and informal mining companies and a hydroelectric company (CELEPSA), whose main seasonal reservoir is located in Paucarcocha, a lagoon located in the community of Tanta, close to the Cañete River source. Uses, agreements and disputes involving these external agents who use hydrological resources have affected the landscape and rules that traditionally regulated the area.

In addition, since 2001, these eight communities are part of the Reserva Paisajística Nor Yauyos Cochas (North Yauyos Cochas Landscape Reserve— RPNYC). The creation of the RPNYC changed the legal and administrative status of the territory, which is currently under the management of the National Service of Natural Protected Areas by the State (SERNANP). This change introduced restrictions on the use of natural resources, and helped to promote tourism development in the area. A landscape reserve is in fact a conservation figure that allows the direct but regulated use of natural resources within a protected territory, under the guidelines of a master plan prepared by the local SERNANP division. The current master plan was developed in line with the medium-term development vision of peasant communities in the reserve, whose interests were represented by local authorities, and includes the vision of other stakeholders in the territory (SERNANP, 2016a). The obligation to protect natural resources, particularly wildlife, is restrictive for some stakeholders, such as livestock herders, and the SERNANP creates discussion spaces to resolve any disputes that may arise.

Within this socio-political and economic context, the Ministry of Environment of Peru (MINAM), and various international organizations, such as the World Wide Fund for Nature (WWF), CARE–Peru, the Peruvian Society of Environmental Law (SPDA), Conservation International and the International Fund for Agriculture Development (FIDA), have been leading the design of a RMHES for the Cañete River Basin since 2010. The objective is to reward upstream peasant communities for the provision of a hydrological service. The mechanism is a reward for those already providing a valued service. Additionally, it aims to help ensure the sustained implementation of adequate grazing practices and conservation of peat bogs, relic forests and natural pastures, to improve ecosystem services delivery in the area. The International Center for Tropical Agriculture (CIAT) carried out a hydrological study to prioritize specific areas in the upper watershed that needed to be targeted to ensure ecosystem service provision (Quintero et al., 2013). CIAT researchers also studied conservation and development priorities of the eight peasant communities upstream to inform about the implementation of the mechanism (Blundo Canto et al., 2016). Currently, the RMHES has reached its inception phase and activities to be implemented are being prioritized by MINAM with the different stakeholders that are affected by the reward mechanism.

The study of competing uses and access to water resources in North Yauyos

The study presented in this chapter is part of a larger research project on conservation and development priorities of peasant communities in North Yauyos (Blundo Canto et al., 2016). This research is based on qualitative and participatory data. The focus is on the socioeconomic reality of upstream communities based on the views and perceptions of peasant families who participated in the research project.

The data were collected through workshops, one per community, in eight communities located in the North Yauyos province in the RPNYC. Each workshop comprised four focus group discussions on different issues; each focus group discussion lasted 1.5–2 hours. Focus groups are particularly useful for exploring the degree of consensus on a given topic (Morgan and Kreuger, 1993). About 10 to 16 people participated in each focus group, with a total of 102 participants, of which 48% were women. Participants were selected to represent the different stakeholders present in the communities, who engaged in different activities, including agriculture and livestock farming, crafting, tourism, health service provision and community administration. A heterogeneous group of participants was chosen in order to represent different views; this approach was ethically appropriate as the communities were accustomed to coming together as a group for public discussion. Each focus group lasted 3 to 4 hours and was audio tape-recorded. The data were complemented with data from semi-structured interviews with local authorities in order to understand the institutional arrangements in each community, the dynamics of land tenure and the main events in the recent history of the communities. All who participated in the study did so freely with prior informed consent. Table 1.1 shows the different types of uses that define social spaces in the upper part of the Cañete River watershed, the stakeholders, issues and demands involved.

Water resources, their uses and struggles among peasant communities

Peasant communities of the upper Cañete River watershed are managed through a political structure based on three institutions: the communal board, the assembly and specialized committees. According to the General Law of Peasant Communities (Law No. 24656) the board governs and administrates the peasant community. It is elected every 2 years by individuals registered in the community registry. The president runs the board and orders the assembly to take governance decisions. The specialized committees complement the functions of the board on specific issues, such as livestock farming, tourism, irrigation, and so on.

All communities, except Tanta, which is over 4,000 m.a.s.l., own farming land, which is mainly used for subsistence agriculture and pasture cultivation. Farmers grow maize and Andean grains such as quinua (*Chenopodium quinoa*) or *tarwi* (lupine *Lupinus mutabilis*), Andean roots and tubers, such as potato (*Solanum tuberosum*), oca (*Oxalis tuberosa*), olluco (*Ullucus tuberosus*) and mashua (*Tropaeolum tuberosum*) and other crops including peas (*Pisum sativum*), beans (*Phaseolus vulgaris*) and barley (*Hordeum vulgare*). Farmed fields are either rain-fed or irrigated. Every community has irrigated areas, and water distribution is managed by a community organization. All irrigation farmers are part of the so-called irrigation committee, which is managed by a board elected by the farmers. They have three main functions: to manage water for irrigation, to resolve conflicts and to plan collective work activities for maintaining irrigation systems. Not all individuals have access to irrigated land; while rain-fed plots are owned by the community and assigned to the peasant families for their use, irrigated plots are usually privately owned. Management decisions over communal land are discussed in the assembly, or in the *ayllu*,[3] while private plots are under the sole jurisdiction of the owner.

---

[3] The *ayllu* is an ancestral form of social organization, constituted by a group of families. It has rights and duties over productive activities and rituals. The communities of Vitis and Huancaya have farming and livestock *ayllus* who possess and rule on part of the land dedicated to these activities.

Table 1.1 Uses, stakeholders, issues of and demands for hydrological resources in the upper Cañete River watershed.

Uses	Stakeholders	Hydrological resources	Issues related to hydrological resources*	Demands
Agriculture	Peasant families from Alis, Carania, Huancaya, Laraos, Miraflores, Tanta, Tomas and Vitis	The Cañete River and tributaries Lagoons Springs	Uncertainty about the rainy season (length and timing) Dry seasons are longer, rainy seasons shorter	Construction and/or improvement of reservoirs or dams Construction and/or improvement of irrigation channels Irrigation systems
Livestock farming	Peasant families from Alis, Carania, Huancaya, Laraos, Miraflores, Tanta, Tomas and Vitis Community committees: Carania, Laraos, Alis, Tomas y Tanta	Lagoons Springs Rivers	Overgrazing; reduced biomass regrowth especially in peat bogs Uncertainty about the rainy season (length and timing) Dry seasons are longer, rainy seasons shorter	Construction and/or improvement of reservoirs or dams Agreements with contaminating private enterprises Implementation of a livestock farming plan (set of rules)
Fish farming Fishing	Individuals from Huancaya, Laraos y Tanta	Lagoons	Reduced number of fish Changes in the volume of flow (rivers) Pollution of lagoons, rivers and springs	Permits to build fish farms Purification of water sources
Tourism and ecotourism	Peasant families from Laraos, Huancaya y Vitis SERNANP and municipalities	Lagoons	Solid waste	Tourism management plan

Mining	Private enterprises in Laraos, Carania, Tomas, Alis	Rivers Lagoons	Pollution of lagoons, rivers and springs Partial or complete depletion of lagoons Appropriation of water resources	Accountability for polluting behavior Agreements between communities and private enterprises
Hydroelectric power production	Private enterprise: Celepsa	Lagoons Rivers	Changes in the volume of flow (Cañete River) Disappearance of flora and fauna Changes in the ecosystem surrounding the dam	Agreements between communities and private enterprises Accountability for changes in the ecosystem

*From the perspective of local communities

In the communities, agricultural issues and conflicts are mostly related to water availability and access. Rain-fed plots are affected by inter-annual variability in the duration and timing of the rainy season, which, according to perceptions of people participating in the study, was more predictable in the past. They perceive that in recent years, the rainy season tends to start later and to be shorter in duration.

However, lower crop production was not only related to water availability, but also to lack of labor due to out-migration and extreme climatic events such as droughts and frost. Participants in the focus groups stated that these events have grown longer and more intense, while glaciers appear to have reduced in size. Moreover, the poor state of irrigation infrastructure has reduced the availability of and access to water for irrigated plots, especially in Alis, Huancaya, Vitis and Carania.

Livestock farming occupies another social space in the hydro-social territory of the upper watershed. Cattle, sheep, and Andean camelids graze in the natural pastures of the reserve, and their feed is complemented with cultivated pastures in irrigated plots in the communities. Selling livestock is often the main income source for families, who sell live animals and their by-products such as suede, wool, and dairy products. The number of animals a family owns depends on their economic resources and labor availability, and on communal rules about the maximum number of cattle, sheep and camelids allowed, which differs by community. Natural pastures are communal lands assigned to local residents by the local administration. Families pay for the right to use them and the payment varies in duration and amount, depending on the community. In Laraos, Tanta, Tomas and Carania, specific sectors of natural pastures are reserved for the exclusive use of communally owned livestock. Privately owned animals are not permitted to graze in these areas; they usually graze in the remaining sectors.

Livestock farming in the communities was extensive and was mostly unmanaged; only a few communities practised rotational grazing, which was partly a result of recent ecosystem-based adaptation interventions by development agencies (Miraflores, Vitis, Huancaya). Focus group participants identified overgrazing as a main problem for natural pastures. According to SERNANP, livestock pressure on natural pastures particularly affects peat bogs, which are important for underground water recharge. Unmanaged grazing in these areas is depleting the condition of pastures and reducing their regrowth capacity. Moreover, local inhabitants noticed an increase in the population of invasive plants that were causing livestock diseases. Finally, unpredictable weather patterns were seen to affect water availability in natural pastures.

Fish farming and river fishing were carried out in lagoons of Carania, Huancaya, Laraos, Tomas and Tanta; and in the rivers of Alis, Carania,

Huancaya, Laraos, Miraflores y Tanta. Some communities established fish farms close to their villages, taking advantage of nearby streams and springs. Fish farming in lagoons was carried out using floating cages. All fish production in the area was for local consumption or for supply to local restaurants. A few individuals (males only) worked on fish farms, which provided them with a good income source. People carried out occasional fishing to complement their household consumption. Focus group participants identified a reduction in the number of trout as an issue. In Carania, participants connected this to reduced water availability, while in other communities such as Alis and Tomas they stated that this might be a result of water pollution by mining activities.

In terms of tourism, the creation of the RPNYC has positioned the area as an emerging tourism site for its undeniable landscape beauty. Access constraints affect this activity, which has been nonetheless growing quickly in the past 10 years. SERNANP and communities such as Huancaya and Laraos are investing in this activity through the creation of tourism committees and associations. While tourism does not represent the main activity in any of the communities, it is an important income source.

Finally, private enterprises are important stakeholders in the hydro-social territory in the upper watershed. Their relationship with the local population is characterized by latent tensions. The main mining companies in the area are the Chumpe processing hub, which has been in operation since 1996 in Alis, the San Valentin mine in Laraos, and the Yauricocha mine in Tomas. The latter was founded in 1920 as a public mining enterprise, which contributed to local economic growth until its privatization, when the interaction between the company and the communities was reduced. Other small mining activities are present in the territory, contributing considerable pressure to hydrological resources.

Finally, the generation of hydroelectric power represents an important water use in the upper Cañete River watershed. In 2006, the El Platanal (Celepsa) hydroelectric plant began operating with the construction of a dam in the Paucarcocha Lagoon in Tanta, close to the Cañete River. The infrastructure was built despite tensions with the local population. Additionally, it has generated latent conflicts with other communities of the upper part of the watershed, which will be described later.

Current issues and demands related to the use of water

Each social space described earlier presents different issues and demands in relation to hydrological resources. Diverse stakeholders have competing needs, which creates tensions over the control of water.

Issues related to agriculture are concerned with water scarcity or lack of access. In terms of water scarcity, the changes perceived in weather patterns particularly affect rain-fed plots, while extreme climatic events appear to becoming more frequent. In Carania, farmers mentioned that even irrigated areas were being abandoned due to water scarcity, given the poor state of water infrastructure and the perceived reduction in the quantity of water available from streams and springs.

Pasture regrowth capacity in natural pastures was also being affected by water scarcity. Fewer natural water troughs, such as springs and water streams in peat bogs, were available. Overgrazing exacerbated these issues: SERNANP and the communities called for the creation of livestock management plans that enforce sustainable grazing practices. However, the dichotomy between communal lands, where rule enforcement is up to the community board and assembly, and private pasture areas, where individuals have complete control over how their resource is managed, reduces the effectiveness of such plans. Wild animals such as pumas, condors and foxes threaten livestock, but are protected by the SERNANP as part of the conservation of natural resources of the reserve; this generates disagreements between the population and SERNANP.

These issues affect the food security of peasant communities in North Yauyos and have, among other causes, caused out-migration of the younger population. This had led to abandonment of plots, which poses a threat to the conservation of agricultural biodiversity and crop genetic resources. Water scarcity also affects the development of crops important in local diets. In such a context, the population is demanding the construction and improvement of irrigation and water catchment infrastructure, and irrigation systems to ensure water availability in their fields. They are demanding an efficient use of water, especially during the dry season.

Issues related to the hydrological resource were different for fish farmers and fishers. Inadequate fish farming practices affected the quality of water where fish farms were installed, and SERNANP aims to incentivize sustainable fish farming in lakes and lagoons by revaluing fish diversity instead of focusing on trout farming. People who carried out fishing activities identified a reduction in the type and number of species and stated that mining activities and the construction of a dam by Celepsa were probably to blame. They demanded greater accountability for private companies and some form of compensation for the livelihood losses they had incurred.

Hydrological and other natural resources in the landscape are attracting tourists and their numbers are growing exponentially every year (SERNANP, 2016a), which poses some concerns for the long-term sustainability of this activity in terms of solid waste and carrying capacity of selected routes. However, tourism is still occasional and concentrated in peak periods, such as on national holidays. Local populations are demanding capacity building to offer improved tourism services. SERNANP is focusing on developing sustainable tourism plans that consider the conservation priorities of the reserve.

In addition to inadequate water provision for agriculture and livestock farming, participants stated that poor water quality was an issue in some rivers, springs and lagoons due to pollution by mining companies. Communities affected by mining activities, such as Alis and Tomas, identified conflicts over their hydrological resources. In Alis, they stated that the Silacocha Lagoon had been driven into its poor condition by the processing activities of the Chumpe plant. The Alis and Tomas Rivers were viewed as being polluted by mining activities in Yauricocha. Unlike other uses, where local populations were directly impacting natural resources, it is external stakeholders who are occupying communal territory and the communities perceive that their actions are depleting the ecosystem services. Although they give payments to local governments under current laws, this retribution does not cover local demands for accountability and benefit-sharing. Peasant families reported reduced water quality and quantity that affected their livelihood, but they stated that they did not have the financial means to make these private companies accountable. The communities called for more accountability by mining companies and more rules for responsible use of water resources and shared benefits.

Finally, the use of water to produce hydroelectric power has also generated conflict. The construction of the dam in the Paucarcocha Lagoon in Tanta provoked tensions within the community as the dam inundated areas that had been used for livestock grazing. Celepsa, the private hydroelectric company, acquired the fields surrounding the lagoon, which were owned by the local population. In Tanta, pasture ownership is individual rather than communal, and while some peasant families wanted to sell, a significant group of families were opposed to this idea. After the construction of the dam, the population perceived changes in the ecosystem, such as lower overall temperature, a reduction in flora and fauna, apart from the reduction in natural pasture area. Additionally, people from Huancaya reported that trout size had decreased since the creation of the Celepsa dam in Tanta. They stated that constant river flow all year round had affected the reproduction and growth cycle of fish, affecting income generation and household consumption. They also reported that flora and fauna had disappeared, as previously the seasonal flow reduction used to favor their presence on the now permanently inundated riverbanks. The affected communities demanded not only accountability by the hydroelectric company, but also improved communication and agreements with communities higher up in the Cañete River catchment.

The importance of disentangling internal and external pressures in the hydro-social territory

The analysis presented in this chapter is based on the conceptualization that competing uses and access to the same hydrological resource can generate competition, superimposition, and eventual alignment, between different interests and use rights within a hydro-social territory (Boelens et al., 2016). Focusing on the social implications of this competitive hydro-social territory, where collective and private uses coexist, we show that this competition is not only internal, but also significantly affected by the action of external agents of change.

In the case of internal claims, it appears that differences in the rights of use and access to some resources generate differences in material well-being within the communities. Access to irrigated plots, the ability to work in fish farming or owning a larger amount of livestock, which are activities that put high pressure on hydrological resources, also imply greater economic well-being. Therefore, stakeholder groups and social spaces within single communities are not homogeneous. The complexity of the structure and dynamics of the micro-politics that result from this heterogeneity require special attention when studying competing uses and access to natural resources. However, it is common for researchers and decision-makers to consider communities as homogeneous communities in relatively close and similar hydro-social territories (Rodríguez de Francisco et al., 2013). Disentangling this heterogeneity within and between communities is necessary to achieve social equity issues. The upper part of the Cañete River watershed is a clear example of this heterogeneity and of the different demands that arise from it. Taking social equity into account would help to avoid the marginalization of vulnerable groups or capturing of benefits by elites when incentive mechanisms for the conservation of natural resources are designed and implemented (Pascual et al., 2014).

The management of natural resources implies rules that regulate their use, sanctions and decision-making structures (Ostrom, 1990). Collective action (Meinzen-Dick and Di Gregorio, 2004) in the communities studied takes form in the activity of community institutions. In the case of tensions with external agents of change, forms of collective action come into play in order to safeguard water resources affected by private companies' operations. As Boelens (2008a) states, water is a source of conflict and collaboration. In the upper Cañete River watershed, conflicts related to hydrological resources are commonly against private companies that use water resources claimed by the local population. When these conflicts against external agents of change arise, internal stakeholder differences appear mitigated and the communities unite in claiming rights against an external agent (such as a mining company or a hydroelectric power plant). The disputes can also transcend the space of a single community and assume a timing dimension (Wolf et al., 2005). Such is the case of the hydroelectric dam that modified not only the space of the communities where it was built, but also affected communities further downstream through changes in the natural river flow. Activities that had been traditionally carried out were altered, such as the case of natural pastures that were inundated to build the dam in Tanta, or the effects on fisher's livelihoods in Huancaya.

Finally, the different stakeholders that use hydrological resources in the upper Cañete River watershed are concerned about the availability of and access to sufficient and adequate water. In such a scenario, the most immediate solution appears to be the construction of efficient irrigation systems, and restoration of ancestral channels of irrigation and reservoirs in order to improve water distribution from catchment to distribution to the plots. As Bigas (2012) notes, one of the main causes of concern related to water scarcity in the medium- and long-term is the perceived increasing unpredictability of weather patterns. If water resources become more scarce, this higher uncertainty could bring about new conflict scenarios (Wolf et al., 2005; Salehyan, 2008; Scheffran et al., 2012). Additionally, future climate change scenarios might accelerate the vulnerability of those spaces in the hydro-social territory that depends exclusively on climatic conditions, such as rain-fed plots and natural pastures. These spaces represent central subsistence activities for peasant communities, and so increasing their resilience and capacity to adapt to change is imperative.

In this panorama of competition and superimposition of different claims, environmental and social change, dialogue and negotiation spaces are needed. Taking into account the different needs and claims of stakeholders within the communities, and the need to achieve concerted decisions between multiple communities, is imperative for equitable and sustainable development in the watershed.

Conclusions

This chapter disentangled the different uses, access, conflicts and demands related to hydrological resources in an Andean watershed of strategic economic importance. Diverging claims over the same water resources compete or superimpose, and answer economic and environmental transformations in the hydro-social territory. The study highlights the importance of understanding the heterogeneity of stakeholders in the hydro-social territory in order to achieve both efficiency and social equity when designing and implementing conservation incentive mechanisms, with special attention to the upper watershed. Peasant communities of the upper Cañete River watershed observe and perceive changes in the availability of water, which they relate to access rights, weather unpredictability and unsustainable uses, among others, by local communities and private enterprises. These changes have resulted in different responses by the communities, including formal agreements, active and passive adaptation, tensions, and especially for the younger generation, out-migration. The concern for water availability for peasant communities is linked to internal and external causes that affect productive activities, which appear to be exacerbated by weather variability and climate change. By taking into account the uses and demands of the stakeholders that affect the provision of hydrological resources, the reward mechanism that will be implemented in the watershed could contribute to the generation of alignment between their needs, improving environmental efficiency and social equity.

Acknowledgments

The authors would like to thank the Ministry of Environment of Perú (MINAM) and the International Fund for Agricultural Development (IFAD), who are implementing a Mechanism to Reward Hydrological Ecosystem Services in the Cañete River watershed. We are also grateful to the National Service of Natural Areas Protected by the State (SERNANP) and the North Yauyos-Cochas Landscape Reserve (RPNYC) for their support in the implementation of the study. We would like to thank the CGIAR Research Program on Water, Land and Ecosystems for financing this study through the project: Finding common ground: Bringing together ecosystem services, agricultural productivity and smallholder livelihoods in watershed planning. We are grateful for the support of the community presidents and authorities who helped to organize the focus groups. Finally, we must say a big word of thanks to the local population who generously shared their time and knowledge to help generate the information for this study.

References

Alvarado Merino, G. 2011. Políticas neoliberales en el manejo de los recursos naturales en Perú: El caso del conflicto agrominero de tambogrande. Seminario Permanente de Investigación Agraria 67–104.

Barrett, C.B., Travis, A.J. and Dasgupta, P. 2011. On biodiversity conservation and poverty traps. Proceedings of the National Academy of Sciences 108(34): 13907–12. Doi: 10.1073/pnas.1011521108.

Bigas, H. (ed.). 2012. The global water crisis: Addressing an urgent security issue. Papers for the InterAction Council 2011–2012. Doi: 10.13140/2.1.2593.8721.

Blundo Canto, G., Cruz-García, G.S., Febres, M.C.T., Cabrejos, P.P. and Quintero, M. 2016. Prioridades de conservación Y desarrollo en las comunidades de nor yauyos. Informe Para El MRSEH de la Cuenca Del Río Cañete. Cali, Colombia.

Boelens, R. 2008a. Rules of the game and the game of the rules: Normalization and resistance in Andean water control. Wageningen University, the Netherlands.

Boelens, R. 2008b. Water rights arenas in the Andes: Upscaling networks to strengthen local water control. Water Alternatives 1(1): 48–65. Doi: urn:nbn:nl:ui:32-366911.

Boelens, R., Hoogesteger, J., Swyngedouw, E., Vos, J. and Wester, P. 2016. Hydro-social territories: A political ecology perspective. Water International 41(1): 1–14. Doi: 10.1080/02508060.2016.1134898.

Budds, J. 2010. Las relaciones sociales de poder y la producción de paisajes hídricos. CENSAT Agua Viva/Amigos de la Tierra, Colombia.

Burneo de la Rocha, M.L. and Chaparro Ortiz de Zevallos, A. 2010. Poder, comunidades campesinas e industria minera: El gobierno comunal y el acceso a los recursos en el caso de michiquillay. Anthropologica XXVIII(28): 85–110.

Damonte Valencia, G.H. 2015. Redefiniendo territorios hidrosociales: Control hídrico en el valle de ica, Perú (1993–2013). Cuadernos de Desarrollo Rural 12(76): 109. Doi: 10.11144/Javeriana.cdr12-76.rthc.

Defensoría del Pueblo. 2015. Conflictos sociales y recursos hídricos 151.

Diez, A. 2006. Las organizaciones colectivas. Los Recursos Y Los Pueblos Indígenas en El Perú 111–30.

Edelman. 2013. What is a peasant? What are peasantries? A briefing paper on issues of definition. Prepared for the first session of the Intergovernmental Working Group on a United Nations Declaration on the Rights of Peasants and Other People Working in Rural Areas, Geneva, 15–19 July 2013.

Giller, K.E., Leeuwis, C., Andersson, J.A., Andriesse, W., Brouwer, A., Frost, P., Hebinck, P., Heitkönig, I., van Ittersum, M.K., Koning, N., Ruben, R., Slingerland, M., Udo, H., Veldkamp, T., van de Vijver, C., van Wijk, M.T. and Windmeijer, P. 2008. Competing claims on natural resources: What role for science? Ecology and Society 13(2). Doi: 10.1016/j.biocon.2005.10.047.

Gleick, P.H. 1993. Water and conflict: Fresh water resources and international security. International Security 18(1): 79–112.

Hendrix, C.S. and Glaser, S.M. 2007. Trends and triggers: Climate, climate change and civil conflict in sub-Saharan Africa. Political Geography 26: 695–715. Doi: 10.1016/j.polgeo.2007.06.006.

Hoogesteger, J. 2012. Transforming social capital around water: Water user organizations, water rights, and nongovernmental organizations in Cangahua, the Ecuadorian Andes. Society & Natural Resources 1920: 1–15. Doi: 10.1080/08941920.2012.689933.

International Water Management Institute. 2007. Water for food, water for life: A comprehensive assessment of water management in agriculture. Earthscan, London and International Water Management Institute, Colombo.

Meinzen-Dick, R.S. and Di Gregorio, M. 2004. Collective action and property rights for sustainable development. 11. 2020 Vision Focus Briefs. Washington DC.

Millennium Ecosystem Assessment. 2003. Ecosystems and Human Well-being: A Framework for Assessment. Island Press, Washington DC.

Morgan, D.L. and Kreuger, R.A. 1993. When to use focus groups and why. pp. 3–19. *In*: Morgan, D.L. (ed.). Successful Focus Groups. Sage, London.

Oré, M. 2011. Las luchas por el agua en el desierto iqueño: El agua subterránea y la reconcentración de tierras y agua. pp. 423–434. *In*: Budds, M., Cremers, R. and Zwarteveen, L. (eds.). Justicia Hídrica: Acumulación, Conflicto y Acción Social. IEP, Lima.

Ostrom, E. 1990. Governing the Commons: The Evolution of Institutions for Collective Action. Cambridge University Press, Cambridge. Doi: 10.1017/CBO9780511807763.

Ostrom, E., Burger, J., Field, C.B., Norgaard, R.B. and Policansky, D. 1999. Revisiting the commons: Local lessons, global challenges. Science 284(5412): 278–282. Doi: 10.1126/science.284.5412.278.

Panfichi, A.I. and Coronel, O. 2011. Conflictos hídricos en el Perú 2006–2010: Una lectura panorámica. pp. 393–422. *In*: Boelens, M., Cremers, R. and Zwarteveen, L. (eds.). Justicia Hídrica: Acumulación, Conflicto y Acción Social. IEP, Lima.

Pascual, U., Phelps, J., Garmendia, E., Brown, K., Corbera, E., Martin, A., Gomez-Baggethun, E. and Muradian, R. 2014. Social equity matters in payments for ecosystem services. BioScience 64(11): 1027–36. Doi: 10.1093/biosci/biu146.

Quintero, M., Tapasco, J. and Pareja, P. 2013. Diseño E Implementación de Un Esquema de Retribución Por Servicios Ecosistémicos Hidrológicos En La Cuenca Del Río Cañete. Lima.

Raworth, K. 2012. A safe and just space for humanity: Can we live within the doughnut? Oxfam Discussion Papers. February 2012.

Rivera, D., Godoy-Faúundez, A., Lillo, M., Alvez, A., Delgado, V., Gonzalo-Martín, C., Menasalvas, E., Costumero, R. and García-Pedrero, A. 2016. Legal disputes as a proxy for regional conflicts over water rights in Chile. Journal of Hydrology 535: 36–45. Doi: 10.1016/j.jhydrol.2016.01.057.

Rodríguez de Francisco, J.C., Budds, J. and Boelens, J. 2013. Payment for environmental services and unequal resource control in Pimampiro. Society & Natural Resources 26(10): 1217–1233. Doi: 10.1080/08941920.2013.825037.

Rodríguez de Francisco, J.C. and Budds, J. 2015. Payments for environmental services and control over conservation of natural resources: The role of public and private sectors in the conservation of the Nima watershed, Colombia. Ecological Economics 117: 295–302. Doi: 10.1016/j.ecolecon.2014.05.003.

Salehyan, I. 2008. From climate change to conflict? No consensus yet. Journal of Peace Research 45(3): 315–326. Doi: 10.1177/0022343308088812.

Scheffran, J., Brzoska, M., Kominek, J., Link, P.M. and Schilling, J. 2012. Climate change and violent conflict. Science 336: 869–871. Doi: 10.1126/science.1221339.

SERNANP. 2016a. Plan Maestro RPNYC—SERNANP.

Sosa, M. and Zwarteveen, M. 2012. Exploring the politics of water grabbing: The case of large mining operations in the Peruvian Andes. Water Alternatives 5(2): 360–375.

Sukhdev, P. 2009. Costing the earth. Nature 462(7271): 277. Doi: 10.1038/462277a.

Vela-Almeida, D., Kuijk, F., Wyseure, G. and Kosoy, N. 2016. Lessons from Yanacocha: Assessing mining impacts on hydrological systems and water distribution in the Cajamarca region, Peru. Water International 41(3): 426–446. Doi: 10.1080/02508060.2016.1159077.

Vera Delgado, J. 2015. The socio-cultural, institutional and gender aspects of the water transfer: Agribusiness model for food and water security. Lessons learned from Peru. Food Security 7(6): 1187–1197. Doi: 10.1007/s12571-015-0510-5.

Wolf, A.T., Kramer, A., Carius, A. and Dabelko, G.D. 2005. Managing water conflict and cooperation. pp. 80–99. *In*: State of the World 2005: Redefining Global Security. The Worldwatch Institute.

Wunder, S. 2015. Revisiting the concept of payments for environmental services. Ecological Economics 117: 234–243. Doi: 10.1016/j.ecolecon.2014.08.016.

CHAPTER 2

Socio-environmental Issues Related to Mineral Exploitation in the Andes

Douglas Aitken,[1,*] *Alex Godoy-Faundez,*[1] *Oscar Jaime Restrepo-Baena,*[2] *Diego Rivera*[3] and *Neil McIntryre*[4]

Introduction

The Andes mountain range is home to some of the world's largest and most important mineral reserves such as copper (590 Mt of resources), molybdenum (20 Mt), silver (250,000 t) and gold (13,000 t) (Cunningham et al., 2008). These resources have brought considerable wealth to many of these areas and have helped in alleviating poverty. Yet mining remains a controversial activity, which in some cases has divided society and has caused large scale environmental impacts (Paredes, 2016). This chapter investigates the socio-environmental impacts of mineral exploitation in selected Andean countries with a particular focus on water resources. Specific issues such as local hydrology, community engagement and technological advancements are investigated and discussed with respect to each country.

Mineral resources exist in each of the countries that contain sections of the Andes mountain range, countries with some of the largest reserves are Chile, Peru and Colombia (Cunningham et al., 2008). The scale of industry and the

[1] Facultad de Ingeniería, Universidad del Desarrollo, Av. Plaza 700, San Carlos de Apoquindo, Las Condes 7610658, Chile.
[2] Departamento de Materiales y Minerales, Facultad de Minas, Universidad Nacional de Colombia, Carrera 80 # 65-223 – Nucleo Robledo, Medellín, Colombia.
[3] Laboratory of Comparative Policy in Water Resources Management, Departamento de Recursos Hídricos, Facultad de Ingeniería Agrícola, Universidad de Concepción, Chillán 3812120, Chile.
[4] Centre for Water in the Minerals Industry, Sustainable Mineral Institute, University of Queensland, Australia.
* Corresponding author: D.aitken@SmiiceChile.cl

environment in which mining operates varies considerably between these three countries. Chile has the most mature and developed mining industry, and mining has traditionally been a cornerstone of the national economy, with the industry providing about 11.2% of the total Gross Domestic Product (GDP) in 2014 (Fundacion Chile, 2017). The majority of mining exports in Chile are copper, which in 2014, accounted for 37.6% of total global production (Fundacion Chile, 2017). Molybdenum, gold and silver production also contribute to the GDP but are less significant than copper (Quirland and Leclerc, 2013). The vast majority of copper mining in Chile occurs in the northern regions, particularly the region of Antofagasta, which contains some of the world's largest copper mines and accounts for around 56% of Chile's total production (Arena Minerals, 2015). Figure 2.1 displays the location of copper mines in Chile using spatial data of all Chilean copper mines obtained from Albers (2012). The environment in which many of the mines operate is one of the most arid in the world, with extremely low levels of precipitation and consequently low water availability (Aitken et al., 2016). With reducing levels of water availability and increasing demand, productivity of some operations has been affected with a consequent loss in earnings (Onstad and O'Brien, 2015). Furthermore, the existence of other industries and agriculture has led to conflicts and competition regarding water rights and overuse of freshwater resources (Rivera et al., 2016; Aitken et al., 2016). The situation is becoming more serious with predictions of lower water availability in the future and greater use in all sectors (Valdés-Pineda et al., 2014). It is extremely important for the mining industry in Chile, therefore, to hasten the transition to a more sustainable industry, particularly with respect to its water management and relationship with other water users.

In Peru, the mining industry is worth less to the national GDP than in Chile (in 2013 mining, oil and gas accounted for around 4.8%), it is, however, an important industry as it accounts for 55.2% of all exports (Ernst and Young, 2016). The mining industry is highly diversified with the production of copper, gold, zinc, tin, molybdenum, iron, cadmium alongside various other metals, the industry also supports around 210,000 jobs directly each year (PWC, 2013). The locations of mines in Peru is well distributed throughout the country as can be observed in Fig. 2.2 where locations of Peru's copper, zinc, silver and gold mines are displayed using spatial data from the USGS (USGS, 2014). The mining industry is extremely competitive in Peru with comparatively low operating costs, low energy prices and a legal environment favorable to mining (PWC, 2013). Nevertheless, despite the considerable economic productivity of the mining industry in Peru, there exists a strong social movement against the industry as a result of environmental impacts, mismanagement of projects, poor wealth distribution, corruption and violence against protestors (Paredes, 2016). In 2014, the Ombudsman's Office identified mining as being the primary cause of 50% of the 270 conflicts in Peru since 2004 (Paredes, 2016). The

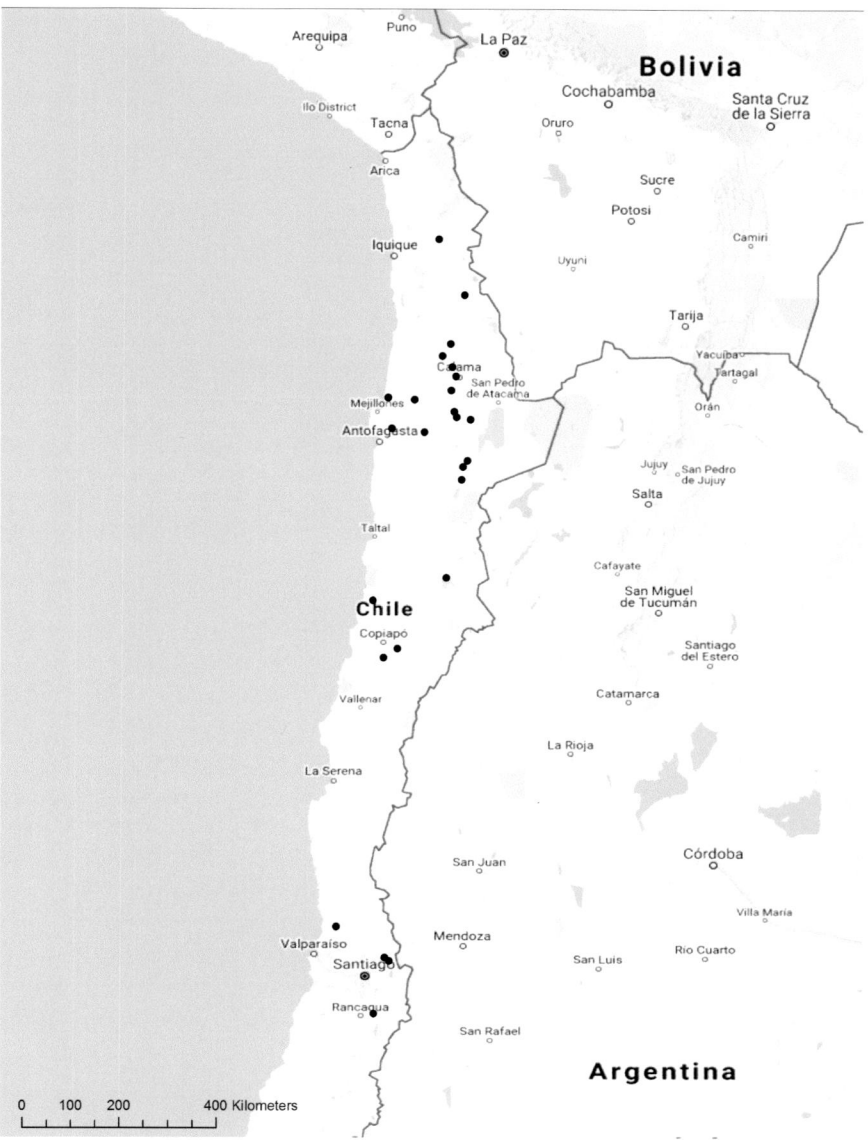

Fig. 2.1 The locations of medium to large scale copper mines in Chile using spatial data from Albers (2012), Map layer: Google.

reasons behind these conflicts are diverse, including: perceived damage to livelihood from large scale projects and the associated environmental damage, increased economic inequality and in many of the cases, the depletion and contamination of local water bodies. The issues related to water resources in Peru are often the most divisive causes between

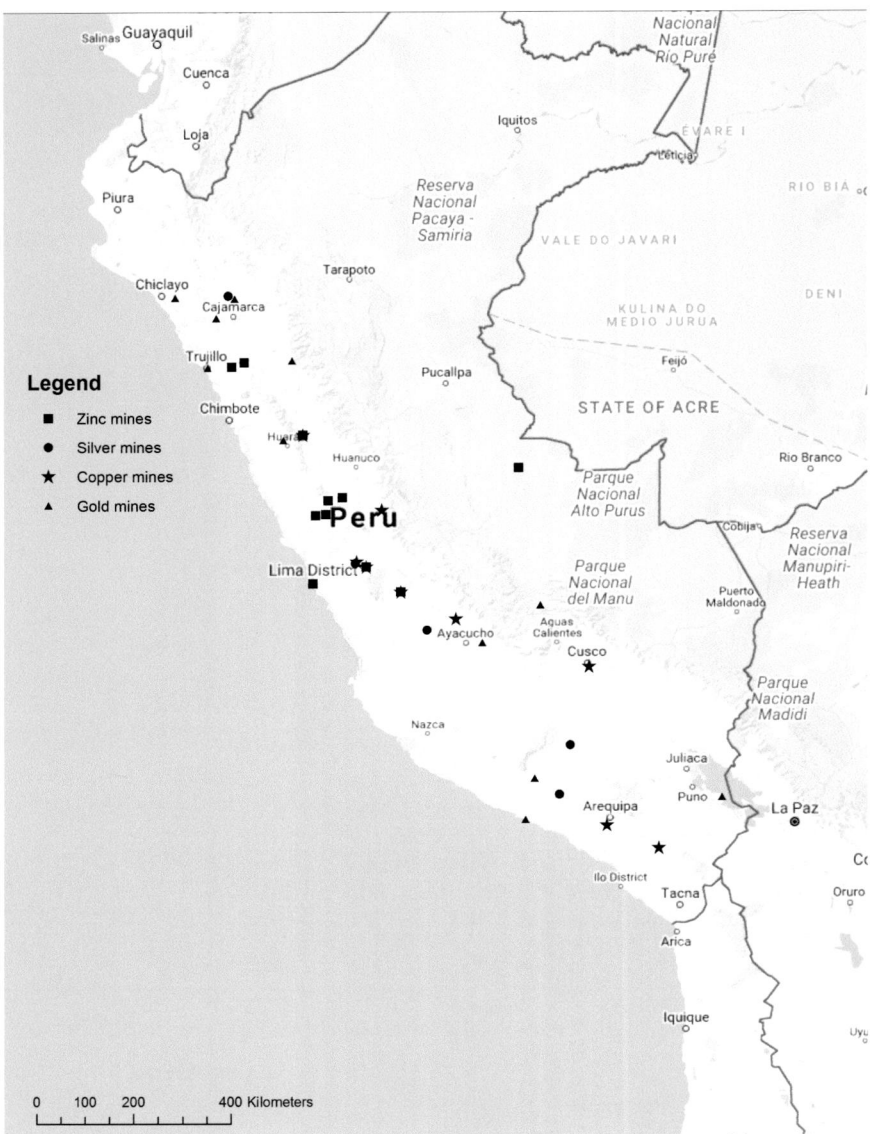

Fig. 2.2 Locations of copper, silver, zinc and gold mines in Peru using spatial data from the USGS (USGS, 2014), Map layer: Google.

communities and companies as water has an extremely high importance in the Andean communities both for practical reasons and also spiritual.

The situation in Colombia is quite distinct from Chile and Peru. Mining contributes a comparatively small percentage to the national GDP (2.1%) (Rojas, 2015). Yet with much of the industry operating illegally, there exists a great potential for the industry to develop into a considerable economic provider. The majority of mining in Colombia is for coal but gold mining located mostly around the foothills of the Andes mountain range is becoming increasingly important (Castellanos et al., 2016; Bustamante et al., 2016). Figure 2.3 displays the location of the United States Geological Society

Fig. 2.3 The locations of USGS listed gold mines in Colombia (USGS, 2014), Map layer: Google.

(USGS) listed gold mines, there are many other illegal operations around the same area and within the whole Andean range but their exact locations are difficult to obtain. According to the government's Financial Inspector, Edgardo Maya Viallzon, in 2015, around 80% of gold mining in Colombia was taking place illegally (Maya Villazon, 2015), other sources record the figure to be as high as 86% (Rettberg and Ortiz Riomalo, 2016). Under such circumstances, economically, the country is not benefitting to its full potential and furthermore, with illegal operations disregarding environmental regulations, contamination from the industry is causing extensive damage to the health of workers, local residents and the environment. Contamination arises mainly from the use of mercury for the amalgamation of gold, the mercury is subsequently vaporised and released to the environment without, in most cases, thought for the subsequent damage caused (Garcia et al., 2015). The release of mercury from artisanal mines has been identified as the main cause of health defects in many communities living close to operations and the cause of considerable destruction of aquatic habitats (Garcia et al., 2015; Cordy et al., 2015, 2011). It is highly important for Colombia's continued development to begin regulating the large illegal mining that is taking place to combat human rights abuses, environmental degradation and increase public revenue.

This chapter considers four particular areas related to socio-environmental issues, firstly, socio-hydrology, water pollution and related issues are investigated for each region, state regulations and company culture are then considered, the area of community engagement is examined followed by a section on technological solutions. Some issues are of more importance to a certain country than the others and in that case may be expanded more for that region.

Socio-hydrology, Water Pollution and Related Issues

Severe water pollution in Colombia

Unlike in the arid zones of Chile and Peru, Colombia has little issue regarding the availability of water and water stress, the more pressing issue with regards to the mining industry is the pollution entering the water cycle and its impacts upon human and ecological health (Cordy et al., 2011, 2013; Garcia et al., 2015). Small scale gold mining, which is the greatest production method of gold in Colombia, is the cause of considerable wastewater discharge related issues in the main gold mining areas (Antioquia, Chocó). Indeed, it is estimated that only 3% of Colombia's gold mines are in possession of an environmental license (Güiza Suárez and Aristizabal, 2013). Given the harmful nature of the mining practices and materials, the impacts of mismanagement are great. In 2014, it was reported that thousands of residents were forced from their homes in the state of Chocó as a result of mercury contamination in their water source and lawsuits were filed claiming the death of 37 children as a result of the contamination (McKenzie, 2014).

Artisanal gold mining operations in Colombia generally rely upon gold extraction using the application of mercury to the crushed ore to produce a gold-mercury amalgam. The amalgam is then heated allowing much of the mercury to be vaporised leaving the gold in a relatively concentrated form. This process has an efficiency of about 10% in which 10 g of mercury is used to produce 1 g of gold, 9 g of the mercury is released directly into the environment without adhering to the amalgam, the remaining 1 g is released following vaporisation (Güiza Suárez and Aristizabal, 2013). According to Cordy et al. (2015), of the mercury added to the processing system, 46% is lost in tailings and 4% is lost through vaporisation. The vaporisation normally takes place in 'gold shops', amalgam processing workshops which tend to be located in urban centers, the release of mercury into the air has the potential to greatly impact the health of the workers and the local population. There have been a number of studies conducted to identify the concentrations of airborne mercury around gold shops in various towns in Colombia where gold extraction takes place. The region of Anitoquia was recorded as the world's largest emitter of mercury pollution from artisanal gold mining in 2010, with an average of 92 tonnes of mercury emitted (Garcia et al., 2015). With respect to air quality standards, the World Health Organisation considers an annual average mercury concentration of $0.2\ \mu g/m^3$ to be tolerable with a value of $1\ \mu g/m^3$ considered hazardous to human health (Cordy et al., 2015). In five cities within Antioquia, mercury levels were found to range from $0.02\ \mu g/m^3$ up to $1,000\ \mu g/m^3$ with a value of $10\ \mu g/m^3$ being common in residential areas (Cordy et al., 2011). Exposure to mercury vapor can have serious short term and long term consequences. Exposure to high levels of mercury vapor $(1,200,000 + ng/m^3)$ can be fatal, lower levels can cause chest pains, dyspnoea, impairment of pulmonary functions, long term neurological and renal degeneration among other illnesses (Cordy et al., 2013; Bustamante et al., 2016). Several health assessments were conducted by Garcia et al. (2015) in the town of Segovia to determine mercury levels in the urine of residents. The results from 37 residents in 2013 demonstrated that 43% of those tested had normal levels of mercury toxicity ($< 5\ \mu g/g$ creatinine), 22% fell within an abnormal level considered an alert category (Drasch et al., 2002), 19% were at a level considered to require action (20–50 $\mu g/g$ creatinine) and 16% were found to have extremely high and dangerous levels ($> 50\ \mu g/g$ creatinine).

The proportion of mercury that is released within the mine tailings also poses a substantial threat to local water bodies. Deposited mercury is capable of entering the food chain rapidly either in soil or in water (Güiza Suárez and Aristizabal, 2013). In soils, the mercury can be absorbed by plants which can then damage the functioning of herbivorous animals that consume the plants (Azevedo and Rodriguez, 2012). Without containment infrastructure in place, mercury can very easily travel to water bodies causing direct impact to people who are abstracting water for drinking and also through the bio-accumulation within organisms living in the area. Pinedo-Hernández et al. (2015) conducted a study investigating the speciation and bioavailability of mercury in gold mining areas of Colombia, the study found higher levels of mercury from sampling stations receiving water downstream from mining areas, the exchangeable concentrations of mercury were determined to be low but considered high risk due to the capacity to bio-accumulate in organisms. Similarly to the inhalation of mercury vapors, the consumption of mercury contaminated water and fish can lead to severe organ and neurological damage, and in extreme cases, death (Yard et al., 2012). The issue of water contamination from mining in Colombia clearly must be addressed, various solutions regarding regulation and technological advancement will be discussed later.

Water stress and contamination in Peru

Peru's water resources vary considerably depending upon location, there are three main watersheds within the country: the Atlantic, the Pacific and the Titicaca watershed (Lavado Casimiro et al., 2012). The Atlantic watershed, which contains Peru's section of the Amazon rainforest contains the vast majority of the total water resource volume of the country, accounting for 97.8%. The Pacific watershed, which contains the majority of the Andes mountain range accounts for 1.7% of the total available freshwater, and the remaining 0.5% is located in the Titicaca watershed (Eda and Chen, 2010). The distribution of water resources is problematic in Peru as the majority of the population (60.4%) live in the Pacific watershed which is also the area with the greatest water demand from agriculture, industry and mining. Mining accounts for only a small proportion of total water use in Peru, in 2010–2011, this value was 1.4% compared to a value of 86.8% for agriculture. In the Atlantic watershed, the figure is highest at 5.4%, in the Pacific it is 0.7% and in the Titicaca watershed it is 1.5% (Eda and Chen, 2010). In Fig. 2.2 it is possible to observe that mining is widespread throughout the country, at least on a north-south axis, the larger mine sites are located mainly in the highlands of the Andes. The severity of water stress with respect to mine sites depends greatly upon local conditions. Those sites located in the Atlantic Basin are unlikely to be as affected by low water availability as mine sites located in the Pacific Basin. Areas which have been reported as having particular issues regarding scarcity are the southern regions of Tacna and Arequipa (Budds and Hinojosa, 2012). It has been alleged that the mining industry in the south has had a strongly adverse impact upon water availability for local communities and ecological services (Budds and Hinojosa, 2012).

This particular issue has been the cause of many community-company conflicts in Peru over the past several decades. Peru has witnessed a huge increase in mineral exploration and extraction from 1990 through to the present day. From 1990 to 1997, mineral exploration was recorded as having increased by 2,000% (Jaskoski, 2014). From 2001 to 2007, the number of mining claims increased from around 1,000 to 8,000, increasing the areal claims from under 500 hectares to 3,500 hectares. The government has since gone further to encourage investment, between 2006 and 2011, 99 decrees were passed to break up community land for the exploitation of natural resources. As a result, over 50% of mining concessions encroach upon areas containing agricultural communities (Jaskoski, 2014) who often end up in conflict with the operating mining companies, particularly over the use or contamination of local water bodies. In March 2012, 162 active conflicts between companies and communities were registered in Peru, 117 of these were related to the exploitation of natural resources (Jaskoski, 2014). A number of these conflicts have received considerable international attention due to their scale and the impact upon the country as a whole (Hill, 2015; Schipani, 2016). An interesting case study is the Tia Maria project in the Moquegua region which is explained in Case study 2.1.

Case study 2.1. The Tia Maria Project

The Tia Maria (Fig. 2.4) project is a copper mining project in the southern Arequipa region of Peru with a proposed 120,000 tonnes per year production of copper (Kozak, 2015). The operating company conducted three public audiences between 2007 and 2009, the first before the development of the EIA and the other two during its development (Jaskoski, 2014). Opposition to the project arose in the local community as a result of the plans to use local freshwater sources for the mines requirements. The EIA developed by the company proposed the use of water from existing waterways, underground sources or from desalination but heavily backed the use of river water. Despite protests, the operating company was slow to shift their final position from backing freshwater use to developing plans to implement desalination. During this time, a non-binding referendum was held in Cocachacra in which 90% of voters opposed the project (Spillan et al., 2011). The decision to use desalinated seawater was eventually proposed but by which point the protesters and the main opposition group declared that the project must be terminated (Jaskoski, 2014). The government continued to back the project, in 2010 however, a review of the EIA was conducted by the United Nations Office for Project Services which found considerable deficiencies in the study. Protests escalated and clashes with the police left three people dead, as a result, the project was finally cancelled.

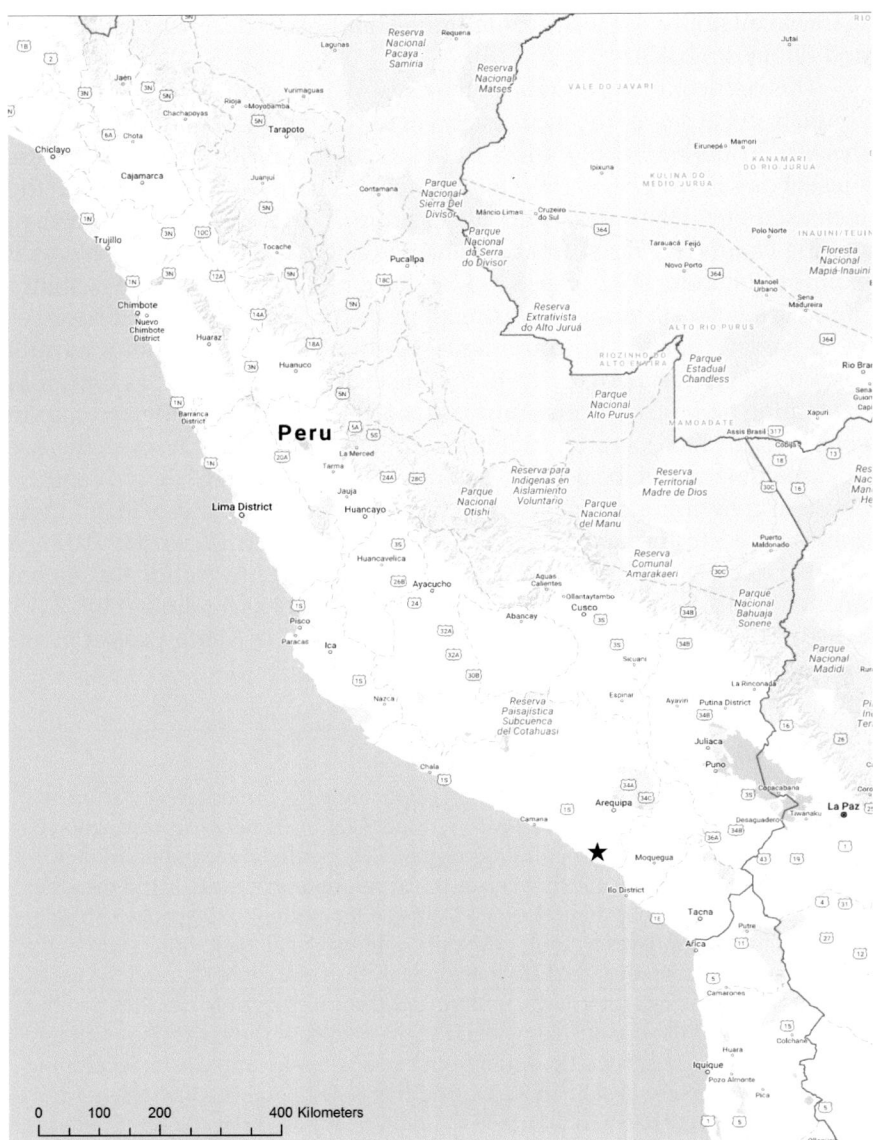

Fig. 2.4 Location of the Tia Maria mining project, Map layer: Google.

The Tia Maria project is by no means an irregularity in Peru, in June of 2015, according to the Defensoria del Pueblo (a state sponsored organization to protect citizen's rights), of the 116 active socio-environmental conflicts 72 cases (62%) were directly related to mining activities. Similarly to the Tia Maria project, many of these conflicts have resulted from water issues. Another large proposed mining operation, Conga, has faced social pressure against its development due to the lack of faith in the mining company to deliver the pledged water diversions to the local communities. With these fears unalleviated, protests took place that saw a forceful response from riot police prompting the government to declare a state of emergency (Jaskoski, 2014). After the escalation of the conflict it became too late for concerns to be addressed due to the deep divisions between stakeholders. Further work had been conducted on revised water strategies however the project has since been suspended (Kozak, 2015).

Contamination of water bodies from the mining industry is another dominant issue in Peru that can be linked to many of the current conflicts. Contamination from the mining industry was more problematic in the past when environmental regulations were less stringent although issues remain as state monitoring is often lacking (Budds and Hinojosa, 2012). Water monitoring studies downstream of Peru's largest gold mine in the Cajamarca area have shown water quality that has failed to meet company and government standards (Bebbington and Bury, 2009). Water contamination is largely blamed on leaching of acids and metals (acid mine drainage), the dumping of tailings (Budds and Hinojosa, 2012), and has been implicated in the high contamination levels of some of the most polluted rivers in Peru in most of the Pacific watershed (Monge, 2016). Furthermore, residents and companies tend to use different methods of water quality measurement, companies use quantitative measurement and residents are more vernacular and qualitative. Naturally, this difference leads to mistrust as residents often do not understand the scientific methods employed by the industry and companies may believe that residents exaggerate their experiences (Bebbington and Bury, 2009). According to Bebbington, the mining industry released around 13 billion m^3 of effluent into Peru's water courses annually prior to 2008. In July of 2008, a state of emergency was declared at a mine site near Lima due to fears that the tailings dam could leak arsenic, lead and cadmium into the capital's water supply. Naturally, the impact of the mining industry on water quality in Peru remains a considerable issue in the public domain and source of tension between communities and companies.

Severe water stress in Chile

In Chile, the majority of mining occurs in the arid and relatively sparsely populated areas in the northern regions which can be observed in Fig. 2.1. Despite the low population density, however, the impacts upon local communities and the environment are high due to the very low levels of precipitation, water storage and runoff flows (Aitken et al., 2016; Valdés-Pineda et al., 2014). The mining industry, in general, has provided Chile with largely positive economic impacts and is one of the reasons why Chile has a higher GDP and a more stable economy than many of its Latin American neighbors (Fundacion Chile, 2017). As a result, mining is generally seen in a positive light by the Chilean population as a producer of wealth and employment. In the CSIRO 2014 study of Chilean attitudes towards mining, it was found that mining was seen by the majority surveyed as a "necessary economic pillar for Chile" (Moffat et al., 2014). Recently, however, there has been an increase in conflicts between communities and mining operations. Rivera et al. (2016) investigated the change in water related conflicts in Chile over the past 30 years. In the 1980s very few disputes were registered throughout Chile, the number has increased considerably up until present day with disputes present in the far south to the far north. The northern regions have witnessed a particularly noticeable increase in disputes, the region of Antofagasta for example accounts for around 10% of total water rights conflicts despite having a very small population relative to the central regions (Rivera et al., 2016). Many of these conflicts relate to the change in water rights ownership which in many cases has changed from agriculture to mining alongside the banning of new groundwater rights from the Government Water Agency (DGA in Spanish) (Rivera et al., 2016).

With regards to the regional water balances, the northern regions of Chile suffer from an extreme imbalance between water demand and availability. According to the World Bank (2011), the water availability per person per year in Antofagasta is 52 m^3, in the Atacama region it is 208 m^3 and in the far northern region of Arica and Parinacota it is 854 m^3. For comparison, the national average value is 53,953 m^3/hab/yr. The regions of Arica and Parinacota, Antofagasta and Atacama each have rivers flowing within the regional boundaries, the flow rates of the rivers, however, are extremely low, values of mean annual flow in Arica and Parinacota range from 0.3 to 0.6 m^3/s, 0.4 to 2.43 m^3/s in Antofagasta and 0.8 to 1.85 m^3/s in the region of Atacama. Furthermore, there have been indications that rainfall has been decreasing over the past century (Burton et al., 2005) and future predictions suggest that there will be significant reductions in Altiplano precipitation over the coming century (Minvielle and Garreaud, 2011).

In the country as a whole, agriculture is by far the greatest user of water at around 77.8% whilst mining accounts for about 7.2% (Valdés-Pineda et al., 2014). Nevertheless, as the majority of mining takes place in the north of the country, a high proportion of the water use in the regions with very low availability is consumed by the mining industry. Aitken et al. (2016) calculated that in the region of Antofagasta, consumption from the mining industry accounted for around 64% of total consumption with agriculture accounting for about 15%. The high consumption from the mining industry puts strain upon other users and also on ecological services, particularly since much of the water is sourced from 'fossil ground water' with an extremely low recharge rate (Foster and Louks, 2006). Indeed, the competition between the mining industry and local communities and industries has been covered in the international media (Jarroud, 2013; Moskvitch, 2012; Onstad and O'Brien, 2015). Despite the mining companies operating, in most cases, in accordance with government legislation, the negative coverage can be highly damaging to large companies particularly when the opinion of local and national populations transition to a more anti-mining stance. Aside from the negative publicity, mining operations are also at risk of even lower levels of freshwater availability which in periods of drought can greatly impact the productivity of operations, this has already been observed in the case of the Los Bronces mine during a recent drought period (Onstad and O'Brien, 2015).

Regulations and Company Culture

Inadequate regulations and illegal mining in Colombia

In 2011, the National Mining Agency and the Colombian Geological Society were created alongside the Deputy Minister of Mines to oversee and regulate activities, superseding the Mines and Energy Ministry (USGS, 2015). With respect to environmental management, at the commencement of a project, mining companies are required to submit environmental impact assessments (EIAs) of their proposed projects which has been part of Colombian legislation since 1993 (Toro et al., 2010). The EIA must contain identification of environmental impacts, an environmental management plan, zoning of environmental management measures, a monitoring program, a contingency plan and an abandonment and final restoration plan (M.e.r., 2016). Public consultation with indigenous communities is also mandatory if the project is expected to affect their territory.

One major criticism of the environmental management of the mining industry in Colombia is that no EIA or environmental study is required for the exploration stage and furthermore, disclosure to local communities is not required (Alvarez, 2013). Local communities are often only aware of the project at a later date which makes it difficult for communities to appeal against the project leading to conflict which in some cases can end in violent incidents (Alvarez, 2013). In a study of the effectiveness of EIA in Colombia, Toro et al. (2010) used a review model which evaluated the EIA process in Colombia using 16 different criteria. The fundamental criticisms in the study were:

- The lack of competent professionals working in administration of environmental licenses and mining being exempt from EIA requirements before 2010 (the year the analysis was published)
- The absence of methodological guidelines for identification and evaluation of significant impacts, and the absence of a standard for weighing environmental factors
- Strategic environmental assessments being considered optional
- Public participation is only mandatory for indigenous communities and is therefore discriminatory against the general public
- No law ensuring projects are monitored for adherence to the EIA
- Incentives for conducting an EIA are lacking and rehabilitation bonds are not enforced meaning that there is no guarantee of environmental management being applied

Suggested potential improvements are:

- An increase in specific guidelines for the EIA process, ensuring that those conducting EIAs have the necessary knowledge and skills via registration and development of a generic methodology
- Analysis of alternative scenarios necessary for all projects
- Ensuring public participation is increased through open consultation groups and publication of plans in local media
- Making an environmental insurance policy available for projects to ensure that environmental commitments are kept and incentives (economic benefits) are available for conducting EIAs.

In Colombia many of the environmental regulations require improvement, yet as much of the gold mining industry is illegal, these regulations are still not being adhered to and therefore, a first priority is to engage with and legalize those mining illegally. According to the last mining census in 2010, 63% of mining operations were operating illegally, the value was 86% for gold mining operations (Rettberg and Ortiz Riomalo, 2016; Bustamante et al., 2016). Indeed, it has been well documented that recently, many of the illegal organizations cultivating and trafficking drugs in Colombia have switched to the illegal mining of gold as a result of a government crackdown on drug production and high gold prices (Rettberg and Ortiz Riomalo, 2016). This increase in illegal mining has consequently led to a growth in violence within gold producing municipalities, despite an overall reduction in homicide rates countrywide. To address the issue of violence, human rights abuses and mercury contamination, it is necessary that artisanal miners obtain licences and are regulated to ensure safe and environmentally benign operating practices. Nevertheless, to legalize mining in areas that are extremely remote, influenced by illegal organizations and suspicious of government intervention poses considerable difficulties. Illegal miners do not adhere to environmental regulations for a number of reasons: they are not well regulated locally, they are uninformed about the consequences of their actions, they do not have adequate capital to invest in alternative processing methods and in many cases do not have the knowledge to operate alternative processes (Güiza Suárez and Aristizabal, 2013). So far government interventions in illegal gold mining in Colombia have met with little success. Persecution of illegal miners has led to many mines being closed and 3,400 artisanal miners being incarcerated during the Santos presidency (Güiza Suárez and Aristizabal, 2013). This method of dealing with illegal mining can have profoundly negative impacts upon communities and families that are dependent upon small scale mining. It is more important, therefore, that the government engages positively with those involved in artisanal mining to support local communities in a manner which ensures efficient and environmentally benign operations alongside the eradication of human rights abuses. Providing the government or local authority can gain the trust of the miners, it may be possible to increase the number of mining licences and provide guidance on methods to mine in a more efficient manner with reduced impacts. Furthermore, the government can provide financial support for miners to implement new techniques to avoid the use of mercury in processing which will be discussed later in the chapter. The benefit to the government will be increased tax revenues from the newly legal miners, reduced environmental damage and an improved relationship with local communities. On the part of the government, this strategy will require considerable investment, in resources to engage with mining communities, development of guidelines for miners, investment in communities to purchase technologies to process gold without mercury and personnel to ensure adherence and provide constant communication between the authorities and communities. Nevertheless, it is likely that such investment could provide a long term economic and environmental return.

A laissez-faire attitude to environmental regulations in Peru

In Peru, mining is regulated by the Ministry of Energy and Mines, the Geological Mining and Metallurgical Institute, and the National Authority of Environmental Certification. For a mining project to be given approval it is necessary that the operating company produces an environmental impact assessment, a step introduced in 1990 as a condition of international financial institutions (Li, 2009). EIA procedures were only approved legally in 2009, mining companies must now, therefore, adhere to certain procedures for conducting an EIA. In brief, the environmental impact assessment must contain a summary of the project, proposed actions, a baseline study (current environmental conditions), identification of environmental impacts for the whole duration of the project, environmental management, contingency, closure, public participation, monitoring and control plans, project scheduling and budget. Throughout the process, public hearings are also mandatory and for affected indigenous people, they have the right to prior consultation. The purpose of the EIA is that projects that are deemed to have an overly negative impact upon the environment or local communities are unable to proceed or are required to adapt their project plans. Nevertheless, there has been considerable criticism of their implementation in project development in Peru. One of the major criticisms is the difficulty in disputing the approval of EIAs and the leniency of the regulating agency. Indeed, up until 2009, according to Li (2009) only one project had been halted as a result of public opposition, however, even in this case the reason was reported as being a financial not an environmental issue by the Ministry of Energy and Mines.

One common theme in problematic mining projects in Peru is inadequate EIAs which requires improvement both by companies undertaking the studies and by the regulating government agency. The current attitude to the EIA is that it is simply a necessary hurdle with the acceptance of a project a for gone conclusion. It is therefore important that EIAs are scrutinized more closely by the regulating agency to ensure that the depth and quality of the study is acceptable and if lacking, revisions are demanded. With a higher level of scrutiny and the possibility of projects to be put on hold for inadequate studies, the industry should react by conducting higher quality studies that in the long term will ensure improved project outcomes. A criticism of the EIA format in Peru is that project designs do not require adjustment following community complaints (Jaskoski, 2014). To avoid situations in which conflict and violence erupts between communities and companies it is necessary for the interests and concerns of the local communities to be of primary importance in the planning of a project and also in development of the EIA.

Progressing beyond the national standards in Chile

Many of the socio-environmental issues surrounding water management and mining in Chile are related to water scarcity and the overuse of resources by the industry. Much of the issue dates back to the creation of Chile's 1981 Water Code in which water rights can be freely bought and sold on the open market (Rivera et al., 2016). The concept of the code was that water would be used for the most economically productive industries as these would be the industries with sufficient capital to invest in the rights. Naturally, this has driven rights away from agriculture and small communities, and concentrated rights with large mining companies. There is little protection of human rights regarding water availability in Chile, particularly with respect to international standards (Cavallo, 2013). Indeed, the Chilean constitution is at odds with several international obligations with respect to the protection of indigenous people, particularly regarding water. Unsurprisingly, the use of freshwater within areas inhabited by indigenous communities has caused some friction between the industry and the communities, as often, indigenous communities do not recognize ownership of water resources, considering it to belong to all and in some cases to have a spiritual value (Camacho, 2012; Prieto, 2015; SDSG, 2010; IIED, 2002). There are various interesting case studies in Chile demonstrating strained community-company relations. One such case is the mining operation of Chuquicamata and the local population of Chiu Chiu in the region of Antofagasta (Camacho, 2012). The commodification of the water in the area dates back to 1983 in which informal management systems were transformed into a formal model of private ownership restricting access to measureable flow rates (Prieto, 2015). Given that mining provides a far higher economic return than the agricultural practices of the local community, the water was redistributed to local mining operations via the water market (Prieto, 2015). Since the redistribution of the water rights, the community has reported a considerable change in the local water resources with wetlands turning to areas of dust. This change has led to a decline in local agriculture, the herding of livestock and crop cultivation (Prieto, 2015). According to Camacho (2012) the leaders of the Chiu Chiu community feel aggrieved at their water allocation, yet as the law stands there is little the community can do to access increased water rights. A larger and more high profile case study in Chile related to water management and the industry attitude to environmental regulations is that of the Pascua Lama project which is explained below.

Case study 2.2. The Pascua Lama Project

The Pascua Lama (Fig. 2.5) project is a gold mining operation located high in the Andes mountains above the Huasco valley in the Atacama region. The project is located between Chile and Argentina, and according to BarrickGold, the operating company, the reserves of gold amount to around 18 Moz and 718 Moz of silver, with the mine expected to have a life span of over 25 years (Barrick, 2009). The operating company was granted rights to survey the mine location in 2001 and preparations were made to begin construction in 2009. In 2009, the estimated project budget was US$ 3 billion (Urkidi, 2010). The project has since been plagued by environmental violations and unsatisfactory environmental impact assessments (Urkidi, 2010). Many of the issues have been related to the impact upon three glaciers (Toro 1, Toro 2 and Esperanza) that feed the Huasco valley with snowmelt. The population of the Huasco province is 66,491, with many of those living in the valley being directly dependent upon agriculture which is fed by snowmelt from the Andean glaciers likely to be affected by the project (Urkidi, 2010). Issues arose between the mining company and the local population when the impact upon the glaciers was not mentioned in the first environmental impact study. After consideration the operating company developed a plan to mitigate impacts by relocating the affected glaciers, this was however, eventually rejected by the National Environmental Commission (CONAMA). In 2006, modifications to the project were finally approved with the condition that the project could not cause the glaciers to retreat, be displaced, undergo destruction or physical intervention (Cavallo, 2013). In 2009, it was discovered by the Operational Control Committee that compliance was breached, water had been extracted from a non-authorized point and dust from operations was leading to enhanced melting of the glaciers. In 2012, the Latin American Water Tribunal submitted a verdict on the case of the Pascua Lama project and water damages (Cavallo, 2013). The States of Chile and Argentina were prompted to declare a moratorium on the project. The project was temporarily put on hold by the Court of Appeal in Copiapo in 2013 as a result of the environmental destruction, the project remains in a state of cessation. The Pascua Lama project is a clear example where both state regulations and company engagement has failed to address the needs of the local community which has resulted in economic loss, unnecessary environmental damage and generation of mistrust in the community and the country.

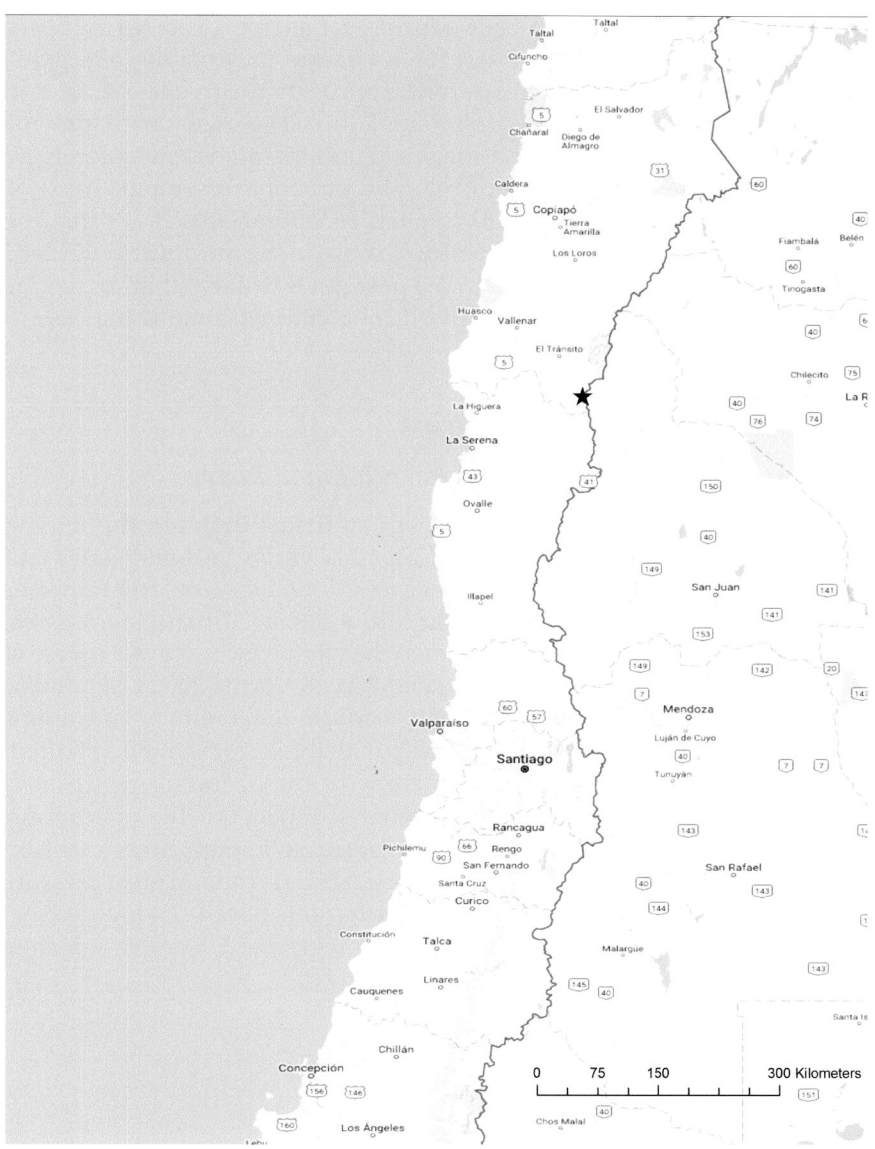

Fig. 2.5 Location of the Pascua Lama mining project, Map layer: Google.

For new projects, the case study of Pascua Lama provides a cautionary example of the need for a change in attitude, alongside a thorough and highly transparent approach to investigating and communicating the likely environmental impacts of new developments whilst considering both international and national standards. Alongside project proposals, by law, environmental impact assessments must be conducted to identify potential impacts of operations. It is often the case, however, that companies perform these to the minimum standard which may have been accepted in the past but now face a higher degree of scrutiny from NGOs and the public as demonstrated in the case of Pascua Lama.

Community Engagement and Trust

Endemic mistrust in the Colombian mining industry

The level and positivity of engagement and trust between the mining industry and local communities in Colombia varies between states and communities, on the whole however, relations between communities and industry can be considered poor (ABColombia, 2015). In many cases, communities are highly dependent upon the mining industry as a source of employment, for example, it is estimated that around 314,000 Colombians work directly in artisanal mining and medium scale mining operations (ABColombia, 2015). Yet, the relationship with the sector is often a strained one due to the violence inflicted by many of the operating groups and the considerable social and environmental impacts of the operations. Additionally, the government's push to increase foreign investment by granting mining rights to multinational companies is putting great pressure on communities living in the areas in which rights have been granted. An interesting case study is that of the Chocó region of Colombia which is explained below.

Case study 2.3. The Region of Chocó

Chocó is a remote region on the west coast of Colombia (Fig. 2.6), it has a low population density and is extremely biodiverse, the region contains about 65% of all Colombian bird species (ABColombia, 2015). It is estimated that 95% of the population are indigenous or Afro-descendant. Chocó is rich in natural resources, the exploitation of minerals began in the 1980s as miners arrived using rudimentary extraction techniques to recover gold and platinum. The industry quickly became an important source of income for illegal organizations, paramilitary and guerrilla groups (Güiza Suárez and Aristizabal, 2013). The arrival of these groups severely limited the ability of the local communities to continue with traditional mining techniques and much of the population became economically dependent upon working in mining camps but continue to face high levels of human rights abuses within the camps. Furthermore, with regards to multinational corporations (MNCs), it has been estimated that around 41% of the Chocó region is subject to mining concessions, 75% of which have been granted to MNCs. The potential for conflict is great, as 96% of the region belongs to Afro-descendants or indigenous peoples (ABColombia, 2015). Indeed, there have been several recent protests and court cases against the mining concessions (Acosta, 2014). There is, however, very little protection for the indigenous and Afro-descendant communities in Colombia, particularly as the government is promoting foreign direct investment and the auctioning of large areas that have been identified as mineral rich. As a result of the weak government support and legislation, mining companies (both at small and large scales) have a poor record with regards to community engagement and protection of their human rights. There is considerable evidence from Chocó that forced displacement is common, as well as physical violence and intimidation. It has also been recorded that often those who oppose the industry can face death threats and the possibility of being murdered. Communities are hesitant to protest against the industry due to the indifference displayed by the state and the potential for violence inflicted by security forces that have in the past killed protestors. Aside from the more violent aspects of the industry, the environmental negligence of companies has greatly affected the environment that many communities are dependent upon for their source of food. High mercury levels in water and fish has led to considerable damage to the health of residents in certain cases in Chocó (ABColombia, 2015).

Fig. 2.6 Location of the Chocó region, Map layer: Google.

The extremely poor community engagement and human rights issues in Colombia are mainly a result of the lack of enforcement from state regulators, regulation on such matters is weak and enforcement is limited. It is therefore a government responsibility to first of all ensure that mining operations are operating without violation of human rights which will require a robust overhaul of the current industry in which operators must obtain licences and be regularly audited. The required investment in such a scheme will undoubtedly be great but necessary for the protection of citizen's human rights. To address the social and environmental problems, outside pressure from non-governmental organizations and mineral buyers can also have considerable impact. Customers of mined products have the ability to demand products that are mined without the violation of human rights and with minimal environmental impacts. Those managing the operations will therefore be pressured into complying with demands to continue operations.

Building trust and cooperation in Peru

In Peru, mining companies have a mixed reputation among the population, the communities in close proximity to operations, however, tend to have a particularly negative opinion (Paredes, 2016; Armstrong et al., 2014). The manner in which mining companies engage with local communities is often criticized as a result of the lack of understanding and respect shown to communities, particularly indigenous communities. Water management is one area that is particularly important for all stakeholders and is an area which strong co-operation and agreement is necessary if the industry is to mature. In Peru, it has been reported that there is often a considerable lack of trust between communities and mining companies as a result of the manner in which companies interact with them. Armstrong et al. (2014) looked at the responses of local communities to two mining projects in Peru. In both cases, the communities made clear that water resources were the most pertinent issue due to its importance for agriculture and everyday living. Communities expressed concerns about overuse of water from mining and the possibility of contamination of water supplies. The researchers recorded residents explaining that they blame contaminated water for health issues and instead they buy water from the local town. Many of the residents felt that the mining operations were not being transparent about the impacts the operations are having upon local water supplies. The communities also showed high levels of scepticism to the company's approaches to community engagement. It is generally perceived that companies do not take engagement seriously and rarely go further than the law requires. In the records of Armstrong et al. (2014) it is regularly mentioned that companies provide insufficient time for communities to understand the project, reports are criticized as being oversized and overly technical, the fact that many communities speak Quechua is overlooked and many important impacts are simply not included or poorly explained.

In the case of Peru, it is likely that many of these aforementioned situations could be largely avoided through a change in the manner of community engagement by the industry. It is important primarily that companies approach local communities at the earliest possible stage of a project to detail proposals and ask for feedback with any doubts regarding the project. During this first phase, communities should be approached with respect regarding their local values, knowledge and customs (Armstrong et al., 2014). Companies must take measures to understand what is important to the community and if communities feel that an area of importance is under threat, the company should devise a suitable solution in conjunction with the community. It should also be a responsibility of the government to provide information regarding the social and cultural practices of local communities, with this information, companies can therefore conduct prior research to understand what concessions will be most effective. Often companies make decisions based on their own values of what communities require such as the development of supermarkets and modern houses but fail to understand what is considered valuable by the community. Furthermore, such offerings are often treated as simple bribes by communities, which can cause division with some residents willing to accept such gifts while others refuse. Early dialogue is therefore important to understand how the community can live peacefully alongside the operation and in which ways the company can improve local living standards without condescendingly simply offering money or gifts based on western style values. In some cases, local residents are grateful for infrastructure developments that are made by the companies but often wish to have more input with respect to developments by the company.

Many of the complaints in Peru are also concerned with the lack of transparency of the mining companies, particularly with regards to water use and water contamination. It is essential for developing good relations that companies are upfront about the water requirement of a mine site and any potential impacts related to discharge. The company can then suggest potential solutions which can be assessed by the community. In many of the interviews of community members (Armstrong et al., 2014), the members demonstrate a desire to understand more about the projects and appear willing to embrace the idea provided their views are respected and the necessary adjustments are implemented. Given that water is such a widespread and important concern for almost all communities in the Andes, it is important that companies detail where water will be extracted from, the flow that extracted, how much will be left for the environment and local consumption and what strategies the company is planning to undertake to minimize the impact to the community. With respect to water contamination, the company must be extremely clear about what contamination is possible, the likelihood of such events and the potential impacts. They must provide a robust plan to minimize contamination and emergency plans to deal with any unplanned releases.

In Peru, the mining industry, communities and the country as a whole have suffered as a result of poor environmental regulation and the indifference of mining companies. With a change in attitude, a strengthening of regulation and improved engagement with local communities, the industry can become more productive, profitable and impacts upon the population can be reduced with community-industry relationships becoming more positive and progressive.

Developing a gold standard in community relations in Chile

As mentioned earlier, despite community relation issues being smaller scale and less violent than other Latin American countries, poor community relations remain a problem for the sector in general. In Chile, due to the increasing power of smaller communities resulting from the rise of social media and interest from the population as a whole, mining companies are under considerable pressure to minimize their impacts upon local communities and indeed work proactively alongside the communities. It is extremely important that companies in Chile develop a more positive and deeper attitude towards their community relationships as the opinions of local communities will continue to strengthen and impact project outcomes.

Many of the solutions that are necessary for Chile are those that have been discussed with respect to Peru, the industry in Chile is, however, in a position to go further and potentially develop into a gold standard for the industry. It is important for the industry to ensure that the basic engagement is conducted well, such as providing transparency regarding impacts of operations and an outlet for continual dialogue but also to achieve more by enhancing local areas, creating high skill jobs and developing educational programs with a longer term vision.

In the mining industry in general, it is unusual for skilled employment to be sourced from local communities, as in many cases, the local skill levels tend to be low. An important step for the industry would be to increase the training of the local population which would allow them over time to work in high skilled positions within the project and transfer those skills elsewhere following the project closure. It is an area that has been discussed in the Roundtable Debates of the 2016 Mining Report published by Fundación Chile (Fundacion Chile, 2017) by the industry's business leaders. It is not solely the responsibility of the industry but should be developed in conjunction with the Ministry of Education to ensure young people have the capacity to further develop their skills to work in the industry, particularly from remote communities. Providing young people have the necessary basic skills, they can then be enrolled within a company training program to further develop. Alongside the need to develop educational programs, in the Round Table report, the importance of including communities in the project supply chain was mentioned, this would ensure that local businesses are used, which include any service from catering to maintenance of machinery and construction.

Another area for the industry to focus is local infrastructure, a contentious issue with local communities is the development of mining towns with poor infrastructure, a 'fly in, fly out' workforce and many social issues. The industry can improve such scenarios through investment in leisure facilities available to workers and the local community, these could be parks, swimming pools and golf courses, etc. The outcome would be an area more suitable to develop tourism, attract mine workers to live more permanently there, both of which could also benefit the local economy.

It is important in the case of Chile that mining companies go beyond what is currently expected of them, once experience is gained on how to develop strong relationships, similar techniques can be used for new projects. Fostering a good reputation will greatly reduce the possibility of negative project outcomes which has caused considerable damage in certain cases in Chile. With an improved track record, communities will likely be more receptive to new projects.

Technological Solutions

Implementing basic technological improvement in Colombia

In Colombia, alongside working with mining communities to legalize and regulate operations, the implementation of technical solutions to reduce environmental impacts is of prime importance. As mentioned, the greatest environmental issue in artisanal mining is the use of mercury in gold extraction. Mercury is used because it is a cheap material, it is easily available, simple to use and suits small scale operations (Saldarriaga-Isaza et al., 2015). Aside from the environmental issues, however, the use of mercury is an extremely inefficient method to recover gold with an efficiency rate of about 10%. The reasons for this are the poor equipment used to mix the ore and mercury, and the restriction in gold particle size which must be around 0.07 to 1.5 mm (Hylander et al., 2007). It is necessary that the use of mercury in gold mining is eventually phased out due to its highly destructive nature. Small steps are first necessary, however, to improve the current technologies. Increasing the safety of use and reducing environmental impacts must be the first priority before complete replacement of mercury based extraction. One method for reduced use of mercury is through the implementation of ore concentration to maximize gold content, thus minimizing material inputs. Gravity separators or centrifuges can be used for this purpose (UNEP, 2011). Furthermore, the size of the particles is extremely important, high efficiency milling should be incorporated along with sieving to produce a consistent feed. Advanced techniques such as shaking tables and flotation cells can also be used to increase the capture of gold particles from the ore (UNEP, 2011).

With respect to vaporization, emissions can be greatly reduced through the use of closed circuits (retorts) which can reduce mercury emissions by 75 to 95% over open burning or through the use of fume hoods with water boxes to trap the mercury which are capable of capturing about 80% of emissions (UNEP, 2011). The United Nations Industrial Development Organisation (UNIDO) tested various techniques to reduce mercury emissions in Segovia, Colombia (Cordy et al., 2013). The techniques were: education regarding the economic benefits of reducing mercury use, installing vapor filters and retorts, supporting the conversion to cyanidation tanks and improving the working environment. They found that those processing centers that participated in the project reduced mercury use by half, many of the new processing operations were operating mercury-free and mercury vapor concentrations in the town were reduced by almost 50% despite an increase in gold production. To completely phase out mercury use, replacement techniques such as direct smelting and cyanidation can be implemented. Direct smelting is highly energy intensive and can only be conducted on a very small scale, cyanidation is therefore a more practical option that is now fairly common in larger operations, although cyanide is also highly toxic and requires careful management. Cyanidation allows the leaching of gold from the ore followed by absorption with carbon and the eventual recovery via electrowinning. The process offers a highly efficient and cheap method to recover gold using a degradable compound that can be oxidized, after its use, in reaction tanks (Botz et al., 2016). Nevertheless, if gold recovery with cyanide is to be promoted, robust guidelines must be made and enforced with regards to the wastewater treatment to avoid damaging discharges. Furthermore, the reprocessing of mine tailings previously processed with mercury must be avoided due to the extremely potent environmental impacts (UNEP, 2011). To start to reduce the use of mercury and eventually replace it, the Colombian government must address the need to engage with illegal miners and allow them to obtain licences and therefore be regulated. The government can then provide the support and assistance that the miners require to operate in a more environmentally sustainable manner by providing financial incentives and organizing the sharing of expensive equipment.

Reducing and replacing freshwater use in Chile and Peru

Water management issues in both Chile and Peru are often related to lack of availability, here the possibility of reducing and replacing freshwater use for both countries will be considered. To increase water use efficiency it is necessary to consider where in the operation the greatest losses occur. In a copper mining water balance study, Gunson et al. (2012) calculated the greatest losses (89%) in a copper sulfide processing operation to be in the tailings storage facility due to the entrainment, seepage and evaporation of the embodied water. The second greatest source of water loss was determined to be water sprayed on roads for dust suppression at 9%. The authors considered the implementation of various strategies to reduce losses such as synthetic road dust suppressants, tailings thickening, evaporation covers and ore pre-concentration among others. A combination of solutions was implemented to the model and a maximum water use reduction of 74% was calculated. Pre-sorting and the filtration of tailings provided 88.4% of the savings and dust suppression accounted for 9.8%. There is a lack of literature studying the economics of water consumption solutions, Aitken et al. (2016) calculated the most cost effective solution to be road dust suppression due to the low costs relative to the high savings potential. With respect to dewatering of the tailings, the use of primary sedimentation with the inclusion of flocculation was determined to be the most cost effective option, however the greatest water savings were calculated to be from filtration of the tailings.

A further option to combat issues associated with freshwater use is the use of seawater within, a solution currently being pioneered by several operations in Chile (COCHILCO, 2008). Uptake of seawater remains relatively low, however, due to impacts upon copper recovery, potential damage to equipment and the high cost to transport and treat the water. There are examples of mining operations in Chile which are currently using raw seawater in operations, the Michilla and Esperanza operations are such examples. Construction of Esperanza was completed in 2010, the seawater is pre-treated with electrochlorination and filtration with the addition of anti-corrosion chemical. Following use in flotation, the water is recycled to be reused in the process with the addition of lime to control the pH. The mine also operates a small (2,400 m^3/day) desalination plant to provide desalinated water for domestic use and final washes. In the Michilla operation, the operators have incorporated the use of seawater for leaching of ore and agglomeration. The seawater undergoes sand filtration prior to use. The high chlorine content in the leaching process resulting from the seawater requires demineralised water to be added.

The alternative to using raw seawater is the implementation of desalination allowing the water to be used as effectively as freshwater. Despite the high costs of the desalination, its use is becoming more commonplace as downstream processes do not require modification and it can completely supplant the need for freshwater. The large Escondida copper mine for example has used desalinized seawater for its sulfur bioleaching plant since 2006 (COCHILCO, 2008). The reverse osmosis plant produces 535 litres/s which is mainly consumed in the concentrator plant 3,160 meters above sea level. The water is transported using an aqueduct along a distance of 176 km. The seawater is pre-treated with sulfuric acid to allow flocculation of organic matter via flotation. Reverse osmosis then removes the salts which end up in the brine which is returned to the sea after running through a turbine to generate electricity. The operators of Escondida mine are currently developing a considerably larger desalination plant and conveyance system to provide the mine's complete water use through desalinised water. The final production rate will be 2.5 m^3/s, the project is expected to be completed in 2017 at a cost of around US$ 2 bn (BHP Billiton, 2013).

The majority of copper mines in northern regions of Chile and the south of Peru are located inland from the coastline in the Andes mountain range at a high altitude requiring large conveyance distances and high pumping costs. Despite these cost barriers to implementation, the uptake of systems to use and treat seawater is expected to increase greatly as a result of the pressure on companies to reduce their use of freshwater. In Chile, for example, the total water consumption in the Chilean mining industry is expected to increase from 14.7 m^3/s in 2015 to 24.6 m^3/s in 2025, with the majority of the increase to be made up by seawater (COCHILCO, 2014). Indeed, there is currently a bill being debated in the Chilean Government to ensure that all mines with a water consumption above 150 L/s substitute freshwater with seawater (Espinosa Monardes, 2013).

A combination of greater water use efficiency and seawater use is most likely the optimal option for the future of mining in Chile and Peru. The replacement of freshwater with seawater would obviously greatly reduce the impacts of freshwater use but the economics will require careful analysis as well as the impacts generated by using seawater such as the greater energy requirements and the direct impacts of extraction and waste generation. If these mining regions can successfully implement new technologies to largely replace freshwater economically and with minimal impact, social and environmental issues could be greatly reduced.

Conclusions

The mining industry will continue to be of primary economic importance to many Latin American countries and for Chile, Peru and Colombia, the progression of socio-environmental management is essential. In Chile and Peru, the socio-environmental issues revolve mainly around low water availability and conflicts between large mining operations and local communities. In Chile, despite robust environmental standards, there remain cases of friction between communities and operations where the communities feel disregarded by the mining companies. The industry must ensure that their relations with communities improve through collaborating honestly, with transparency and with long-term community development. The industry must take measures to reduce water consumption, a particular area of friction, which can be achieved by greater water use efficiency or replacement with seawater. The industry in Peru faces many of the same issues as in Chile but with more severity with respect to community relations. Environmental regulations remain lenient towards industry and require strengthening to protect the environment and the rights of many of the local communities affected by large mining operations. Similar to Chile, the industry must take a considerably more respectful approach to working with local communities. It is necessary that companies understand the issues considered important by those affected by their operations and are willing to work collectively to find solutions. In Colombia, the mining industry is gaining importance nationally, particularly for gold extraction. The majority of the industry, however, operates illegally causing widespread environmental damage and serious impacts to human health. The first key move for the government is to engage with those operating illegally and to allow transition to a legal and regulated framework. Additionally, support needs to be provided to encourage the implementation of techniques to first reduce the consumption of mercury and secondly to phase out its use. Educational support should be provided to local mining communities alongside financial assistance to provide new technologies, which can greatly minimize the current socio-environmental impacts. It is imperative that the mining industry and governments of each country give these issues the attention and resources required to implement solutions that will deliver the necessary change. Chile has the opportunity to develop an advanced industry that can become a global leader in environmental management giving the industry yet more economic importance. In Peru, if the industry can improve its engagement with local communities and provide more local employment opportunities whilst working to improve water use and contamination issues with regulators, the industry would have the potential to expand more sustainable, bringing high revenues and local benefits.

References

ABColombia. 2015. Fuelling Conflict in Colombia: The Impact of Gold Mining in Choco. London, UK. ABColombia. Available at: http://www.abcolombia.org.uk/downloads/ABC-Choco_mining_report_V7_Screen.pdf. (Accessed: 20/04/2017)

Acosta, Luis Jaime. 2014. Colombian Court Orders Mining Companies to Return Land to Natives. Reuters. Available at: http://www.reuters.com/article/colombia-mining-idUSL2N0RQ30V20140926. (Accessed: 15/10/2017)

Aitken, Douglas, Diego Rivera and Alex Godoy Faúndez. 2016. Cost-effectiveness of strategies to reduce water consumption in the copper mining industry. *In*: Gecamin. Water in Mining 2016: Proceedings of the 5th International Congress on Water Management in Mining, 18–20 May 2016, Santiago, Chile. Santiago, Gecamin.

Aitken, Douglas, Diego Rivera, Alex Godoy-Faúndez and Eduardo Holzapfel. 2016. Water scarcity and the impact of the mining and agricultural sectors in chile. Sustainability 8(2): 128. Doi: 10.3390/su8020128.

Albers, C. 2012. Coerturas SIG Para La Enseñaza e La Geografie En Chile. Temuco: Universidad de La Frontera. www.rulamahue.cl/mapoteca. (Accessed: 12/04/2017)

Alvarez, Juan Diego. 2013. Big industry small rules: The mining industry in Colombia. Sortuz: Oñati Journal of Emergent Socio-Legal Studies 5(1).

Arena Minerals. 2015. Regional Exploration in Chile. January. Available at: http://www.arenaminerals.com/_resources/presentations/Arena_Pres_January_2015.pdf. (Accessed: 09/07/2017)

Armstrong, Rita, Caroline Baillie, Andy Fourie and Glevys Rondon. 2014. Mining and Community Engagement in Peru: Communities Telling Their Stories to Inform Future Practice. International Mining for Development Centre. Available at: https://im4dc.org/wp-content/uploads/2014/09/Mining-and-community-engagement-in-Peru-Complete-Report.pdf. (Accessed: 23/05/2017)

Azevedo, Raquel and Eleazar Rodriguez. 2012. Phytotoxicity of mercury in plants: A review. Journal of Botany 2012: 1–6. Doi: 10.1155/2012/848614.

Barrick. 2009. Building Pascua-Lama. Available at: http://www.barrick.com/files/doc_presentations/2009/05.07.2009%20-%20Pascua%20Lama%20Go%20Ahead%20Webcast.pdf. (Accessed: 10/04/2017)

Bebbington, A.J. and Bury, J.T. 2009. Institutional challenges for mining and sustainability in Peru. Proceedings of the National Academy of Sciences 106(41): 17296–301. Doi: 10.1073/pnas.0906057106.

BHP Billiton. 2013. BHP Billiton Approces Escondida Water Supply Project. Available at: http://www.bhpbilliton.com/investors/news/BHP-Billiton-Approves-Escondida-Water-Supply-Project. (Accessed: 20/05/2017)

Botz, M.M., Mudder, T.I. and Akcil, A.U. 2016. Cyanide Treatment. Gold Ore Processing, 619–45. Elsevier. Doi: 10.1016/B978-0-444-63658-4.00035-9.

Budds, Jessica and Leonith Hinojosa. 2012. Restructuring and rescaling water governance in mining contexts: The co-production of waterscapes in Peru. Water Alternatives 5(1): 119–137.

Burton, Ian, Bo Lim, Erika Spanger-Siegfried, Elizabeth L. Malone and Saleemul Huq. 2005. Adaptation Policy Frameworks for Climate Change: Developing Strategies, Policies, and Measures. Cambridge, UK; New York: Cambridge University Press.

Bustamante, Natalia, University of Queensland, Neil McIntyre, University of Queensland, Juan Carlos Díaz-Martínez, Universidad Nacional de Colombia, Oscar Jaime Restrepo-Baena and Universidad Nacional de Colombia. 2016. Review of Improving the Water Management for the Informal Gold Mining in Colombia. Revista Facultad de Ingeniería Universidad de Antioquia, no. 79. Doi: 10.17533/udea.redin.n79a16.

Camacho, Francisco Molina. 2012. Competing rationalities in water conflict: Mining and the indigenous community in Chiu Chiu, El Loa Province, Northern Chile: Competing rationalities in water conflict, Chile. Singapore Journal of Tropical Geography 33(1): 93–107. Doi: 10.1111/j.1467-9493.2012.00451.x.

Castellanos, Angélica, Pablo Chaparro-Narváez, Cristhian David Morales-Plaza, Alberto Alzate, Julio Padilla, Myriam Arévalo and Sócrates Herrera. 2016. Malaria in gold-mining areas in Colombia. Memórias Do Instituto Oswaldo Cruz 111(1): 59–66. Doi: 10.1590/0074-02760150382.

Cavallo, Aguilar Cavallo. 2013. Pascua lama, human rights, and indigenous peoples. Goettingen Journal of International Law 5(1): 215–249. Doi: 10.3249/1868-1581-5-1-cavallo.

COCHILCO. 2008. Best Practices and Efficient Use of Water in the Mining Industry. Comision Chilena del Cobre. Santiago, Chile. Available at: http://www.cochilco.cl/descargas/english/research/research/best_practices_and_the_efficient _use_of_water.pdf. (Accessed: 20/05/2017)

COCHILCO. 2014. Proyección de Consumo de Agua En La Minería Del Cobre 2014–2025. Comision Chilena del Cobre. Santiago, Chile. Available at: http://www.mch.cl/wp-content/uploads/sites/4/2015/01/Informe-Proyeccion-consumo-de-agua.pdf. (Accessed: 20/05/2017)

Cordy, Paul, Marcello M. Veiga, Ibrahim Salih, Sari Al-Saadi, Stephanie Console, Oseas Garcia, Luis Alberto Mesa, Patricio C. Velásquez-López and Monika Roeser. 2011. Mercury contamination from artisanal gold mining in Antioquia, Colombia: The World's highest per capita mercury pollution. Science of The Total Environment 410-411: 154–60. Doi: 10.1016/j.scitotenv.2011.09.006.

Cordy, Paul, Marcello Veiga, Ben Crawford, Oseas Garcia, Victor Gonzalez, Daniel Moraga, Monika Roeser and Dennis Wip. 2013. Characterization, mapping, and mitigation of mercury vapour emissions from artisanal mining gold shops. Environmental Research 125: 82–91. Doi: 10.1016/j.envres.2012.10.015.

Cordy, Paul, Marcello Veiga, Ludovic Bernaudat and Oseas Garcia. 2015. Successful airborne Mercury Reductions in Colombia. Journal of Cleaner Production 108: 992–1001. Doi: 10.1016/j.jclepro.2015.06.102.

Cunningham, Charles, Eduardo Zappettini, Waldo Vivallo, Carlos Mario Celada, Jorge Quispe, Donald Singer, Joseph Briskey, David Sutphin, 1 Mariano Gajardo, Alejandro Diaz, Carlos Portigliati, Vladimir Berger, Rodrigo Carrasco and Klaus Schulz. 2008. Quantitative Mineral Resource Assessment of Copper, Molybdenum, Gold, and Silver in Undiscovered Porphyry Copper Deposits in the Andes Mountains of South America: U.S. Geological Survey Open-File Report 2008–1253. United States Geological Survey. Reston, Virginia: United States.

Drasch, G., Böse-O'Reilly, S., Maydl, S. and Roider, G. 2002. Scientific comment on the German human biological monitoring values (HBM Values) for Mercury. International Journal of Hygiene and Environmental Health 205(6): 509–12. Doi: 10.1078/1438-4639-00178.

Eda, Laura E. Higa and Weiqi Chen. 2010. Integrated water resources management in Peru. Procedia Environmental Sciences 2: 340–48. Doi: 10.1016/j.proenv.2010.10.039.

Ernst and Young. 2016. Peru's Mining & Metals Investment Guide 2015/2016. Lima, Peru: Ernst and Young. Available at: http://www.rree.gob.pe/promocioneconomica/invierta/Documents/MiningGuide_2015_2016.pdf. (Accessed: 22/05/2017)

Espinosa Monardes, Marcos. 2013. Establece La Desalinización Del Agua de Mar Para Su Uso En Proceso Productivos Mineros Boletín N° 9185–08. Camara de Disputados. Available at: https://www.camara.cl/sala/verComunicacion.aspx?comuid=10478&formato=pdf. (Accessed: 22/05/2017)

Foster, Stephen and Daniel Louks. 2006. Non-Renewable Groundwater Resources: A Guidebook on Socially-Sustainable Management for Water-Policy Makers. Paris, France: UNESCO. Available at: http://unesdoc.unesco.org/images/0014/001469/146997E.pdf. (Accessed: 23/05/2017)

Fundación Chile. 2017. Chile 2016 Mining Report. Fundación Chile. Santiago, Chile. Available at: http://fch.cl/wp-content/uploads/2015/12/II_MiningReport_Chile2016_S.pdf. (Accessed: 23/05/2017)

Garcia, Oseas, Marcello Veiga, Paul Cordy, Osvaldo Suescun, Jorge Martin Molina and Monika Roesser. 2015. Artisanal gold mining in Antioquia, Colombia: A successful case of Mercury reduction. Journal of Cleaner Production 90: 244–52.

Güiza Suárez, Leonardo and Juan David Aristizabal. 2013. Mercury and gold mining in Colombia: A failed state. Universitas Scientiarum 18(1). Doi: 10.11144/Javeriana.SC18-1.mgmc.

Gunson, A.J., Klein, B., Veiga, M. and Dunbar, S. 2012. Reducing mine water requirements. Journal of Cleaner Production 21(1): 71–82. Doi: 10.1016/j.jclepro.2011.08.020.

Hill, David. 2015. What Is Peru's Biggest Environmental Conflict Right Now? Available at: https://www.theguardian.com/environment/andes-to-the-amazon/2015/jun/08/tia-maria-perus-biggest-environmental-conflict-right-now. (Accessed: 15/08/2016)

Hylander, Lars D., David Plath, Conrado R. Miranda, Sofie Lücke, Jenny Öhlander and Ana T.F. Rivera. 2007. Comparison of different gold recovery methods with regard to pollution control and efficiency. CLEAN—Soil, Air, Water 35(1): 52–61. Doi: 10.1002/clen.200600024.

IIED. (ed.). 2002. Breaking New Ground: Mining, Minerals, and Sustainable Development: The Report of the MMSD Project. Sterling, VA: Earthscan Publications.

Jarroud, Marianela. 2013. Mining and Logging Companies Leaving All of Chile without Water. The Guardian. Available at: http://www.theguardian.com/global-development/2013/apr/24/mining-logging-chile-without-water. (Accessed: 15/08/2016)

Jaskoski, Maiah. 2014. Environmental licensing and conflict in Peru's mining sector: A path-dependent analysis. World Development 64: 873–83. Doi: 10.1016/j.worlddev.2014.07.010.

Kozak, Robert. 2015. Southern copper temporarily suspends copper project amid protests. The Wall Street Journal. Available at: http://www.wsj.com/articles/major-peru-mine-sparks-nationwide-protests-1431726237. (Accessed: 20/02/2017)

Lavado Casimiro, Waldo Sven, Josyane Ronchail, David Labat, Jhan Carlo Espinoza and Jean Loup Guyot. 2012. Basin-scale analysis of rainfall and runoff in Peru (1969–2004): Pacific, Titicaca and amazonas drainages. Hydrological Sciences Journal 57(4): 625–42. Doi: 10.1080/02626667.2012.672985.

Maya Villazon, Edgardo. 2015. El 80% de la mineria de Colombia es ilegal segun la Contraloria. Available at: http://www.bluradio.com/108039/el-80-de-la-mineria-de-colombia-es-ilegal-segun-la-contraloria. (Accessed: 18/01/2017)

McKenzie, Victoria. 2014. Mercury Poisoning Adds to Humanitarian Crisis in West Colombia: Govt. Colombia Reports. Available at: http://colombiareports.com/colombians-forced-homes-lack-water/. (Accessed: 18/01/2017)

M.e.r. 2016. Countries and Profiles: Colombia. Utrecht, The Netherlands: Netherlands Commission for Environmental Assessment. Available at: http://www.eia.nl/en/countries/sa/colombia/eia. (Accessed: 18/01/2017)

Minvielle, Marie and René D. Garreaud. 2011. Projecting rainfall changes over the South American Altiplano. Journal of Climate 24(17): 4577–83. Doi: 10.1175/JCLI-D-11-00051.1.

Moffat, Kieren, Naomi Boughen, Airong Zhang, Justine Lacey, David Fleming and Kathleen Uribe. 2014. Chilean Attitudes towards Mining: Citizen Survey—2014 Results. Canberra, Australia: Commonwealth Scientific and Industrial Research Organisation. Available at: http://www.csiro.au/en/Research/Mining-manufacturing/CSIRO-Chile/Chilean-attitudes-to-mining. (Accessed:23/05/2017)

Monge, Carlos. 2016. Water Management, Environmental Impacts and Peru's Mining Conflicts. New York, US: National Resource Governance Institute. Available at: https://resourcegovernance.org/blog/peru%E2%80%99s-troubled-mining-sector-civil-unrest-copper-conflict-and-watersheds. (Accessed: 19/01/2017)

Moskvitch, Katia. 2012. War for Water in Chile's Atacama Desert: Vines or Mines? London, UK: BBC. Available at: http://www.bbc.com/news/business-17423097. (Accessed: 23/06/2016)

Onstad, Eric and Rosalba O'Brien. 2015. Drought in Chile Curbs Copper Production, to Trim Global Surplus. Reuters. Available at: http://uk.reuters.com/article/copper-drought-chile-idUKL5N0VY2X920150225. (Accessed: 23/06/2016)

Paredes, Maritza. 2016. The glocalization of mining conflict: Cases from Peru. The Extractive Industries and Society 3(4): 1046–1057. Doi: 10.1016/j.exis.2016.08.007.

Pinedo-Hernández, José, José Marrugo-Negrete and Sergio Díez. 2015. Speciation and bioavailability of mercury in sediments impacted by gold mining in Colombia. Chemosphere 119: 1289–95. Doi: 10.1016/j.chemosphere.2014.09.044.

Prieto, Manuel. 2015. Privatizing water in the Chilean Andes: The case of Las Vegas de Chiu-Chiu. Mountain Research and Development 35(3): 220–29. Doi: 10.1659/MRD-JOURNAL-D-14-00033.1.

PWC. 2013. Mining Industry: Doing Business in Peru. Lima, Peru: Pricewaterhouse Coopers. Available at: https://www.pwc.de/de/internationale-maerkte/assets/doing-business-in-mining-peru.pdf. (Accessed: 23/05/2017)

Quirland, Andres and Max Leclerc. 2013. Study of the Mining Sector and Business Opportunities for Swiss Companies. Zurich, Switzerland: OSEC, Switzerland Global Enterprise. Available at: http://web.swisschile.cl/wp-content/uploads/2015/05/ESTUDIO-MINERO-2013.pdf. (Accessed: 23/05/2017)

Rettberg, Angelika and Juan Felipe Ortiz Riomalo. 2016. Golden opportunity, or a new twist on the resource-conflict relationship: Links between the drug trade and illegal gold mining in Colombia. SSRN Electronic Journal. Doi: 10.2139/ssrn.2719686.

Rivera, Diego, Alex Godoy-Faúndez, Mario Lillo, Amaya Alvez, Verónica Delgado, Consuelo Gonzalo-Martín, Ernestina Menasalvas, Roberto Costumero and Ángel García-Pedrero. 2016. Legal disputes as a proxy for regional conflicts over water rights in Chile. Journal of Hydrology 535: 36–45. Doi: 10.1016/j.jhydrol.2016.01.057.

Rojas, Carolina. 2015. Mining Sector in Colombia—Moving Forward. Bogota, Colombia: National Mining Agency of Colombia. Accessed at: https://www.anm.gov.co/sites/default/files/DocumentosAnm/anm_ladu_final.pdf. (Accessed: 18/01/2017)

Saldarriaga-Isaza, Adrián, Clara Villegas-Palacio and Santiago Arango. 2015. Phasing out mercury through collective action in artisanal gold mining: Evidence from a framed field experiment. Ecological Economics 120: 406–15. Doi: 10.1016/j.ecolecon.2015.04.004.

Schipani, Andres. 2016. Mining Taps Deep Reserves of Rage in Peru. Financial Times. Available at: https://www.ft.com/content/758b7caa-00e0-11e6-ac98-3c15a1aa2e62. (Accessed: 19/01/2017)

SDSG. 2010. Report: Current Issues in the Chilean Mining Sector. Colorado, USA: Sustainable Development Strategies Group. Available at: http://www.sdsg.org/wp-content/uploads/2010/02/10-10-08-CHILE-REPORT.pdf. (Accessed: 23/05/2017)

Spillan, John, John Parnell and Cesar Antunez De Mayolo. 2011. Exploring crisis readiness in Peru. Journal of International Business and Economy 12(1).

Toro, Javier, Ignacio Requena and Montserrat Zamorano. 2010. Environmental impact assessment in Colombia: Critical analysis and proposals for improvement. Environmental Impact Assessment Review 30(4): 247–61. Doi: 10.1016/j.eiar.2009.09.001.

UNEP. 2011. Reducing Mercury Use in Artisanal and Small-Scale Gold Mining: A Practical Guide. Nairobi, Kenya: United Nations Environment Programme. Available at: https://wedocs.unep.org/handle/20.500.11822/11524. (Accessed: 23/05/2017)

Urkidi, Leire. 2010. A glocal environmental movement against gold mining: Pascua–Lama in Chile. Ecological Economics 70(2): 219–27. Doi: 10.1016/j.ecolecon.2010.05.004.

USGS. 2014. Mineral Resources On-Line Spatial Data. Reston, Virginia: United States Geological Survey. Available at: https://mrdata.usgs.gov/mineral-operations/find-minfac.php. (Accessed: 20/05/2017)

USGS. 2015. 2013 Minerals Yearbook. Reston, Virginia: United States Geological Survey. Available at: https://minerals.usgs.gov/minerals/pubs/mcs/2013/mcs2013.pdf. (Accessed: 23/05/2017)

Valdés-Pineda, Rodrigo, Roberto Pizarro, Pablo García-Chevesich, Juan B. Valdés, Claudio Olivares, Mauricio Vera, Francisco Balocchi, Felipe Pérez, Carlos Vallejos, Roberto Fuentes, Alejandro Abarza and Bridget Helwig. 2014. Water governance in Chile: Availability, management and climate change. Journal of Hydrology 519: 2538–67. Doi: 10.1016/j.jhydrol.2014.04.016.

World Bank. 2011. Diagnostico de La Gestion de Los Recursos Hidricos. Departamento de Medio Ambiente y Desarrollo Sostenible Region para America Latina y el Caribe. Available at: http://www.dga.cl/eventos/Diagnostico%20gestion%20de%20recursos%20hidricos%20en%20Chile_Banco%20Mundial.pdf. (Accessed: 23/05/2017)

Yard, Ellen E., Jane Horton, Joshua G. Schier, Kathleen Caldwell, Carlos Sanchez, Lauren Lewis and Carmen Gastañaga. 2012. Mercury exposure among artisanal gold miners in madre de dios, Peru: A cross-sectional study. Journal of Medical Toxicology 8(4): 441–48. Doi: 10.1007/s13181-012-0252-0.

CHAPTER 3

Waters of Andean Indigenous Peoples
Ancestral Rights and the Neutralization of their Claims

Amaya Álvez,[1,] Verónica Delgado,[1]*
Fernando Ochoa[2] and Carla Cid[3]

Introduction: Ancestral Waters as Part of International Human Rights Standards

Indigenous people inhabited Chile long before the Spanish invasion, led by conquistador Diego de Almagro, in 1536. Indigenous people of the region had their own customary laws, languages, religions and traditions. The conquest and colonization of Latin America, and in particular the one pursued in Chile (1536–1810), annulled any alternative legality to that imposed by the Spaniards (Anghie, 2004). Only recently has there been a move towards "the recognition of indigenous people as political subjects and not merely as the object of a politics dictated by others; that is, as subjects with rights to control their own institutions and self-define their own destinies" (Yrigoyen, 2004; Álvez, 2017).

[1] Associate Professor of Law, Faculty of Legal and Social Sciences, University of Concepción, Barrio Universitario s/n, Concepción, Chile.
   Email: vedelgado@udec.cl
[2] Researcher, Water Research Center for Agriculture and Mining (CRHIAM), University of Concepción, Barrio Universitario s/n, Concepción, Chile.
   Email: fernando.ochoa.udec@gmail.com
[3] LL.B., Faculty of Legal and Social Sciences, School of Law, University of Concepción, Barrio Universitario s/n, Concepción, Chile.
   Email: carlacid@udec.cl
* Corresponding author: aalvez@udec.cl

Benedict Kingsbury (Kingsbury, 1998), defends the idea of 'indigenous peoples' as a legal category that provides them great normative power as subjects of contemporary international law. He claims that this principle of international law could also be partially applied to them using the language of human rights as legal entitlements based on the similarities to the required elements: a territory, a population, the existence of a governmental institution and legal international relationships (Kingsbury, 1998). In summary, among the international instruments and institutions applicable in Chile that address protection of indigenous land and the water resources located within it are (for a more complete analysis, Rojas, 2014) the United Nations Charter,[1] the Universal Declaration of Human Rights,[2] the International Covenant on Civil and Political Rights,[3] the International Covenant on Economic, Social and Cultural Rights,[4] the United Nations Declaration on the Rights of Indigenous People,[5] ILO Conventions 107 and 169,[6] the Machu Picchu Declaration on Democracy,[7] Agenda 21[8] and the Andean Charter for the Promotion and Protection of Human Rights of 2002.[9] In addition, in the Inter-American System, the American Convention on Human Rights[10] and the work of the Inter-American Court of Human Rights[11] should be noted.

As a concrete example, there has also been an evolution towards the protection of the environment as natural heritage. For example, the Convention on Biological Diversity (CBD, 1992) focuses its attention on indigenous peoples in order to protect their interests in biological resources aiming to consecrate a sustainable use of traditional lifestyles to biological diversity and encourage the equitable sharing of the benefits arising from

---

[1] Signed June 26, 1945 in San Francisco, entering into force in October of the same year, signed by Chile in 1945 as a founding member.

[2] Adopted December 10, 1948 in Paris, France, with Chile voting in favor.

[3] Covenant adopted by the UN General Assembly on December 16, 1996 and signed by Chile on the same date.

[4] Promulgated by the UN General Assembly on December 19, 1966 and signed by Chile on September 16, 1969.

[5] Adopted in New York on September 13, 2007, with Chile voting in favor.

[6] Convention of 1989, ratified by Chile in 2008.

[7] Passed on July 29, 2001, in the presence of the presidents of Argentina, Brazil, Chile, Costa Rica, Panama, Paraguay, the Dominican Republic, Uruguay, the Prince of Asturias and the Secretary General of the Andean Community; non-binding.

[8] Enacted during the United Nations Conference on Environment and Development in 1992, and signed by Chile during the conference.

[9] Adopted by the Andean Presidential Council in Guayaquil, Ecuador, on July 26, 2002, without the participation of Chile; non-binding.

[10] Also called the 'Pact of San José,' promulgated by Decree 873 on January 5, 1991.

[11] Created by the OAS in 1959, convening for the first time in 1960.

the utilization of such knowledge (CBD S. 8 J). Also relevant is the adoption of the AKWÉ KON guidelines regarding water in 2000.[12] These are voluntary guidelines for the execution of cultural, environmental and social impact assessments regarding developments proposed to take place on, or which are likely to impact, sacred sites and lands and waters traditionally occupied or used by indigenous and local communities.[13] In Chile these consultation processes are in a primitive state, very unfortunately they are neither binding, not fully transparent.[14]

Sources of Ancestral Indigenous Rights

Self-determination in the case of indigenous peoples requires the recognition of collective rights, the existence of alternative legalities and a general political space in which to make decisions. For example, in reference to ancestral indigenous rights, or indigenous title, we can identify at least three possible sources of legitimacy (Aguilar Cavallo, 2005):

(A) Legal recognition of the indigenous people by the state in which their territory is located. Indigenous peoples as a unified legal category have never been recognized in any of the previous Chilean constitutions. In fact, the Chilean state was formed in 1810 under the premise of excluding those who inhabited the country prior to the conquest by the Spanish Empire.

(B) Recognition of treaties between colonizers or states and indigenous peoples. The Spanish Empire signed treaties with indigenous peoples, though they were not regarded as independent nations in their relationship with the Spanish Empire. Moreover, authors like Bartolomé Clavero (Clavero, 2008) have argued for the renewed binding value of past treaties between states and indigenous peoples. Shortly after Chile achieved independence from Spain, the new republic was recognized by the Mapuche nation through the Treaty of Tapihue (1825).

(C) Recognition of indigenous people's legal rules, mostly through customary laws. The paradigm of uniform rights gradually started shifting in the 1990s. Many Latin American countries, Chile among them, started taking legal and political measures to recognize indigenous rights as they were transitioning to democratic regimes (Molina and Yañez, 2008).

---

[12] Decree 1963, published in the Official Gazette on May 16, 1995, ratified the Convention on Biological Diversity in Chile.

[13] The Akwe Kon guidelines, online: https://www.cbd.int/doc/publications/akwe-brochure-en.pdf.

[14] Information about the consultation process on areas of biodiversity protection can be found here: consultaindigena.mma.gob.cl.

Characteristics of Ancestral Indigenous Rights in International Law

Ancestral indigenous rights have some common characteristics in the recognition granted by international human rights law:

(A) Collective character: The communal nature of life in various groups, who share not only a common space but also a way of life, involves a close relationship with the collective and the land. This is addressed by Article 13 of ILO Convention 169, which states that governments shall respect the "special importance for the cultures and spiritual values of the peoples concerned of their relationship with the lands or territories, or both as applicable, which they occupy or otherwise use, and in particular the collective aspects of this relationship." It is also recognized in various rulings by the Inter-American Court of Human Rights,[15] which has been consistent on the protection of collective rights, ruling that the concepts of property and possession in indigenous communities can have a collective meaning, in the sense that land ownership "is not centered on an individual but rather on the group and its community" and that "both the private property of individuals and communal property of the members of the indigenous communities are protected by Article 21 of the American Convention."[16] With an evolutionary interpretation particular to human rights treaties, and in accordance with the provisions of Article 29 of the Convention, the court indicated that the right to collective property falls within the assumptions of Article 21 of the American Convention, since "failing to recognize specific versions of the right to use and enjoyment of property that emanate from the culture, practices, customs and beliefs of each people would be equivalent to maintaining that there is only one way of using and enjoying property and this, in turn, would make the protection granted by Article 21 of the Convention meaningless for millions of individuals."[17]

---

[15] Case of the Sawhoyamaxa Indigenous Community v. Paraguay. Merits, Reparations and Costs. Judgment of March 29, 2006. Series C No. 146, para. 119 and Case of the Yakye Axa Indigenous Community v. Paraguay. Merits, Reparations and Costs. Judgment of June 17, 2005. Series C No. 125, para. 136.

[16] Case of the Yakye Axa Indigenous Community v. Paraguay. Merits, Reparations and Costs. Judgment of June 17, 2005. Series C No. 125, para. 143; Case of the Kuna Indigenous People of Madungandí and Emberá Indigenous People of Bayano and Their Members v. Panama. Judgment of October 14, 2014. Series C No. 284, p. 79; Case of the Mayagna (Sumo) Awas Tingni Indigenous Community v. Nicaragua, para. 148; and Case of the Xákmok Kásek Indigenous Community v. Paraguay, para. 85.

[17] Case of the Sawhoyamaxa Indigenous Community v. Paraguay, para. 120, and Case of the Xákmok Kásek Indigenous Community v. Paraguay, para. 87.

(B) Exercise over a broad physical space: This refers not only to land but instead a broad concept of territory or environment, along with the resources found within it. Article 13.1 of ILO Convention 169 recognizes the "special importance for the cultures and spiritual values of the peoples concerned of their relationship with the lands or territories," acknowledging the worldview of the communities; it defines indigenous territories as covering the "total environment of the areas which the peoples concerned occupy or otherwise use." Article 14 establishes the obligation, in appropriate cases, to take measures to "safeguard the right of the peoples concerned to use lands not exclusively occupied by them, but to which they have traditionally had access for their subsistence and traditional activities," broadening the concept of land usually recognized by legislation. In this regard, the Inter-American Court has resolved that the protection of natural resources falls within the protection of the right to property established in Article 21 of the American Convention. Article 27 of the International Covenant on Civil and Political Rights establishes, in a similar manner, the right of persons belonging to ethnic, religious or linguistic minorities to enjoy their own culture in community with other members of their group. The United Nations Declaration of 2007 states that indigenous peoples have the right to "own, use, develop and control the lands, territories and resources that they possess by reason of traditional ownership or other traditional occupation or use, as well as those which they have otherwise acquired."[18]

(C) Historical and immemorial title: This refers to the customs of indigenous peoples, behaviors repeated by the majority of the community over time with the conviction that doing so is obligatory. This customary law, consists of standards developed and transmitted orally that change according to the evolution of the imperative values in each indigenous community. Thus, from a dynamic perspective of identity and ethnicity, it is incorrect to refer to ancestral customs; instead, it is necessary to analyze the modern customs of indigenous people (Aravena, 2000). We disagree with the previous statement, since customary norms are amenable to adapting to practices recognized as binding by the particular community and are called ancestral because they are customs that have long been maintained in the community. Article 8 of ILO Convention 169 states that these norms must be taken into consideration when applying remaining national laws. What is

---

[18] United Nations Declaration on the Rights of Indigenous Peoples, Article 26.2.

necessary to affirm is that the recognition of the cultural pre-existence of indigenous peoples also assumes acceptance of their law as a cultural product. The main consequence of ancestral indigenous law being an immemorial title is its unrecorded nature, since it is founded in the uses and customs that indigenous people have practised on their land since ancestral times, including the pre-Columbian era. For this reason, the Inter-American Court of Human Rights pronounced in August 2001, in *Mayagna (Sumo) Awas Tingni Community v. Nicaragua*, "Indigenous peoples' customary law must be especially taken into account for the purpose of this analysis. As a result of customary practices, possession of the land should suffice for indigenous communities lacking real title to property of the land to obtain official recognition of that property, and for consequent registration," a position that has been reaffirmed in subsequent rulings, in which, following its line of reasoning, the court has stated, (1) "traditional possession of their lands by indigenous peoples has equivalent effects to those of a state-granted full property title," (2) "traditional possession entitles indigenous people to demand official recognition and registration of property title" and (3) that the state must "delimit, demarcate and grant collective title over lands" to members of indigenous communities.[19]

Particularities of the Ancestral Rights of Andean Indigenous Peoples

Colonized people are forced to act within a frame; therefore, the proposal here is to look to indigenous knowledge as a source to articulate a new model of society. With specific regard to water, the Andean region presents a distinctive irrigation system that brought together people and their attachment to the land and water resources in a decentralized institutional framework (Boelens et al., 2009). It is important to research available information regarding indigenous legal traditions because, as pointed out by Yrigoyen (2004), "the colonial reality put the native peoples in a condition of political subordination, economic exploitation and cultural devaluation."

---

[19] Case of the Moiwana Community v. Suriname. Preliminary Objections, Merits, Reparations and Costs. Judgment of June 15, 2005. Series C No. 124, para. 209; Case of the Mayagna (Sumo) Awas Tingni Community v. Nicaragua, para. 151 and 153; Case of the Xákmok Kásek Indigenous Community v. Paraguay. Merits, Reparations and Costs. Judgment of August 24, 2010. Series C No. 214, para. 109; and Case of the Kuna Indigenous People of Madungandí and Emberá Indigenous People of Bayano and Their Members v. Panama. Judgment of October 14, 2014. Series C No. 284, p. 40.

If we research the normative system prior to the arrival of the Spanish Empire, we discover that the Aymara people's development was possible through a system of irrigation channels for agricultural use that descended from the Andes. Due to its vital role in their lives, water played a part in most ceremonies and rituals in addition to being a vital element for agriculture and human survival. On the other hand, the Atacameño people have inhabited the driest desert in the world, the Atacama Desert, since 9000 BC (Gonzalez, 2005). They also developed highly complex irrigation systems through floods that allowed them to cultivate in areas called *ayllus*. The population settled in the *ayllus*, developing art offerings for their ancestors at local cemeteries (Mamani, 2005).

Authors have highlighted the level of sustainable development in terms of water resources achieved by Andean indigenous peoples as custodians of a hydraulic culture of surprising technology, with water as one of the foundations of their existence and culture (Cuadra, 2000). It is even asserted that in no other part of the world has such importance been placed on water, nor has it been so successfully used for extractive and ritual purposes since before the arrival of the Spanish conquistadors (Gentes 2001, 2002 and 2004).

Constitutional and Legal Situation of Ancestral Water Rights in Chile

The 1980 Constitution and the 1981 Water Code created a water market system. Some authors have realized that the neoliberal water modernity project does not allow plurality of water systems or management modes (Boelens, 2009). There is a more progressive interpretation of the 1980 Constitution under which ancestral rights over water of indigenous peoples in Chile is constitutionally guaranteed. Indeed, the founding document establishes in Article 19, section 24, paragraph 11 that *"rights of private citizens over waters, recognized or created in conformity with the law, shall grant proprietorship to the owners thereof."* From this rule, it is inferred that the rights of private citizens over waters can have various origins, those granted by an act of concession by authority and those recognized in accordance with immemorial or ancestral use that involves the existence of an indigenous custom that complies with the parameters described earlier: collective, exercised in a broad territorial space and through which an indigenous title can be adduced (Atria and Salgado, 2016). That said, within these 'recognized' rights for authors such as Vergara Blanco (1998) are:

(a) Customary uses such as those recognized in D.L. 2.603 of 1979.

(b) Minimum or limited uses of sources such domestic wells, springs that originate, flow and terminate within the same property, minor lakes and water found during minding work.[20] According to the author, there are also other old rights recognized by current law:

(1) Those accorded to agricultural communities.[21]

(2) Rights acquired by prescription.[22]

(3) Those accorded to Andean indigenous communities by Article 64 of Law 19.253 of 1993 on the Protection, Promotion and Development of Indigenous People.

The importance of these customs being recognized as rights over watersheds associated with the constitutional protection they enjoy, which is enshrined in Article 19, section 24, paragraph 11.

The doctrine in Chile, proposed by Daniela Bravo (Rivera Bravo, 2013) in her work on customary water uses and rights, among others, concludes that they have not been sufficiently studied in the case of ancestral uses of indigenous waters. The author understands 'recognized' rights as those protected by legislation, even if they originated under old regulatory systems and regardless of whether they are registered (Rivera Bravo, 2013).

The Water Code of 1981 aimed to record every 'transaction' carried out regarding rights over water, beginning, of course, with its constitution. The granting of use rights is done through concessions issued by the General Water Directorate (hereafter DGA) or in exceptional cases by the president. The procedure is strictly regulated by the Water Code[23] and complemented, in matters not covered, by Law 19.880, which establishes the foundations of administrative procedures. The authority cannot refuse to grant rights such that if all legal requirements are fulfilled it must issue the administrative act, which in turn will be recorded as a public document and registered in the appropriate Water Registry of the Real Estate Registrar and in the Public Registry of Water Use Rights of the Public Water Registry (hereafter CPA, for its acronym in Spanish) administered by the DGA.[24]

Despite the establishment of the administrative registration process, a significant number of users have continued making use of water without

---

[20] Article 20, subsection 2 of the Water Code: "With the exception of use rights over water corresponding to springs that originate, flow and terminate within the same property as well as water of lakes not navigable by vessels of more than 100 tons, small lakes and swamps located within one property and that over which use rights created in favor of third parties are not registered as of the effective date of this Code. The ownership of these use rights beings, by the mere operation of the law, to the riparian landowner."

[21] Article 54 bis D.F.L. 5 of 1968.

[22] Article 310 3 of the Water Code of 1981.

[23] Article 140 et seq. of the Water Code of 1981.

[24] Article 150 of the Water Code of 1981.

any registration, a situation which various title clearing or regularization systems established in the transitory articles of the Water Code are intended to remedy. A clear example is found in one of the main water consumer categories in our country, agricultural irrigators, who usually form associations to distribute water on the basis of time/fees, with this system creating true feelings of ownership (Sandoval, 2015). In these situations, the law has had to adapt to reality and either grant them legitimacy or risk becoming a dead letter. Our law has not been oblivious to this reality, as shown by D.L. 2.603 of 1979, which explicitly recognizes the user of the land as the holder of water rights.[25]

In addition, when these rights are 'recognized,' the jurisdictional act that clears them has the characteristic of being a deed of declaration, that is, one that is limited to establishing a pre-existing situation, which entails certain determinant considerations regarding the water distribution system, as Vergara Blanco (1998) reminds us: the distribution of water is subject to its origin, in this case the use of the person claiming the right; thus, the characteristics of the water right will be limited to establishing the prior effective use.

The Water Code of 1981 has undergone numerous reforms, the most important of which occurred in 2005 with Law 20.017, which again recognized customary uses, in this case of groundwater, in Transitory Article 4,[26] granting a period of six months in which to make recognition requests, which was subsequently extended to one year by Law 20.099. In addition, already in 1992 Law 19.145 had been enacted, which modified the Water Code to limit groundwater exploitation and exploration in aquifers that feed meadows and *bofedales* (high-altitude peatlands) in northern Chile. Thus, recognizing the existence of saturated or special protection areas, Law 20.411, promulgated on December 29, 2009, banned the creation of rights in determined areas, a prohibition that did not extend to indigenous people or indigenous communities that fulfill the requirements of Transitory Article 5 of Law 20.017.

---

[25] Article 7: "The use right holder will be presumed to be the owner of the property who is currently using said rights. In the event that the preceding regulation is not applicable, the use right holder will be presumed to be the person currently effectively making use of the water."

[26] Article 4: "The General Water Directorate will allocate permanent water use rights over groundwater with flows of up to two liters per second from the First to Metropolitan regions, both included, and up to four liters per second in the remaining regions for wells that were built before June 30, 2004. Requests must be presented no more than six months after the effective date of the present law." Subsection 2 of the same article states that "The regularization indicated in this article shall not be a requirement for the use of groundwater as indicated in the first subsection of Article 56 of the Water Code." Article 56 refers to drinking water and other domestic uses on one's "own land."

In addition, Article 6 of Law 20.017 of 2005 aimed to facilitate the regularization of water intended to supply the population in rural areas through the rural drinking water system by establishing a series of requirements. For the purposes of the creation of the respective use right in the name of the Rural Drinking Water Committee, records must be presented that prove ownership of the property in the name of the respective committee or the authorization of its owner, requirements not imposed on indigenous communities.[27] This public policy of prohibiting the creation of new water use rights in determined areas—except for 'disadvantaged' groups—was debated and declared constitutional by the Constitutional Court in Case 2512–2013,[28] as it was considered possible to establish differences in access to water resources on the basis of public interest.

These regulations demonstrate the importance that, even today, more than three decades after the passage of the Water Code, is placed on the uses and customs of water by users. In addition, the modifications of the requirements involving indigenous people or communities shows the existing consensus regarding heightening the level of protection of their rights as a way of restoring part of their autonomy and quality of life, which had been diminished by third party acquisition of rights to the waters of their natural environments (Ramirez and Yepes, 2011). These regulations decreed by Law 20.017 of 2005 were extended to the entire indigenous population of the country regardless of their geographical area, even though the regulation mentioned in Article 64 of Law 19.253 was limited to the Andean zone, that is, the XV, I, II and III regions.

Among the proposals presented in the current draft that discusses modifications to the water system, it has been proposed to add to the Water Code an Article 4 bis that decrees: "*Water is a national good of public use. Thus, its ownership and use belong to all the inhabitants of the nation. Water has, among others, environmental, survival, ethnic, productive, scenic, landscape, social and land use planning functions. It is the duty of the State to guarantee all inhabitants access to the functions stated in the previous subsection.*"[29]

If approved, this law would recognize the ethnic dimension of water, and once again open the discussion on the level of protection that it should be granted. Yáñez (2010) criticized the project of law (Bulletin 7543-12) because the draft does not pronounce on communities' ancestral uses of

---

[27] Article 6 of Law 20.017.

[28] Constitutional Court, Alcalde Villalón, Julia, v. DGA (2013), cons. 2.

[29] Proposal presented in Bulletin 7543-12 on March 17, 2011. Available at: https://www.camara.cl/pley/pley_detalle.aspx?prmID=7936. [Date consulted: February 19, 2016].

water (as they are not included in the list of 'priority uses'), nor does it extend the special protection set forth in Article 64 of Law 19.253 to the other ethnic groups in our country.[30]

Under the idea of recognizing indigenous customary law, the current Water Code reform neither includes ancestral waters within those that take precedence for 'clearing' nor considers the creation of an administrative process that is adapted to the indigenous vision of natural resource administration, without imposing criteria unfamiliar to their culture to prove ancestral uses. In addition, the draft seeks to redesign the legal nature of the water use right, granting it a temporary character,[31] which would allow the state to allocate sufficient water to protect human consumption and other essential local development, environmental and territorial uses; however, the proposal neither modifies nor regulates the currently established neoliberal system, the water market, which would impede allocation to the non-competitive sectors of the country. It represents a partial advance in that the system is attenuated, although not substantially modified.

Rights Accorded to Andean Indigenous Communities

There is a special protection over indigenous water rights in Law 19.253 in Title VIII, paragraph 2, which establishes "particular complementary dispositions for the Aymaras, Atacameños and other indigenous communities of the north of the country." The institutions and mechanisms in place to meet this objective are specified between Articles 62 and 65. It begins by establishing to whom these dispositions apply, understanding that Aymaras are the indigenous people who belong to Andean communities located mainly in the I Region (currently the XV Region of Arica-Parinacota and Tarapacá) and Atacameños are the indigenous people who belong to communities located mainly in towns of the interior of the II Region, as well as, in both cases, the indigenous people originally from these communities. Article 62 ends by stating that the dispositions shall apply to other Andean communities such as the Quechuas and Collas, including the Diaguitas.[32]

---

[30] Yáñez [Date consulted: February 19, 2016].

[31] Artícle 5 bis 2, proposed in Bulletin 1543-12: "For the fulfillment of the provisions in the previous article, the General Water Directorate may create water reserves over natural sources of water.

Regarding said reserves, the Directorate may grant private citizens temporary use concessions for the development of the functions indicated in the second subsection of Article 4 bis."

[32] Law 20.117 promulgated on September 8, 2006.

It is Article 64 that recognizes ancestral rights of communities over the water located in their territories.[33] In this way, the legal right of Andean communities to certain waters is configured, recognizing the ancestral use of water resources, the legal basis of which is indigenous customary law (Aylwin et al., 2013). Thus, the law recognizes a *de facto* situation whose foundation lies in the fact that indigenous peoples have used waters belonging to said spaces since pre-colonial times, be it for human or animal use or irrigation, making clear that the right exists because it is recognized by law, even though it is not recorded in the corresponding registry for purposes of certainty.

Another element is the concept of 'community lands' used in Article 64, which, with an eye toward protection, cannot be reduced to the existence of a title of ownership; this concept was inspired by ILO Convention 169, fully applicable in our country since September 2009.[34] Accordingly, Part II, "Land," leads us to an interpretation of the concept that includes lands that are not exclusively occupied by indigenous peoples, but to which they have traditionally had access for their survival and traditional activities. In Article 13.2, it indicates that the term 'lands' "shall include the concept of territories, which covers the total environment of the areas which the peoples concerned occupy or otherwise use," a directive that the Supreme Court has supported in various rulings, entirely protecting the heart of indigenous society.[35] This notwithstanding, we are obligated to mention that these precepts fall within those considered 'not self-executing' by the constitutional court,[36]

---

[33] Art. 64: "The waters of the Aymara and Atacama communities shall be especially protected. Water located within community lands, such as rivers, canals, irrigation ditches and springs, shall be considered property and community use assets, without prejudice to rights registered by third parties pursuant to the General Water Code. No new water rights over lakes, ponds, springs rivers and other aquifers that supply the water of the various indigenous communities established by this law will be granted without previously guaranteeing the normal supply of water to the affected communities."

[34] Decree 236 of the Ministry of Foreign Affairs, published in the Official Gazette on October 14, 2008.

[35] Francisca Linconao v. Forestal Palermo (2009) and Alejandro Papic Domínguez v. Aymara Chusmiza and Usmagama Indigenous Community (2009).

[36] The Constitutional Court has held that these precepts fall within the provisions of the convention that are not self-executing, Case 309 of 2000: "The court states that said disposition imposes on the State the obligation to recognize the possession and property rights of 'interested peoples' over 'the lands that they have traditionally occupied,' but also with respect to real estate belonging to non-indigenous third parties that said peoples are interested in using. The foregoing, inasmuch as no. 2 imposes said obligation regarding lands that 'the interested peoples have traditionally occupied.' In addition, Article 13 no. 2 of the Convention considerably broadens the concept of 'lands.' From the foregoing it is concluded that the State assumes the obligation of expropriating lands at the request of a collective entity and for its own benefit, which in the judgment of the petitioner violates the conceptual framework of expropriation in Chile. Thus, Article 14 violates Article 19, no. 24, second and third subsections, of the Constitution."

since in its judgment, brooding the concept of community lands implies that the state assumes the obligation to expropriate lands at the request of a collective entity for its own benefit, which violates the conceptual framework of expropriation in Chile, an interpretation that strikes us as excessive since what is sought is not expropriation based on the claims of indigenous people but rather full recognition of their rights and perhaps a right to pass through or use temporarily those lands.

This proves especially relevant in terms of water, since we cannot avoid the fact that due to the nature of the resource it commonly flows through various territories, which may or may not be registered in the name of the indigenous community. The ruling of November 25, 2009 on Case 2840-2008 handed down by the Supreme Court proves illuminating in this sense, as it affirms that there is *"no doubt that the recognition of rights that Article 64 makes in favor of the Aymara and Atacameño communities refers not only to water located in properties registered to the community, but also to water that, although located on property registered to third parties, supplies the indigenous collective, since what this regulation attempts to protect is, essentially, the water supply of said indigenous communities, which is achieved only by the application of the rule under consideration in the way that is has been understood by lower court judges, which this court shares. With this interpretation that guarantees the water supply of the indigenous communities in question, the objectives expressed in the Presidential Message that inspired the current Indigenous Law are met...."* With this ruling, the court had in mind the purpose of the law, which is to supply the communities in order to foster their repopulation, a situation made difficult by the water scarcity that has led indigenous people to move to cities in order to improve their quality of life. The aforementioned 'community lands' precept also includes a restriction on the recognition of ancestral rights, understanding that regularization should encompass community uses and not those of private individuals, even if they belong to one of the mentioned ethnic groups, which is reinforced by Transitory Article 3 of 19.253 and a factual argument stemming from the custom on which the law is founded: Andean indigenous peoples have traditionally been organized as communities, making collective use of resources as seen in activities such as ceremonies, shepherding and their particular agricultural activity.

Law 19.253, to ensure the ownership of ethnic groups over ancestral waters, decrees that the water rights of the communities are without prejudice to rights that third parties have registered in accordance with the Water Code, that is, even when these third parties have registered water use rights pursuant to legal regulations, the resource is not prevented from remaining the property of the community, which has proved essential in times of conflict, when third parties, in accordance with the law, have subsequently registered rights over water over which indigenous peoples have ancestral rights. Thus has been the case with various emblematic rulings in our country that have established the foundations of the issue. The Supreme Court ruled in this way in Case 2840-2008 of November 25, 2009, *Alejandro Papic Domínguez v. Aymara Chusmiza and Usmagama Indigenous Community*, confirming the ruling of the Iquique Court of Appeals by finding in favor of the Aymara-Atacameño Community, recognizing their use rights over the water of the spring known as the Socavón de Chusmiza on the basis of the presumption of Article 64 of Law 19.253 and dismissing the claims brought by Alejandro Papic Domínguez on behalf of Agua Mineral Chusmiza S.A.I.C. The court stated that *"the appealed judgment has been limited to regularizing pre-existing rights upon verifying that the water claimed by the indigenous community has been used by it in the terms established in the second subsection of Transitory Article 2 of the Water Code, customary rights that, furthermore, precede those of the plaintiff company. In this case new administrative rights are not being granted; rather, the use of water resources since time immemorial, reflected by positive acts of possession such as the construction of the adit from which the water emanates, storage ponds, the canal, the farming terraces and the human settlement itself as reflected by the modest houses that constitute the two villages of Chusmiza and Usmagama, is being judicially recognized. The purpose of the procedure used is that once the customary use is recognized, it be considered a right, which, once regularized, can be entered in the corresponding national state registry, which will allow the survival of the indigenous community on its ancestral land, reasons for which the appeal under consideration is dismissed.*[37]*

In this same regard, and even more relevant for being the first finding in favor of ancestral indigenous ownership over water, was the ruling handed down on March 22, 2004, in Case 986-2003, which confirmed the findings of the Antofagasta Court of Appeals. The court recognized the rights of the Toconce Atacameño community on the basis of their use since time immemorial, noting that it was sought only to normalize and not create the right, as it emanated from the law, and therefore rejected the appeal of ESSAN S.A., Empresa de Servicios Sanitarios de Antofagasta, in whose name the water use rights in dispute were registered. *"In consequence, the court cannot but conclude that the process of the often-cited Transitory Article 2 of the Water Code allows the regularization but not the creation of rights, since those that are normalized existed previously and their possession is not in dispute, since they emanate from the law. It is a procedural regulation that aims to regulate the manner of registration of a right that, as occurs in this case, is not registered, but legally recognized. To achieve regularization, it is important to determine, in*

---

[37] Case Popic Domínguez vs. Aymara Chusmiza and Usmaqama indigenous community (2009) Supreme Court ROL 2840-2008.

the relevant processing stage, if the petition fulfills the requirements indicated in said regulation and if the claimed flow is effectively used in the manner and terms set forth by the legislature."[38]

In this sense the court confirms the purport of the regulation, understanding that the legislature is not creating rights but rather recognizing a pre-existing situation founded in ancestral use; in other words, it stems from indigenous customary law, and therefore precedes any other subsequently created title or right over water.

The second subsection of Article 64 establishes that no new rights will be granted over lakes, ponds, springs, rivers and other aquifers that supply the water possessed by indigenous communities without first guaranteeing the normal supply of water to the communities. It is a sort of positive discrimination that aims to secure the community's supply, producing a true freeze on new rights. The same law is responsible for establishing the institutions that must work on the special protection enacted in Article 64 of the indigenous law; transitory Article 3 indicates that CONADI and the DGA will sign an agreement for the protection, creation and re-establishment of water rights of ancestral property of the communities of the north of the country. This is without prejudice to the fund decreed in the same law in Article 20, which is dedicated to "creation, regularization or purchase of water rights" of all indigenous communities of the country, prohibiting their transfer for 25 years after their registration.

In our domestic law there is much tension between individual private property rights and the recognition of the ancestral rights of indigenous communities; however, as an exception, the law expressly recognizes the ancestral property of indigenous peoples with regard to rights of recognition of the waters of the Aymara and Atacameño communities and other ethnic groups of the Andean region, as justified, in accordance with parliamentary discussion, by the water scarcity in the area. This can be read in the text of Transitory Article 3 of the indigenous law: *"Likewise, the Corporation and the General Water Directorate will establish an agreement for the protection, creation and reestablishment of water rights of ancestral property of the Aymara and Atacameño communities pursuant to Article 64 of this law."*

---

[38] *"The government is concerned about the situation of the communities of the Norte Grande of the country that are affected by the dispute over water resources. It believes it to be of great importance that this law, along with the modifications to the Water Code that have been presented to Congress, regulate these resources in a manner that allows human life to continue in the towns and villages of the North of the country. It believes that it would be a grave mistake for the population of the North to be concentrated only in three large coastal cities, with the interior abandoned as a consequence of faulty planning regarding the water resources that are fundamental for the development of human life."* History of the legislative process of the Indigenous Act Law 19253 (1993) Toconce Atacameño Community vs. ESSAN SA (2004) Supreme Court ROL 986-03.

For its part, Article 64 of Law 19.253 does the same and understands water as an asset of use and community property, dispositions that defend the legitimacy of customary uses as sources of water rights for northern communities, stemming their subordination before national law, even against rights that are conceded to private citizens by an act of authority.

Characteristics of the Ancestral Rights Accorded to Andean Communities

The characteristics of ancestral rights are not specifically indicated, with the laws providing generic guidelines that indicate the elements to be considered when establishing ancestral indigenous ownership. In the examination of these elements we rely on the anthropological study of Cuadra (2000), who indicates that the existing regulations contain the descriptions necessary to recognize ancestral ownership of water:

(A) Geographical area: In defining who is understood as Aymaras and Atacameños, Article 62 of the indigenous law includes the regional element as a central part of their identity, stating that "Aymaras are the indigenous people who belong to Andean communities located mainly in the I Region and Atacameños are the indigenous people who belong to communities located mainly in the towns of the interior of the II Region." Here we include the XV Region of Arica and Parinacota, giving us the territorial limits we must adhere to when establishing ancestral ownership. This territorial division of the 1970s could obviously be objected to, but as that topic exceeds the scope of this chapter and those territories effectively correspond to those in which the Andean communities have typically resided, we will not address it. Second, we have the territorial element presented by Article 64 of Law 19.253, which indicates that rights over water located "in community lands" will be recognized, a declaration that has led to dilemmas between private citizens and communities, especially regarding its scope in accordance with international treaties.

(B) Customary use: The law has expressly recognized ancestral property rights, in this case based on uses and customs dating to the distant past, which spring from the practices of indigenous peoples carried out in accordance with their enduring worldview, their relationship with the community, Pachamama, religious rites, etc. The assumption is thus that there is no granted title of ownership as such, since the basis of possession comes from indigenous customary laws, but that this lack of registration does not prevent the title of the communities over the waters, understanding that it is "without prejudice" to the rights that third parties have registered in accordance with the Water Code, such that the source is practice and the eventual declaration is meant only to achieve registration.

(C) Community use: The basis of communal water use comes from the purport of the law granting ancestral water rights to the Aymara and Atacameño 'communities'; thus, what is sought is protection of the waters that supply or serve the purposes of the collective and not the interests of individual irrigators, even if they belong to these ethnic groups. It merely recognizes the reality of indigenous peoples, who conceive of community progress as part of their identity, along with the incompatibility of privatizing natural resources. In the case of water, common use for shepherding and agriculture is the norm.

There are some coincidences between the essential characteristics of indigenous title in international laws and the rulings of the Inter-American Court of Human Rights and the rights accorded by our legislation regarding ancestral indigenous title over the waters of the Aymaras, Atacameños and other Andean communities. The main coincidence is the protective duty of the state, including with respect to territory, the element that has been most disputed but nonetheless recognized in accordance with international treaties in force in domestic jurisdiction.

The response to the question of whether ancestral waters are protected in Chile is affirmative in the legal sphere, although with important limitations. They are recognized as assets of use and community property integrated into the private concept of property and the use rights recognized in the Water Code of 1981, and indigenous peoples' own system of internal administration of commons, interpreted as part of their worldview, is not acknowledged. In a recent fieldwork done by Amaya Álvez in September 2015 with Aymara people in Putre, we were struck by how customary water models survive to this day in a subtle way, including rules for dividing the water between community members, deciding when someone broke the water governance rules and determining how to punish that member of the community.

Furthermore, protection of waters and recognition by virtue of ancestral use favors only Andean ethnic groups, going beyond the concept of 'territory' applied in the international system and benefitting only the communities that belong to the I, II, III and XV regions, excluding more than 80% of the indigenous population of the country, which belongs to other ethnic groups, without any substantial justification. Both legally and factually they fulfill the requirements of common, historical and immemorial use, but do not live amid water scarcity, a factor that in international standards is not a determining requirement for ancestral recognition. Even when an ethnic group does live amid water scarcity, as do the Rapa Nui people, protection of ancestral waters is not extended (Oyazún et al., 2011; Gaete, 2012; Zelada, 2013).

Praxis regarding Regularization of Ancestral Rights over Community Waters

Part of the matter that this investigation aims to address is related to indigenous practice regarding water resources and the manner in which it is transformed by the legislature's regulation of it. As previously detailed, there is no special procedure instituted by law that focuses on the recognition of water use rights of indigenous communities on the basis of their ancestral use. Today, following administrative practice, the regulations of the Water Code directed at the "regularization" of use rights of water that is being used by people other than its owners, unregistered rights and those applying to water extracted for personal use from a natural source, that is, according to Transitory Article 2 of the Water Code, are applied by analogy.

The legislation provides for a series of obligations toward communities, among which are the execution of agreements[39] aimed at the creation and re-establishment of rights to ancestral water in the terms of Article 64 of Law 19.253. When this law originally came into force, the lawyers that brought the first proceedings knew that what was sought was precisely judicial recognition of these rights and not an administrative concession. However, they assumed that among the existing formal procedural rules that applied to use and possession and in general came from an acquisitive prescription of rights was that decreed in Transitory Article 2 of the Water Code (to 'regularize' rights) and made this situation equivalent to that of the indigenous peoples (Cuadra, 2000).

Most of the national doctrine—especially after the ruling on the case of *Toconce v. ESSAN*, which demonstrated that regarding the right to ancestral property claimed by the community, this procedure from Transitory Article 2 does not regularize but rather creates rights, with the court having only to establish the content and characteristics of the water right—has not questioned the use of this procedure (Yáñez and Molina, 2011). Rivera (2013), in any case, has highlighted the need to discuss if it is necessary and appropriate for indigenous communities to have to 'regularize' their right (which is legal) in accordance with Transitory Article 2 and explains that it would be sufficient to initiate a declaratory action to have the court declare and specify the ancestral water rights recognized. Rivera recognizes, however, that the usual practice by the Chilean judiciary is to employ the aforementioned process.

---

[39] Transitory Article 3, subsection 2 of Law 19.253: "Likewise, the Corporation and the General Water Directorate will establish an agreement for the protection, creation and reestablishment of water rights of ancestral property of the Aymara and Atacameño communities pursuant to Article 64 of this law."

This investigation, while framed within this same line, not only demonstrates that it is incorrect to apply this regularization process from a formal point of view, but also that it gives rise to a series of practical difficulties regarding the collectivist vision that indigenous peoples have of natural resources, which are incoherently remedied.

Public policy developed since 1993, despite intentions, has denied recognition to the alternative legality of indigenous peoples and instead has stipulated "adaptation" to Transitory Article 2. Thus, for example, with respect to the Atacameño people, petitions for recognition must to be collective, which has been solved by filing them through indigenous associations; in addition, the communities' sources of water must comprise all of the sources used by the respective community, with the formation of different associations per channel absolutely prohibited. Another obstacle is that in the worldview of the Atacameños ancestral rights over water extending from the source of each river or spring to all of the land through which the water flows. This is inconsistent with the existing legal system that separates land from water, which claimants attempted to overcome by requesting the greatest possible flow (Cuadra, 2000). This is partially based on a model that does not allow a different logic regarding water rights and natural resources, one in harmony with an indigenous worldview. Indigenous peoples, in order to not see their rights, in this case regarding water, even further diminished, have had to adapt to the imposed processes that do not correspond to their collective vision of natural resources and which have attempted to absorb their communities' internal administration of water into non-indigenous institutions.

The clearing process imposed on indigenous peoples subjects them to requirements and therefore tests that directly correspond to neither their worldview nor the elements that make up their water-related ancestral customs. In addition, it opens the door for third parties with claims to the water to oppose the regularization because it does not meet the demands of Transitory Article 2 of the Water Code of 1981, which are not provided for by the indigenous law in any of its dispositions, broadening the right enshrined in a specific through a general law that is not in line with the spirit of indigenous peoples' claims. Among the most controversial requirements are that five years of uninterrupted use preceding 1981 have been exercised and that the use have been carried out peacefully, free of secrecy or violence and without recognition of ownership by others.

Water serves as a powerful example to show the effects of externally imposed cultural and political values. Scholars have noted how the neoliberal water modernity project prohibits the existence of a plurality of water rights, water identities and management modes (e.g., Boelens et al., 2010). Here we have an example of how it is possible to curb indigenous customary law through procedural means.

Project on Judicialization of Water Resources, 2013–2014

A complementary investigation carried out by the same CRHIAM researchers addressed the concern over judicial creation of water rights in Chile stemming from recent practice (2013–2014), especially in key matters such as the legal regime of access, in the context of the concern over the difference between legal titles over water resources and the reality, in terms of both hydrology and infrastructure, in which these aspects operate in the country. This work presented the results of an empirical study of all of the rulings handed down by civil courts of first instance in Chile during 2013 and 2014, compiled and analyzed within the framework of the FONDAP project that created the Water Research Center for Agriculture and Mining (CRHIAM, for its acronym in Spanish) at the University of Concepción. This systematization effort made it possible to identify some practices of each of the country's courts, particularly those that contribute to the configuration of water rights at a local level as well as their reconfiguration through the case-by-case approach applied by the courts (Ochoa et al., 2017).

In the specific matter addressed by this work, the judicialization data illustrate the described problem. In 2013, of a total of 1,008 rulings, indigenous communities figured in 15 of them (14 regularizations and 1 protection). In addition, a claimant appeared individually in at least one case, invoking indigenous status.

In 2014, of a total 845 rulings, 499 were on regularization cases, nine of which involved an indigenous community claimant.

One of the most interesting aspects that could be inferred from the judicialization project was the use of the concept of "ancestral waters" with a meaning very far from that described as part of indigenous customary law earlier in this work. The judicialization project found that in 2013 the

Table 3.2 2013 regularization proceedings.

Claimant	Number
Rural drinking water system	4
Indigenous community	14
Indigenous person (as a natural person)	1
Natural person	241
Legal person	70
Estate	6
Various claimants	7
Other public entity	5
Total	351

Table 3.3 2014 regularization proceedings.

Claimant	Number
Rural drinking water system	6
Indigenous community	9
Water users' association	1
Legal person	64
Natural person	405
Estate	9
Various claimants	4
Total	499

Table 3.4 Use of the term "ancestral waters" in 2014 rulings.

Winning party	Number
Indigenous community	4
Natural person	10
Estate	1
Legal person	2
Civil corporation	1
Rural drinking water system	1
Total	19

term "ancestral" was used in 29 water-related rulings, 27 of which were regularizations, one a protection judgment and one a punitive proceeding. However, in the majority of the cases the term "ancestral" was expressed in a proceeding that only involved Chileans claiming water use rights.

Communities' uses of ancestral waters fall within the rights accorded *ipso jure* in our legislation, both by Law 19.253 and the Constitution, such that the proceeding has the objective only of legal certainty, not creating or constituting rights, since they pre-exist from ancient times as customs and laws of the indigenous communities—as has been understood and declared by national courts—and is in agreement with the characteristics of ancestral title in international human rights law. As there is no procedure dictated by law or by administration following the provisions of Transitory Article 3 of the indigenous law, the provisions of Transitory Article 2 of the Water Code, which, as we have mentioned, modifies the requirements determined by Article 64 of the indigenous law and is adapted to neither the characteristics of a title of ancestral property nor the obligations of states[40] to the communities, are applied supplementarily.

Part of the current problem that this work aims to highlight is the unsuitability of existing laws and procedures regarding recognition of customary law of indigenous peoples. To this it should be added that the Water Code reform currently in process in the National Congress has proposed ending this transitory system and an adequate substitute process has yet to be laid out.

The recognition granted by the law is directed only at customary use by the Andean indigenous communities of our country; thus, it aims to protect the uses or practices carried out by the members of the respective indigenous people with the conviction that doing so means obeying an obligatory norm that is binding for all of the members of the community. To this end, it is necessary to demand that the collective nature be proved through the structure of the indigenous community, that immemorial or long-standing use by the community be demonstrated through evidence recognized by the people themselves, such as oral traditions, and that the geographic character of the exercise of the right over water involve the entirety of the environment in which the water resources are used, with special emphasis on the place from which they spring. It does not seem reasonable to also demand the fulfillment of the requirements prescribed in Transitory Article 2 of the Water Code if it is understood that the right over water has arisen by the mere operation of the law in accordance with its ancestral use.

Legal Alternatives for the Recognition of Ancestral Waters

Detailed treatment of the procedures that adequately guarantee the parameters of ancestral use of water discussed in this chapter are beyond its objectives. This notwithstanding, we wish to at least outline two legal alternatives to the current procedure and indicate the reasons that they seem more respectful of indigenous customary law.

We believe that one alternative for recognition is filing a declaratory action, moving away from the clearing process decreed by the Water Code. If the water use rights of the communities are not registered, in

---

[40] I/A Court H.R. Case of the Moiwana Community v. Suriname. Preliminary Objections, Merits, Reparations and Costs. Judgment of June 15, 2005. Series C. No. 124, para. 209; Case of the Mayagna (Sumo) Awas Tingni Community v. Nicaragua, para. 151 and 153, and Case of the Xákmok Kásek Indigenous Community v. Paraguay. Merits, Reparations, and Costs. Judgment of August 24, 2010. Series C No. 214, para. 109. Case of the Kuna Indigenous People of Madungandí and Emberá Indigenous People of Bayano and Their Members v. Panama. Judgment of October 14, 2014. Series C No. 284, p. 40.

this case it is possible, with a declaratory action brought before the court corresponding to the commune in which the ancestral waters are located, to request through a claim for declaratory judgment that a right be declared created in favor of the community on the basis of ancestral use and by virtue of the recognition granted in Article 64 of Law 19.253. It is debatable if as part of the ruling the judge must determine the flow, intake point, consumptive or non-consumptive characteristics and whether the exercise of the right is permanent or temporary and continuous, discontinuous or alternating, since in this way the indigenous law of Chile may weaken the capacity of indigenous people to freely define their interests and collective subjectivities and defend them in an autonomous manner. The way in which the community itself manages its water rights must also be acknowledged, respecting its vision and ancestral use. This ruling operates as *res judicata* and its registration must be processed in the Water Registry of the Real Estate Registrar, following the provisions of Article 114 of the Water Code.[41]

It is understood that the claimant must have a legitimate interest (in the ancestral waters) and must also invoke a moral or patrimonial interest, all of which falls within the interests of the communities, which manifest various interests related to water resources, settlement repopulation, the survival of their traditions and a culture connected to water. Among the requirements of the action are proving ownership, a criterion fulfilled in accordance with the indigenous law upon the ancestral waters of the community being found to be within the "recognized uses" which grants the community full ownership over them through a demonstration that the elements that constitute ancestral uses such as indigenous customs required for their ownership to be fully established. The types of admissible evidence are those established in the ordinary procedure of declaratory action such as expert reports, witnesses and inspections that the court can carry out to confirm the use of water for farming terraces, wells, villages near the water, etc., facts that can indisputably verify the uses to which water has been put in the heart of community life.[42] In this particular matter we consider the assumption contained in Article 64 to be of special importance, as it establishes: *"Water located within community lands, such as rivers, canals, irrigation ditches and springs, shall be considered assets of property and community use, without prejudice to rights registered by third parties pursuant to the General Water Code."*

---

[41] Art. 114. "The following must be registered in the Water Registry of the Real Estate Registrar: 7. Final court decisions that recognize the existence of a use right."

[42] The types of evidence indicated by Article 341 of the Code of Civil Procedure are admitted: public and private instruments, witnesses, admissions, court inspections, expert reports and presumptions.

Some have wished to interpret this institution not only as a certainty of ownership but also as a way in which courts order the restitution of the property in dispute (Couture, 2010), an interpretation not sufficiently established in national jurisprudence that could bring about other effects if what is sought is restitution and the invalidation of the registration in favor of a third party (Lathrop, 2011). This brings us to a second assumption, which is if water is registered in favor of a third party, the public law annulment of the administrative act issued by the administration of the state must be requested and a declaratory action subsequently filed. Regarding use rights of water that effectively has owners in accordance with a legal title, indigenous communities, whose rights over water are accorded by the operation of the law, must assert that the DGA has acted outside its jurisdiction in granting use rights to a third party given that the non-recorded nature of indigenous rights over water does not prevent their legal existence, a consideration among those expressly recognized in Article 7 of the Constitution and which also has the advantage of allowing a request for compensation for damages suffered to eventually be presented. The objective of the annulment action is to annul the administrative act, and thus all of the other resolutions that stem from it, while also determining the responsibilities and sanctions defined by the law. Once the controversy is resolved by ordinary courts, in the case that the water has been registered in the name of a third party despite being located in the place in which indigenous peoples have used the water within the framework of ancestral use, it proves appropriate to rule on the declaratory action and the characteristics of the right of the community.

The importance of filing both actions is that the disputed matter, ownership over water, is thus wholly settled, including compensation agreements aimed at resolving, at least partially, the damages caused to the communities through the loss of access to their ancestral waters. The Chusmiza Usmagama Community faced such a situation even after receiving a favorable ruling from the Supreme Court, since the procedure used, Transitory Article 2 of the Water Code of 1981, seeks the regularization of water use rights but neither provides for the indigenous vision of water nor includes the possibility of obtaining monetary compensation for the damage caused.

Conclusion

Water serves as a powerful example to show the effects of externally imposed cultural and political values. Scholars have noted how the neoliberal water modernity project prohibits the existence of a plurality of water rights, water identities and management modes (see Boelens et al., 2010).

The recognition of the right of Andean indigenous communities in our country to the waters to which they have had access since the distant past and through which their ancestral property is acknowledged constitutes one of the great advances of our legislation in terms of protection of natural resources and indigenous rights, even though it is a partial recognition that is not extended to the other ethnic groups of the country. However, mere recognition of rights is insufficient without the legal or administrative mechanisms necessary to exercise them.

As revealed in our preliminary examination, the recognized uses of water in our country have no process that allows indigenous peoples' rights over ancestral waters to be recognized in a way that grants them the public nature necessary to avoid conflicts with third parties, specifically in cases in which the DGA, in the fulfillment of its obligations, grants the claim of a third party. The pronounced neoliberal system of the Water Code of 1981 only worsened the problems of the indigenous communities of our country, who were cast aside and dispossessed of the waters to which they had had access to in favor of users with greater acquisitive power. The passage of the law recognizing the ancestral property of Andean indigenous peoples, the Law on the Protection, Promotion and Development of Indigenous Peoples of 1993, awakened numerous hopes that indigenous customary law would finally have legal recognition.

Through our study and with empirical data on judicialization, it is hoped to answer the questions posed at the beginning of our discussion regarding the elements and requirements that make up the rights over ancestral waters recognized for Andean communities in Article 64 and Transitory Article 3 of Law 19.253. The various dimensions that connect indigenous people with water—their connections with ancestors, spirituality and survival—have been legally recognized, granting *ipso jure* the right over water; therefore, it is understood that the resolution issued by authority is an act of recognition rather than creation of the right, which has important legal consequences: the resolution aims only to establish a pre-existing situation and in the event of conflict, and indigenous communities must prevail by virtue of ancestral ownership.

Although the elements necessary to establish the ancestral rights of communities in order to be recognized by our legislation are not specifically indicated, the law, in Title VIII, second paragraph, "Particular Complementary Dispositions for the Aymaras, Atacameños and Other Indigenous Communities of the North of the Country," provides us with sufficient data to infer them: geographical area, customary use and communal use. The evidence from indigenous peoples must be directed only at proving the three mentioned elements for ancestral ownership to be formally established, with ownership granted by authority to a third party proving unenforceable.

To complement our position, we consider institutions such as a declaratory action to establish a pre-existing situation, in this case ancestral use, and in cases in which authority has granted water use rights to a third party, the advisability of requesting the public law annulment of the improper resolution, which are suggested as an alternative that is consistent with an appropriate legal interculturality and meets the needs of the indigenous communities regarding the protection of their ancestral waters.

Perhaps the study of this matter could foster an advance in legal pluralism in acknowledgment of the existing interculturality in the country, recognition of the ancestral waters of the other indigenous peoples or the recognition of the different sources from which indigenous rights emanate. Such recognition could be a gateway to dialogue that would accord all inhabitants an effective equality in the face of the ethnic, cultural and racial diversity of our country, which, far from being a cause of national fragmentation, dignifies and enriches us both within and without.

Acknowledgments

This research is part of Fondap Project N° 15130015, which created the Water Research Center for Agriculture and Mining (CRHIAM, for its acronym in Spanish) at the University of Concepción. The authors are grateful for this institutional support and the work carried out by undergraduate thesis students Antonia Alfaro, Cristina Benítez, Carla Cid, Fernando Cortez, Javier Peñaloza, Jorge Rodríguez, María Ignacia Sandoval and Nicolás Vidal.

References

Álvez, A. 2016. Constitutional challenges of the south: indigenous water rights in Chile. Another step in the civilizing mission? The Windsor Yearbook of access to justice 33(3): 87–110.

Aguilar Cavallo, G. 2005. El título indígena y su aplicabilidad en el derecho chileno. Ius et Praxis 11(1): 269–295.

Anghie, A. 2004. Imperialism, Sovereignty and the Making of International Law. Cambridge University Press, Cambridge.

Aravena, A. 2000. Derecho consuetudinario y costumbre indígena. La consideración de la costumbre como atenuante o eximente de responsabilidad penal: Informe pericial. Proceedings of the 12th International Conference on Customary Law and Legal Pluralism: Challenges in the Third Millennium. Chile. 147.

Atria, F. and Salgado, C. 2016. La Propiedad, el Dominio Público y el Régimen de Aprovechamiento de las Aguas en Chile. Legal Publishing, Santiago.

Aylwin, J., Meza-Lopehandía, M. and Yáñez, N. 2013. Los Pueblos Indígenas y el Derecho. Editorial LOM, Santiago.

Boelens, R., Guevara-Gil, A. and Panfichi, A. 2009. Indigenous water rights in the Andes: Struggles over resources and legitimacy. J. Water Law 20(5-6): 268–272.

Boelens, R., Getches, D. and Guevara-Gil, A. (eds.). 2010. Out of the Mainstream. Water Rights, Politics and Identity. Earthscan, Oxon.

Clavero, B. 2008. Reconocimiento Mapu-che de Chile: Tratado ante constitución. Revista Derechos y Humanidades 13: 13–40.

Couture, E. 2010. Fundamentos del Derecho Procesal Civil, Book I. Legal Publishing, Santiago.

Cuadra, M. 2000. Teoría y práctica de los derechos ancestrales de agua de las comunidades atacameñas. Estudios Atacameños 19: 93–112.

Gaete Uribe, L.A. 2012. El Convenio N° 169: Un análisis de sus categorías problemáticas a la luz de su historia normativa. Revista Ius et Praxis 18(2): 77–124.

Gentes, I. 2001. Derecho de Agua y Derecho Indígena.-Hacia un reconocimiento estructural de la gestión indígena del agua en las legislaciones nacionales de los Países Andinos. Santiago. Available at: http://www. eclac. cl/drni/proyectos/walir/doc/walir10. pdf.

Gentes, I. 2002. Agua, poder y conflicto étnico. Santiago: CEPAL. Available at: http://www. serindigena.org/archivosdigitales/otros/ensayo_agua_poder_y_conflicto_etnico.pdf.

Gentes, I. 2004. Agua, derechos locales e indígenas y su interacción con la legislación nacional. Estudio de casos de Chile. Available at: http://www. eclac. cl/drni/proyectos/walir/doc/walir33. pdf.

González, J. 2005. Los pueblos originarios en el marco del desarrollo de sus derechos. Estudios Atacameños 30: 79–90.

Kingsbury, B. 1998. 'Indigenous Peoples' in International Law: A constructivist approach to the Asian controversy. American Journal of International Law. 92: 414–415.

Lathrop Gómez, F. 2011. Procedencia de la acción meramente declarativa del dominio en el derecho chileno. Ius et Praxis 17(2): 3–24.

Mamani Morales, J.C. 2005. Los Rostros del Aymara en Chile: El Caso de Parinacota. Plural Editores, La Paz, Bolivia.

Molina, R. and Yáñez, N. 2008. La Gran Minería y los Derechos Indígenas en el Norte de Chile. Editorial LOM, Santiago.

Ochoa, F., Álvez, A., Delgado, V. and Rivera, D. 2017. El acceso al recurso hídrico en la praxis judicial Chilena: Paradojas y malas prácticas. Actas de Derecho de Aguas 6: 5–28.

Oyarzún, S.M. and Aylwin, M.M. 2011. Una visión panorámica al Convenio OIT 169 y su implementación en Chile. Estudios Públicos 121: 133–212.

Ramírez, M.F. and Yépes, M.J. 2011. Geopolítica de los recursos estratégicos: Conflictos por agua en América Latina. Revista de Relaciones Internacionales, Estrategia y Seguridad 6(1): 149–165.

Rivera Bravo, D. 2013. Usos y Derechos Consuetudinarios de Aguas, Su Reconocimiento, Subsistencia y Ajuste. Legal Publishing, Santiago.

Rojas, D. 2014. Análisis Conceptual del Derecho a la Tierra de los Pueblos Indígenas en el Derecho Internacional. School of Law, Pontificia Universidad Católica de Valparaíso.

Sandoval Muñoz, M.I. 2015. Ausencia de la Regulación de Usos Prioritarios de las Aguas en Chile: Cuando el Río (no) Suena es porque Piedras Trae. CRHIAM: School of Legal and Social Sciences, University of Concepción.

Valenzuela, O.O. 2013. El convenio 169 de la OIT: Principales conversaciones acerca de su implementación.

Vergara, A. 1998. Estatuto Jurídico, tipología y problemas actuales de los derechos de aprovechamientos de agua. Estudios Públicos 69: 163.

Vergara, A. 2014. Crisis Institucional del Agua. Descripción del Modelo, Crítica a la Burocracia y Necesidad de Tribunales Especiales. Legal Publishing, Santiago.

Vergara Blanco, A. 1998. Derecho de Aguas, Books I and II. Editorial Jurídica de Chile, Santiago.

Yáñez, N. 2010. Observaciones al proyecto de reforma al Código de Aguas, Boletín 7543. Available at: https://www.camara.cl/pdf.aspx?prmTIPO=DOCUMENTOCOMUNIC ACIONCUENTA&prmID=2539.

Yáñez, N. and Molina, R. (eds.). 2011. Las Aguas Indígenas en Chile, LOM, Santiago.

Yrigoyen, R. 2004. Legal pluralism, indigenous law and the special jurisdiction in the Andean countries. Beyond Law 27: 32–49.

Zelada, L.G. 2013. Comentario jurisprudencial: la consulta a los pueblos indígenas en la sentencia del Tribunal Constitucional sobre Ley de Pesca roles Nºs. 2387-12-CPT y 2388-12-CPT, acumulados. Estudios Constitucionales 11(1): 621–632.

Legislation

National Laws

Political Constitution of the Republic of 1980, Water Code of 1981, Law 19.253 of 5 October 1993, Law 20.017 of 16 June 2005, D.L. 2.603 of 23 April 1979 on modification of Constitutional Act 3, Law 20.411 of 29 December 2009.

International Standards

ILO Convention 169, United Nations Charter, Universal Declaration of Human Rights, International Covenant on Civil and Political Rights, International Covenant on Economic, Social and Cultural Rights, United Nations Declaration on the Rights of Indigenous Peoples, ILO Convention 107, Machu Picchu Declaration on Democracy, the Rights of Indigenous People and the War Against Poverty of 2001, Agenda 21, Andean Charter for the Promotion and Protection of Human Rights of 2002, American Convention on Human Rights.

Jurisprudence

National Jurisprudence

Chusmiza Usmagama Community v. DGA (2005). Santiago Court of Appeals, judgment of June 2, 2005, Case 6848-2002.

Chusmiza Usmagama Community v. DGA (2006). Pozo Almonte Court, judgment of August 31, 2006, Case 1994–1996.

Constitutional review of the draft of the agreement approving ILO Convention 169 on indigenous peoples (2008). Constitutional Court, judgment of April 3, 2008, Case 1050-08.

Draft of the Agreement Approving the Draft of the International Convention for the Protection of All Persons from Forced Disappearances (2009). Constitutional Court, judgment of May 29, 2009, Case 1483-09.

INP v. Muñoz Candia (2009). Supreme Court, judgment of October 29, 2009.

Papic Domínguez v. Aymara Chusmiza and Usmagama Indigenous Community (2007). Iquique Court of Appeals, judgment of July 3, 2007, Case 817-2006.

Papic Domínguez v. Aymara Chusmiza and Usmagama Indigenous Community (2009). Supreme Court, judgment of November 25, 2009, Case 2840-2008.

Request of deputies regarding the unconstitutionality of ILO Convention 169 (2000), Constitutional Court, judgment of August 4, 2000, Case 309-2000.

Request of a group of deputies that the unconstitutionality of Article 1, number 20, letter C, number 3 and number 48 of the draft of the General Fishing and Aquaculture Law contained in Law 18.892 and its modifications be declared (2012). Constitutional Court, judgment of December 21, 2012, Cases 2387-12 and 2388 together.

Thauby Pacheco, Claudio Francisco and others (2006). Supreme Court, judgment of October 3, 2006, Case 2707-2006.

Toconce Atacameño Community v. ESSAN S.A. (2004). Supreme Court, judgment of March 22, 2004, Case 986-03.

Union v. Pinochet and others (2011). Supreme Court, judgment of January 27, 2011, Case 8314-2009.

Valdés v. Irarrázabal (2008). Constitutional Court, judgment of June 10, 2008, Case 943-2007.

International Jurisprudence

Mayagna (Sumo) Awas Tingni Community v. Nicaragua (2000) Inter-American Court of Human Rights, Judgment of February 1, 2000.

Moiwana Community v. Surinam (2005), I/A Court H.R., Judgment of June 15, 2005.

Saramaka People v. Surinam (2007), I/A Court H.R., Judgment of November 28, 2007.

Sawhoyamaxa Indigenous Community v. Paraguay (2006), I/A Court H.R., Judgment of March 20, 2006.

Xákmok Kásek Indigenous Community v. Paraguay (2010), I/A Court H.R., Judgment of August 24, 2010.

Yakye Axa Indigenous Community v. Paraguay (2005), I/A Court H.R., Judgment of June 17, 2005.

CHAPTER 4

Modeling for Management
A Case Study of the Cañete Watershed, Peru

Wendy Francesconi,[1,][*] *Natalia Uribe,*[2] *Jefferson Valencia*[3]
and *Marcela Quintero*[3]

Introduction

The Cañete River watershed located in the central Peruvian Andes, is undergoing hydrological changes due to global rising temperatures, land-use changes and increased water supply demand. At the river's source in the ice-covered mountains at 5,800 m.a.s.l., changes in the landscape are evident given the ever receding snow covered ground. According to aerial photographs of the snowcap mountains, out of the 16 snow peaks that existed in 1962, only 11 remained in 1990 (Cementos Lima S.A.). Exacerbating this situation are the changes in land use occurring at the upper and middle watershed. The watershed's natural habitat functions as a sponge, retaining much of the water that melts from the glaciers or that falls as precipitation (Wiegers et al., 1999). In addition, lakes, puddles, springs and creeks create a water holding network that transports and slows down water losses along the elevation gradient. Yet, changes in ecosystem dynamics affect the water holding capacity of the watershed, leading to more erratic water flows with longer dry periods and flush events.

---

[1] International Center for Tropical Agriculture (CIAT), Av. La Molina 1895 La Molina, Lima, Peru.
[2] UUESCO-IHE, Westvest 7, 2611 AX Delft, The Netherlands.
  Email: n.uribe@unesco-ihe.org
[3] International Center for Tropical Agriculture (CIAT), Km 17 Recta Cali-Palmira, Apartado Aéreo, 6713, Zip code: 763537 Cali, Colombia.
  Emails: j.valencia@cgiar.org; m.quintero@cgiar.org
[*] Corresponding author: W.Francesconi@cgiar.org

These changes in turn cause additional environmental problems and have socioeconomic consequences (Alurralde et al., 2011).

Animal husbandry has been a part of the native culture in the highland Andes areas (Wiegers et al., 1999). Native llama, alpaca, wanaku and vicuñas have been part of the local culture and are integrated into the livelihood strategies at high altitudes as a dominant economic activity (Blundo et al., 2016). Yet changes have been taking place not only in terms of climate, but also in the way farmers in the highland managed their land. Increases in agricultural-related economic activities along with the introduction of cattle ranching in the upper watershed pastures, have been contributing to the diversification of land cover and land-use changes in the region (Valdivia et al., 1996). Understanding these behaviors in agriculture and assessing their impacts on water quantity and quality is vital to make policies that propel changes in behavior in both the upper watershed and those impacted by their land management behaviors at the lower watershed. In the Peruvian Andes, water availability and agriculture go hand in hand. The country's privileged geographical location provides water from the glacier cover mountain tops. However, the implementation of appropriate and integrated water and land resource management is only recently being considered through the development of a political framework by the Peruvian government that allows for the legal retribution for actions that contribute to the conservation of ecosystem services (MINAN, 2014). In light of the threat of climate change, a base line understanding of water dynamics is necessary to begin addressing the current and potential problems in Andean watershed.

To help with the analysis of watershed management, the Soil and Water Assessment Tool (SWAT) has been used extensively and is well documented (Arnold et al., 2012; Moriasi et al., 2007; Francesconi et al., 2016). Within the heterogenic landscape conditions expected from the extreme changes in the elevation gradient in Andean watersheds, the spatially explicit hydrological model can help identify critical areas for water provisioning, soil erosion and others (Quintero et al., 2009). The use of SWAT in the analysis of watershed for the development of payments for ecosystem services has also been common as it can successfully estimate provisioning ecosystem services and generate simulation scenarios for improved management (Francesconi et al., 2016). With the goal of providing science-based information that could contribute in decision making and the implementation of conservation initiatives, water flow dynamics were assessed at the Cañete watershed. In the present study, the calibration and validation of SWAT was conducted for an Andean watershed, to identify the location and contribution to stream flow by different Hydrologic Response Units (HRU), in order to provide spatially distributed information for the analysis of land use and climate change impact on stream flow at the Cañete River.

Methodology

Study site

The study was conducted in the Cañete watershed located at UTM 8'543,759 N–8'676,000 N and 345,250–444,750 E in west central Peru (Fig. 4.11). The watershed and the Cañete River are located in the department of Lima, and has an area of approximately 6,192 Km². The watershed is comprised by the province of Yauyos and Cañete, and to a lower extent, the province of Huarochiri. The watershed contains a population of about 130,000, which declines as elevation increases (Apaclla, 2010). Precipitation also changes drastically with altitude from about 17 mm at the lower watershed to about 1000 mm annually. The Cañete River originates in the Ticllaconcha Lake at an altitude of 4,429 m.a.s.l., and it extends by almost 236 km before ending in the Pacific Ocean. The Cañete River has an average slope of 2%. In the upper areas slopes with steep hills can be 8%, which are smoothen out by the coastal areas in the lower parts of the watershed. The altitudinal gradient of the watershed leads to numerous ecosystem types as well as soils. From subtropical deserts, to shrub lands in the lower mountain areas, to pastures in the highlands, the watershed has been adapted to different agricultural practices.

The Cañete watershed can be divided into three distinct sections: upper, middle and lower. The upper section of the watershed, which ranges between 4,000 to 5,800 m.a.s.l. contains the glaciers. The middle watershed is comprised by the area between 350 and 4,000 m.a.s.l., and the lower section has the smallest extent (4.6% of the watershed) and goes from sea level to 350 m.a.s.l. The lower section contributes to the Cañete valley, and the interbasins of the Omas and Topará Rivers. Agriculture in the Cañete valley is a major economic activity in the watershed. The flattening of the watershed relief has created an area where the accumulation of alluvial nutrient deposits has created adequate conditions for agricultural production. The valley has more than 23,000 ha irrigated for the production of crops such as cotton and maize being the most extensive (27 and 20%, respectively), followed by sweet potatoes, fruit trees and asparagus (INRENA, 2001). Water discharge at the valley is currently 77.5% lowered compared to historical records (55.8 m³/s). Yet, demand for agricultural purposes has increased and is estimated to be about of 75% of the Cañete River stream flow given that 95% of the population depends on this economic activity (Otárola, 2011). The types of crops produced include corn, cotton and cassava, among others. The types of soils at the lower watershed range from alluvial, colluvial-alluvial and residual materials from the *in situ* weathering processes. The main soil orders found in the Canete valley include Fluvisols, Andosols, and Solonchak (ONERN 1970). The predominant soil texture is sandy loam with moderate drainage, an

Ap horizon between 0–25 cm deep. The soils are characterized by having low organic matter (1.47%) and the valley has an average slope between 2–4% (Apaclla, 2010).

SWAT Modeling

The Soil and Water Assessment Tool (SWAT) was developed by the USDA-Agricultural Research Service to model the impact of land use and management on water quality and quantity. The tool is an open source and widely used globally watershed-scale process-based hydrological model developed by the U.S. Department of Agriculture-Agricultural Research Service (USDA-ARS) (Williams et al., 2008). The model version used was the ArcGIS 9.3 software interphase ArcSWAT 2009.93.7b. The model can simulate small, medium and large watershed with varying soil, vegetation, weather and slope conditions over short and long periods of time (Francesconi et al., 2016). The main components of SWAT are hydrology, climate, nutrient cycling, soil temperature, sediment movement, crop growth, agricultural management and pesticide dynamics (Arnold et al., 1998). For modeling purposes, watersheds may be categorized into a number of sub-watersheds or sub-basins. The use of sub-basins in a simulation is particularly beneficial when different areas of the watersheds are dominated by land uses or soils dissimilar enough that it changes the hydrology. The sub-basins are further divided into Hydrological Response Units (HRU). These HRU units are land areas with unique land cover, soil and agricultural management practices. The hydrological cycle as simulated by SWAT is based on the water balance equation, which includes daily precipitation, run-off, evapotranspiration, percolation and return flow components. The surface run-off is estimated in the model using two options (a) the Natural Resources Conservation Service Curve Number (CN) method (USDA-SCS, 1972) and (b) the Green and Ampt method (Green and Ampt, 1911). The percolation through each soil layer is predicted using storage, routing techniques combined with preferential flow model (Arnold et al., 1995). The evapotranspiration is estimated in SWAT using three options (a) Priestley-Taylor (Priestley and Taylor, 1972), (b) Penman-Monteith (Monteith, 1965) and (c) Hargreaves (Hargreaves and Riley, 1985). The flow routing in the river channels is computed using the variable storage coefficient method (Williams, 1969), or Muskingum method (Chow, 1959). The SWAT model uses the Modified Universal Soil Loss Equations (MUSLE) to compute HRU-level soil erosion. The estimation of run-off energy is used in turn to estimate sediment detachment and transportation (Williams and Berndt, 1977). The sediment routing in the channel (Arnold et al., 1995) consists of channel degradation using stream power (Williams, 1980) and deposits in channel using fall velocity. Channel degradation is adjusted using USLE soil erosion tendency and channel cover factors (Betrie et al., 2011).

SWAT Model Inputs

SWAT modeling requires spatially explicit land cover, soil, weather and elevation data. For the Cañete watershed, this information was compiled from various sources (Table 4.5). The Digital Elevation Model (DEM) was extracted from CGIAR-CSI (http://srtm.csi.cgiar.org/). The site provides 90m resolution elevation data for the entire world originally produced by NASA and processed by CIAT to improve the quality of the continuous topographic surfaces for large portions of the tropics (http://srtm.csi.cgiar.org et al., 2008). Physicochemical parameters of soils were taken from national studies of each county and in some cases missing values of bulk density, soil available water content and hydraulic conductivity, were calculated with the Soil Water Characteristics tool (Saxton, 2005) based on percentages of sand, clay and organic matter. To delineate watersheds from the DEM, we manually created outlets in particular locations with the purpose of using most of the weather information for the simulation. An area of 100 hectares was chosen as the threshold for the delineation of watersheds. At the lower basin where the DEM is smoothened, the stream network map was used to force the sub-basins reaches identified in SWAT to follow known stream locations. In the generation of the HRUs, the slope was set to five ranges to get a homogeneously distributed slope map. The combinations among sub-basin, land use, soil and slope allow generated many different HRUs. Given the diverse topographic and biophysical characteristics of the Andean region, a full range of HURs is desired in order to better represent these conditions. The potential evapotranspiration (PET) was simulated using the Hargreaves method (Hargreaves et al., 1985), and the actual evapotranspiration (AET) was calculated based on the methodology developed by Ritchie (1972), which is used by the SWAT model. A more detail description of the model setup process can be found in Uribe and Quintero (2011).

Table 4.5 SWAT data input sources and description.

Data input	Elevation	Land use map	Soils map	Weather stations
Resolution/Scale/Number	SRTM-90m DEM	1:100,000	1:100,000	12
Data Source	CGIAR-CSI	IGN	ONERN-CORLIMA	SENAMHI
Data Format	GeoTiff	ADF ESRI	ADF ESRI	DBF ESRI

Data Analysis

The Natural Resources Conservation Service Curve Number (CN) method (USDA-SCS, 1972) was used to predict the surface run-off. CN2 values were determined based on a previous study in the Colombian Andes where land cover map categories where associated to SWAT's Land Cover codes (IDEAM and Andina, 2007). In this study, common statistical indicators were used to assess the model performance during calibration and validation (Gassman et al., 2007): Nash-Sutcliffe coefficient of efficiency (NSE) (Nash and Sutcliffe, 1970), index of agreement (*d*), Root Mean Square Error (RMSE), and Mean Absolute Error (MAE). Data calibration took place from 1992 until 2000, and validation from 2001 until 2009. The first step in the calibration process was to conduct a sensitivity analysis using stream flow related parameters. To evaluate the impact of the different parameters, a Latin Hypercube-One-factor-At-a-Time (LH-OAT) approach was use, which is an automatic and stratified parameter sensitivity analysis built-in in SWAT. Following, to further improve the model, a manual calibration of individual parameters was conducted by comparing observed and predicted stream flow values. Through this process, a total of 16 parameters were adjusted within the range of values suggested and based on the characteristic at the sub basin level.

Results and Discussion

SWAT Model Calibration

The observed versus the calibrated and simulated discharge values were satisfactory according to Moriasi's (2007) model evaluation standards and performance (Fig. 4.12; Table 4.6). Streamflow trends show simulated values to be underestimated during the calibration period, and overestimated during the validation years (from 2001 on). Model calibration was conducted by primarily modifying parameters related to the aquifer and ground water characteristics (Table 4.7). Parameters such as: initial depth of water in the shallow aquifer (SHALLST), groundwater delay (GW_DELAY), baseflow alpha factor (ALPHA_BF), threshold water depth in the shallow aquifer for flow (GWQMIN), threshold water depth in the shallow aquifer for 'revap' (REVAPMN), deep aquifer percolation fraction (RCHRG_DP), initial groundwater height (GWHT) and specific yield of the shallow aquifer (GW_SPYLD), are used in the estimation of base flow, and therefore directly affect steam flow values from the catchment. Additional parameters related to surface runoff lag time (SURLAG), water loss (soil evaporation compensation factor (ESCO)), snow pack and melt processes (temperature lag factor (TIMP), melt factor for snow (SMFMX and SMFMN), and snow melt base temperature (SMTMP)), were adjusted to improve the overall model calibration values. Compared to other studies, many of these parameters were also used to calibrate SWAT for a central Chilean Andean Basin (Stehr et al., 2009).

The application of SWAT to model watersheds in the Andes has been used primarily to provide information on water provisioning and the impact of climate change, glacier melt and land-use change (Stehr et al., 2009; Stehr et al., 2010a; Stehr et al., 2010b; Uribe and Quintero, 2011; Espinosa and Rivera, 2016). To improve the calibration of mountain areas, SWAT has developed functions and processes such as elevation bands and snow melt parameters to better simulate biophysical and environmental conditions in highlands. A common aspect of modeling Andes watersheds using SWAT is the lack of high resolution and adequate data (Stehr et al., 2009). Among the input data required for SWAT modeling, the most relevant to the Andes highlands will be weather and soil information. Furthermore, when available, data from weather stations does not reflect well the precipitation and temperature conditions found in mountain areas, where conditions can change dramatically within the altitudinal gradient. The same situation will apply to soil input data. Given the array of geological characteristics,

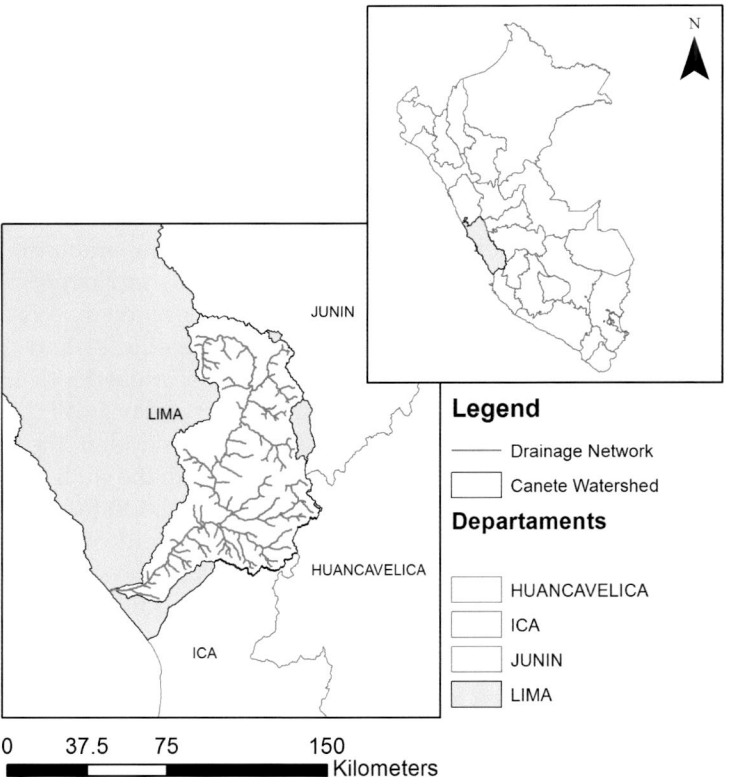

Fig. 4.11 Location of the Cañete watershed in the providence of Lima in Peru. Image depicts main tributaries to the Cañete River from the Andes to the Pacific Ocean.

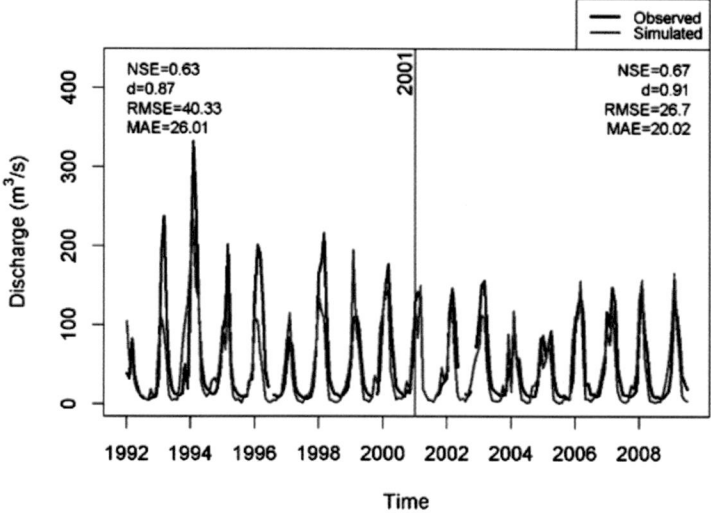

Fig. 4.12 Observed and SWAT simulated values for streamflow (m³/s) at the lower watershed outlet in the Cañete River.

Table 4.6 Calibration and validation model performance evaluation values.

	Calibration period (Jan/92–Dec/00)				Validation period (Jan/01–Jul/09)			
Test	NSE*	d±	RMSE+	MAE^	NSE*	d±	RMSE+	MAE^
Score	0.63	0.87	40.33	26.01	0.67	0.91	26.7	20.02

* Monthly average scores between 0.5 to 1.0 are generally viewed as acceptable levels of performance (Moriasi et al., 2007).

± Scores of 1 indicate a perfect agreement between the measured and predicted values, and values closer to 0 indicate little or no agreement at all (Willmott, 1981).

+ Scores of less than half the observed standard deviation value are considered appropriate (Singh et al., 2004).

^ Values close to zero are considered appropriate (Singh et al., 2004).

soil conditions may be vastly different throughout the watershed, making it difficult to model. To illustrate this point, the Peruvian Andeans holds custody of what is known as the 'rainbow mountain', a barren mountain range where you can see a strip pattern of seven different mineral deposits re-occurring throughout the landscape. The use of elevation bands at the Cañete watershed helped incorporate variation in precipitation and temperature in relation to the altitudinal gradient as has been suggested (Hartman et al., 1999; Ster et al., 2009). Moreover, to address the difficulties posed by the inadequate soil data and glacier melting processes, the calibration of the model required the adjustment of parameters related to the watershed's geological characteristics and snow melting processes.

Table 4.7 Parameter calibration in SWAT for the Cañete Watershed.

Name	Definition	Range	SWAT default	Cañete
ALPHA_BF.gw	Baseflow alpha factor [days]	0–1	0.048	0.01
CH_N2.rte	Manning's n value for main channel	0–0.3	By HRU	By HRU
CN2.mgt	Initial SCS CN II value	20–90	By HRU	By HRU
DEEPST.gw	Initial depth of water in the deep aquifer [mm]	0–3000	1000	500
ESCO.bsn	Soil evaporation compensation factor	0–1	0.95	1
FFCB.bsn	Initial soil water storage expressed as a fraction of field capacity water content	0–1	0	0.8
GW_DELAY.gw	Groundwater delay [days]	0–500	31	25
GW_SPYLD.gw	Specific yield of the shallow aquifer [m^3/m^3]	0–0.4	0.003	0.4
GWHT.gw	Initial groundwater height [m]	0–25	1	25
GWQMIN.gw	Threshold water depth in the shallow aquifer for flow [mm]	0–5000	0	1
PLAPS.suub	Precipitation lapse rate [mm/Km]	0–100	0	200
RCHRG_DP.gw	Deep aquifer percolation fraction	0–1	0.05	0
REVAPMN.gw	Threshold water depth in the shallow aquifer for "revap" [mm]	0–500	1	50
SHALLST.gw	Initial depth of water in the shallow aquifer [mm]	0–1000	0.5	50
SMFMN.bsn	Melt factor for snow on December 21 [mm H2O/°C-day]	0–10	4.5	1
SMFMX.bsn	Melt factor for snow on June 21 [mm H2O/°C-day]	0–10	4.5	3.5
SMTMP.bsn	Snow melt base temperature [CC]	(–5) –5	0.5	2
SURLAG.bsn	Surface runoff lag time [days]	1–24	4	8
TIMP.bsn	Snow pack temperature lag factor	0–1	1	0.5
TLAPS.sub	Temperature Lapse rate [°C/Km]	(–)20 –20	0	–2.7

Water Flow, Climate Change and Land Use

A slight reduction in stream flow can be observed overtime at the lower watershed outlet. The average monthly flow during 1993 to 2000 was 61 m^3/s compared to 48.9 m^3/s from 2001 to 2008. This represents a 20% reduction in flow over a period of 15 years. Influence by ENSO may explain some of these cumulative precipitation differences as the calibration period experience two La Niña events (1998–99 and 1999–2000), as opposed to just a single one during the validation period (2000–01). According to Lavado-Casimiro and Espinoza (2013), increases in precipitation can be expected during La Niña years at the latitude where the Cañete watershed is located in the Andes. However, the average precipitation values during these years was 5% lower than the average precipitation during all the calibration years. If the observed trend continues, we can expect water flow to be reduced to almost half its 1990's values over the next 15 years. Furthermore, greater discharge variability in peak flow can be observed during the validation period (Fig. 4.12), which could lead to flash flooding events or water stress in crop production (Kang et al., 2009; Igbadun et al., 2007).

The slight changes observed in stream flow over time, resulted in the slight overestimation by the model during the validation period. This overestimation seemed to benefit SWAT's model performance during the validation period as the statistical evaluation scores were improved. While the calibration period is characterized by smooth peaks and baselines, there was greater variability during the validation period. Overestimation of surface runoff and/or streamflow may be caused by underestimations in evapotranspiration or subsurface values (Arnold et al., 2012). Yet, despite the overestimations during validation period, the model captured most of the water flow variability compared to the observed data (Krause et al., 2005).

The reductions in flow were positively correlated with reductions in precipitation over time ($r^2 = 0.71$) (Fig. 4.13a). The amount of rainfall at the upper watershed (collected from four weather stations above 3800 m.a.s.l.) was 1039 mm/year during calibration, compared to 904 mm/year during validation. Climate change in Peru is expected to increases precipitation in the coastal areas and the Andes region close to the equator. Towards the south of the country, precipitation variability can be expected. However, reductions in precipitation at high elevations towards the western Andes region are forecasted (Urrutia and Vuille, 2009). The expected reductions in precipitation that are accompanied by increases in temperature at

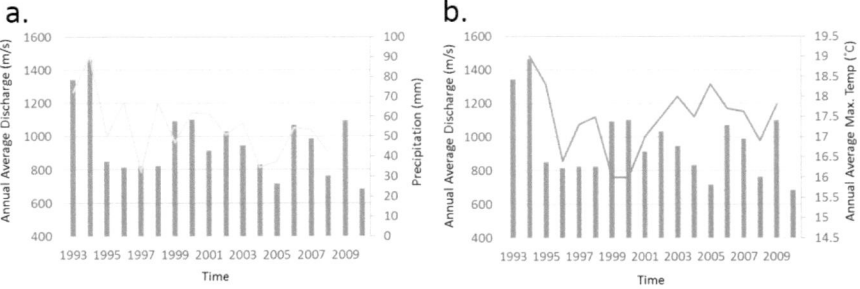

Fig. 4.13 Average annual streamflow values compared to (a) average annual precipitation and (b) average annual max. Temperature values.

high altitudes, have and will continue to negatively impact glaciers and contribute to their mass loss.

Glacier recession and nonrenewable snow melt results in increases in discharge (Mark and Seltzer, 2003; Mark and McKenzie, 2007). Yet, at the Cañete site a trend towards flow reductions is suggested by the data collected and modeled. These results may be portraying a subsequent stage of glacier melt on the hydrological processes in watersheds. Eventually, the contributions by snow melt in high mountain watersheds will diminish as the glacier mass is lost, leading to longer and reduced flows during the dry season, as well as in increases in water flow variability overall (Mark et al., 2010). The Cañete watershed is recognized because it provides year around stream flow given its numerous lakes and glaciers. Yet, the observed increase variability in water flow, especially during the dry season, has been affecting different economic sectors at the lower watershed (Apaclla, 2007). Furthermore, the negative socioeconomic impacts of water reductions are not only estimated in the agricultural sector, but also in other water dependent sectors in the region (Stern and Echavarría, 2013; MINAM, 2013).

According to the results, along with the slight decreases in discharges overtime, the average annual values for maximum temperature seem to increase (Fig. 4.13b). Data for temperature at the upper watershed was available from a single weather station (Huarochiri) located at 3154 m.a.s.l. The average annual values during the calibration period were 0.4°C lower than those during the validation years. While more weather stations and data would be required to more confidently support temperature changes over time at the upper Cañete watershed, the present results provide insight on the potential increases being experienced, which will impact water dynamics throughout the watershed. For the Andes region however, increases in temperature has been reported and forecasted (Valdivia et al., 2013; Valdivia et al., 2010). In response to the expected changes in climate, coping strategies will need to be implemented that target land cover/land use, agriculture and water management.

SWAT modeling can be used to identify the areas that contribute the most to waterflow. Hence, the same areas may be the most vulnerable to changes in climate and land use. Based on the HRU outputs in SWAT, the areas with the highest water yield values were allocated as well as the areas with the highest sedimentation values (Fig. 4.14). The identification of these areas in the watershed allows for the prioritization of potential intervention activities to incorporate agricultural practices and assign conservation zones, which could ensure the watershed's capacity to uptake, evaporate, filter and transfer water resources to the aquifer and to superficial water flow systems. As per the results, a few districts can be identified for the incorporation of conservation activities. Among the 29 districts covered by the watershed delineation area, six of them (Miraflores, Alis, Laraos,

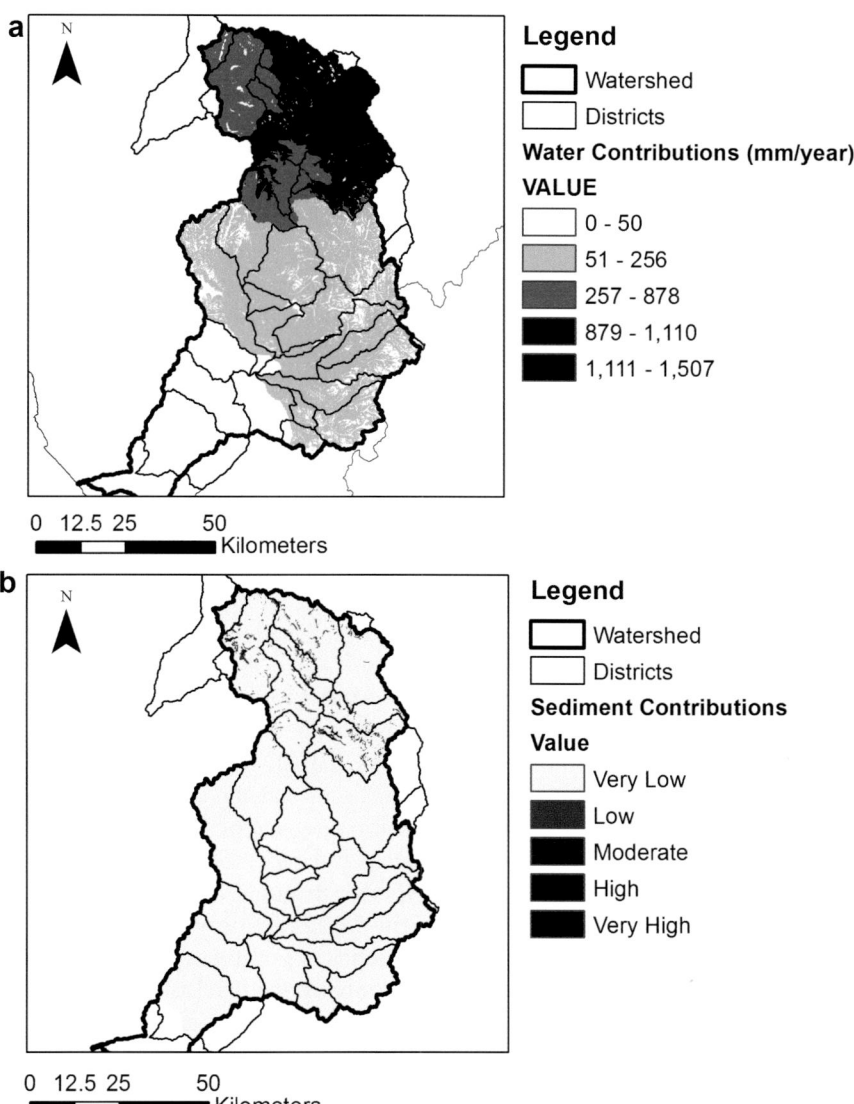

Fig. 4.14 Areas within the watershed that contribute the most to (a) water discharge and (b) sediment yield to stream flow.

Vitis, Thomas and Huancaya), with a combined area of 52,602 ha and with an average water flow contribution of 1507 mm, could be targeted for developing climate change coping strategies (Uribe and Quintero, 2011) (Fig. 4.14a). Subsequent socioeconomic analyses to understand community identified priority conservation practices has been conducted at the Cañete watershed to eventually take into consideration along with SWAT's modeling outputs (Blundo et al., 2016).

The satisfactory calibration and validation of the SWAT model allows us to explore the potential sediment contribution of the landscape to water quality in the Cañete River. The proper estimation of sediment yield values, or any given output in SWAT, requires the calibration and validation of the model for the variable in question (Arnold et al., 2012). Yet, monitoring data on sediments, or any other water quality variable is not easily available in Peru, or Latin America in general. Therefore, we can only speculate and use the stream flow calibrated/validated model to compare the spatial distribution of sediment yield based on the current soil, land use, slope and climatic conditions, and/or use the model to evaluate simulation scenarios comparing relative output values (Arnold et al., 2012). According to the speculative estimation of sediment yield contributions at the HRU level, in addition to the above districts mentioned as important for water contributions, the districts of Laraos and Tanta may be important to prioritize for the incorporation of in soil erosion control strategies (Fig. 4.14b).

Land Use Systems and Policy Design

Previous studies of the Cañete watershed have identified that about 95% of the communities at the upper watershed depend on farming and cattle ranching activates for subsistence (SPDA, 2012). In addition to climate change, anthropogenic changes in the landscape at the upper and middle watershed have contributed to its diminishing capacity to retain water and soil resources. Hence contributing to the reductions in stream flow over time. Among the evolving livelihood strategies for income and in response to climate change, farmers in the Andes are now engaging in crop diversification by introducing greater varieties of potatoes, peas and onions. Consequently, farming activities have not only intensified but also expanded (Valdivia, 2010). Animal rearing has also increased and been transformed in the region. Native and domesticated camelids such as llamas and alpacas traditionally have been a source of income and food security since pre-Inka period (Wieggers et al., 1999). Yet, the introduction of sheep, goats and cattle have modified socioeconomic and landscape management practices at the upper watershed. Increases in the number and size of animals, is prone to cause soil compaction and overgrazing, especially during the dry season when grasses are less available.

In search of increased profits, husbandry diversification seems like a good alternative for the development of resilient livelihoods. However, landscape management dynamics in the Andes is stratified into production zones (Wiegers et al., 1999). In the upper watershed, rain feed agropastoral lands have lost their cropping and grazing potential, in part due to climate change. Hence, a relocation of pastoral practices closer to water catchment areas and into vulnerable ecosystems such as the *bofedales*, which are grass dominated lands with high moisture content due to snowmelt, is now taking place (Wiegers et al., 1999). In addition to identifying priority areas for water provisioning, further analysis of SWAT's outputs at the HUR level can be overlapped with maps depicting areas where water demand is increasing and where land-use changes may present a threat to maintaining stable hydrological dynamics. This will further target the areas for activity implementation reducing cost and increasing effectiveness (Espinosa and Rivera, 2016). Water provisioning as an ecosystem service requires a comprehensive understanding of the natural and social drivers of change in order to help farmers and users throughout the watershed cope with the changes experienced in their environment and become more resilient financially, while not engaging in the continued deterioration of their most productive and vulnerable ecosystems.

The use of hydrological modeling tools to contribute to knowledge-based decision making has the potential of helping meet socioeconomic needs with environmental conservation goals. Faced with climate change threats, the development of sustainable livelihood strategies in Andean watersheds needs to be supported by policy framework that formalize climate adaptation schemes, ecosystem restoration activities, and the incorporation of sustainable agricultural practices by farmers. A payment for ecosystem services framework in Andean Watersheds, is among the mechanisms that could achieve these objectives (Quintero et al., 2009). The results from the hydrological model have been used by the Peruvian Ministry of the Environment to help design a government law that formalizes and promotes a reward mechanism for environmental conservation behavior. Residents and industry at lower and mid watershed can help ensure a stable water supply by supporting the implementation of conservation activities at the upper watershed (MINAM, 2014). The hydrological analysis conducted provided knowledge from which to better understand the impacts of climate change on water provisioning and quality, and served as a research approach in the design of a national policy that required a methodology to prioritize landscape areas for conservation management activities.

Conclusions

SWAT can be an advantageous tool providing scientific information on hydrological processes in the Andean region, where little input data exist and where the need for climate change impact on glacier covered watersheds is needed. SWAT is equipped with capabilities for modeling mountain watersheds such as elevation bands and snow melt hydrological processes. As with other agricultural modeling tools, SWAT is more effective when high quality monitoring data is available. Yet, in Andean watersheds this information may not be always available. If with the available monitoring data calibration and validation of the model is achieved and is reasonable, the outputs can still be informative in providing water accounting and identifying the location of biophysical critical areas for water delivery. This information can help in decision making when needing to target priority areas within the watershed for the incorporation of conservation activities to invest in.

From the observed data, average annual precipitation reductions at the upper Cañete watershed were correlated with a trend towards stream flow reductions at the outlet. Yet, the slight increases in temperature observed at the upper watershed were not associated with increases in stream flow, only what seems to be greater annual variability. Reliant and longer data collection efforts are necessary to thoroughly understand the impacts of climate change on water provisioning. In the meantime, by understanding climate change induced trends in water flow and land-use management, it is possible to develop integrated agricultural management schemes that can increase the financial resilience of farmers throughout the watershed, and in turn, reduce the impacts of farming practices on the environment. The idea is to use agriculture as a tool for conservation and a component in payment for ecosystem services schemes to mitigate climate change impacts. While SWAT can provide information on the biophysical conditions, this is not sufficient to develop an understanding of the problems in climate change affected watershed. In addition to SWAT modeling, socioeconomic research and a participatory approach are required to develop potential solutions that are inclusive and government supported.

SWAT's modeling outputs at the Cañete watershed can help us understand how other similar Andean-Pacific coastline watersheds could be behaving in response to reductions in precipitation and temperature increases, as well as the potential impacts on stream flow by peak precipitation events. Taking advantage of the calibrated and validates model, the tool's additional value lies in its capacity to evaluate distinct land management practices in prioritized sub-watersheds to evaluate appropriate actions to secure stable water provisioning. This would be the subsequent line of modeling required to assist in land management decision making. Yet prior, the identification of the most likely land management activities will need to be determined in a participatory manner in order to develop realistic modeling scenarios to simulate. At the Canete watershed, the current design and implementation of a sustainable (self-funded in time) PES mechanisms piloted by the Peruvian government, will contribute to the identification of such scenarios and hence to the overall formalization of similar initiatives throughout the country.

References

Alexander, L.V., Zhang, X., Peterson, T.C., Caesar, J., Gleason, B., Klein Tank, A.M.G., Haylock, M., Collins, D. Trewin, B., Rahimzadeh, F., Tagipour, A., Rupa Kumar, K., Revadekar, J., Griffiths, G., Vincent, L., Stephenson, D.B., Burn, J., Aguilar, E., Brunet, M., Taylor, M., New, M. Zhai, P., Rusticussi, M. and Vazquez-Aguirre, J.L. 2006. Global observed changes in daily climate extremes of temperature and precipitation. J. Geophys. Res. 111(D5): 1–22.

Alurralde, J.C., Ramirez, E., García, M., Pacheco, P., Salazar, D. and Mamani, R.S. 2011. Living with glaciers, adapting to change the experience of the Illimani project in Bolivia. In Proc. of the XIVth IWRA World Water Congr. Brazil 1–8.

Andrés Doménech, I., García Bartual, R.L., Montanari, A., Segura, M. and Bautista, J. 2015. Climate and hydrological variability: the catchment filtering role. Hydrol. Earth Syst. Sci. 19(1): 379–387.

Apaclla Nalvarte, R. 2007. Estudio de máximas avenidas en las cuencas de la zona centro de la vertiente del Pacifico. Dirección de conservación y planeamiento de recursos hídricos. Ministerio de Agricultura. Autoridad Nacional del Agua. Available online http://www.ana.gob.pe/media/390377/informe%20final%20zona%20centro.pdf.

Arnold, J.G., Srinivasan, R., Muttiah, R.S. and Williams, J.R. 1998. Large area hydrologic modeling and assessment: Part I. Model development. J. Am. Water Resour. Assoc. 73–89.

Arnold, J.G., Moriasi, D.N., Gassman, P.W., Abbaspour, K.C.,White, M.J., Srinivasan, R., Santhi, C., Harmel, R.D., van Griensven, A., Van Liew, M.W. and Kannan, N. 2012. SWAT: Model use, calibration, and validation. Trans. ASABE 55(4): 1491–1508.

Betrie, G.D., Mohamed, Y.A., van Griensven, A. and Srinivasan, R. 2011. Sediment management modelling in the Blue Nile Basin using SWAT model. Hydrol. Earth Syst. Sci. 15: 807–818. Doi: 10.5194/hess-15-807-2011.

Blundo Canto, G., Cruz-García, G.S., Tristán Febres, M.C., Pareja Cabrejos, P. and Quintero, M. 2016. Prioridades de conservación y desarrollo en las comunidades de Nor Yauyos. Informe para el MRSEH de la cuenca del río Cañete. Centro Internacional de Agricultura Tropical (CIAT). Cali, Colombia.

Cementos Lima, S.A. 2001. Estudio glaciológico de la cuenca alta del río Cañete. En: INRENA, DGAS y ATDRMOC. Evaluación y ordenamiento de los recursos hídricos de la cuenca del río Cañete.

Espinosa, J. and Rivera, D. 2016. Variations in water resources availability at the Ecuadorian páramo due to land-use changes. Environ. Earth Sci. 75(16): 1173.

Francesconi, W., Srinivasan, R., Pérez-Miñana, E., Willcock, S.P. and Quintero, M. 2016. Using the Soil and Water Assessment Tool (SWAT) to model ecosystem services: A systematic review. J. Hydrol. 535: 625–636.

Gassman, P.W., Reyes, M., Green, C.H. and Arnold, J.G. 2007. The Soil and Water Assessment Tool: Historical development, applications, and future directions. Trans. ASABE 50(4): 1211–1250.

Green, W.H. and Ampt, C.A. 1911. Studies on soil physics: I. Flow of air and water through soils. J. Agr. Sci. 4: 1–24.

Hargreaves, G.L., Hargreaves, G.H. and Riley, J.P. 1985. Agricultural benefits for Senegal River Basin. Journal of Irrigation and Drainage Engineering 111: 113–124.

IDEAM, C. and ANDINA, O. 2007. Nueva medición de la calidad del agua en los ríos Magdalena y Cauca. Colombia.

Igbadun, H.E., Tarimo, A.K.P.R., Salim, B.A. and Mahoo, H.F. 2007. Evaluation of select crop water production functions for an irrigated maize crop. Agr. Water Manage. 94(1-3): 1–10.

Instituto Nacional de Recursos Naturales (INRENA). 2001. Estudio hidrogeológico del valle de Cañete. Dirección General de Aguas y Suelos. Ministerio de Agricultura y Riego. http://

www.ana.gob.pe/sites/default/files/normatividad/files/estudio_hidrogeologico_
   canete_0_0.pdf.

Jarvis, A., Reuter, H.I., Nelson, A. and Guevara, E. 2008. Hole-filled seamless SRTM data V4,
   International Centre for Tropical Agriculture (CIAT). http://srtm.csi.cgiar.org.

Krause, P., Boyle, D.P. and Bäse, F. 2005. Comparison of different efficiency criteria for
   hydrological model assessment. Adv. Geosc. 5: 89–97.

Lavado Casimiro, W.S. and Espinoza, J.C. 2014. Impactos de El Niño y La Niña en las lluvias
   del Perú (1965–2007). Rev. Brasileira Meteorol. 2: 171–182.

Ministerio de Medio Ambiente (MINAM). 2014. Ley de Mecanismos de retribución por
   servicios ecosistemicos. Ley 30215. Available at: http://www.minam.gob.pe/wp-
   content/uploads/2014/06/ley_302105_MRSE.pdf.

Monteith, J.L. 1965. Evaporation and environment. Symp. Soc. Exp. Biol. 19: 205–234.

Moriasi, D.N., Arnorld, J.G., Van Liew, M.W., Bingner, R.L., Harmel, R.D. and Veith, T.L. 2007.
   Model evaluation guidelines for systematic quantification of accuracy in watershed
   simulations. Am. Soc. Agr. Biol. Eng. 50: 885–900.

Nash, J.E. and Sutcliffe, J.V. 1970. River flow forecasting through conceptual models, Part I: a
   discussion of principles. J. Hydrol. 10(3): 282–290.

Neitsch, S.L., Arnold, J.G., Kiniry, J.R., Williams, J.R. and King, K.W. 2005. SWAT Theoretical
   Documentation Version 2005. Soil and Water Research Laboratory, ARS, Temple Texas,
   USA.

Oficina Nacional de Evaluación de Recursos Naturales (ONERN). 1970. Inventario, evaluacion
   y uso racional de los recursos naturales de la costa. Cuenca del Rio Cañete. Volumen I.
   Autoridad Nacional del Agua. Lima-Peru. Available at http://repositorio.ana.gob.pe/
   handle/ANA/264.

Otarola, E. 2011. Informe final del diseño del esquema de PSA hidrológico de la cuenca del rio
   Cañete. Preparado por MINAM, CIAT, CARE y WWF. http://www.forest-trends.org/
   documents/files/doc_4356.pdf.

Quintero, M., Wunder, S. and Estrada, R.D. 2009. For services rendered? Modeling hydrology
   and livelihoods in Andean payments for environmental services schemes. Forest Ecol.
   Manag. 258(9): 1871–1880.

Quintero, M. and Pareja, P. 2015. Estado de avance y cuellos de botella de los mecanismos
   de retribución por servicios ecosistémicos hidrológicos en Perú. Cali, CO: Centro
   Internacional de Agricultura Tropical (CIAT). 40 p (Publicación CIAT No. 411).

Ritchie, J.T. 1972. A model for predicting evaporation from a row crop with incomplete cover.
   Water Resour. Res. 8: 1204–1213.

Rostamian, R., Jaleh, A., Afyuni, M., Mousavi, S.F., Heidarpour, M., Jalalian, A. and Abbaspour,
   K.C. 2008. Application of a SWAT model for estimating runoff and sediment in two
   mountainous basins in central Iran. Hydrol. Sci. J. 53(5): 977–988.

Saxton, K.E. and Rawls, W. 2005. Soil Water Characteristics: Hydraulics Property Calculator.
   USDA Agricultural Research Service and USDA-ARS, Hydrology and Remote Sensing
   Laboratory. http://hydrolab.arsusda.gov/soilwater/Index.htm.

Singh, J., Knapp, H.V., Arnold, J.G. and Demissie, M. 2005. Hydrological modeling of the
   Iroquois river watershed using HSPF and SWAT. J. Am. Water Resour. Assoc. 41(2):
   343–360.

SPDA. 2012. Viabilidad legal del esquema PSEH Cañete, lineamientos generales para PSEH y
   propuestas legales para canalización de fondos. Preparado por J.L. Capella Vargas para
   Conservación Internacional, Perú.

Stehr, A., Debels, P., ARUMI, J.L., Romero, F. and Alcayaga, H. 2009. Combining the Soil and
   Water Assessment Tool (SWAT) and MODIS imagery to estimate monthly flows in a
   data-scarce Chilean Andean basin. Hydrol. Sci. J. 54(6): 1053–1067.

Stehr, A., Aaguayo, M., Link, O., Parra, F., Romero, O. and Alcayaga, H. 2010a. Modelling the hydrologic response of a mesoscale Andean watershed to changes in land use patterns for environmental planning. Hydrol. Earth Syst. Sci. 14: 1963–1977.

Stehr, A., Debels, P., Arumí, J., Alcayaga, H. and Romero, F. 2010b. Modelación de la respuesta hidrológica al cambio climático, experiencia de dos cuencas del centro-sur Chileno. Tecnolo. y Cienc Ag. 1(4): 37–58.

Stern, M. and Echavarría, M. 2013. Mecanismos de retribución por servicios hídricos para la Cuenca de Cañete, Departamento de Lima, Perú. Mecanismos de retribución por servicios hídricos Perú. Washington, DC: Forest Trends.

United States Department of Agriculture-Soil Conservation Service (USDA-SCS). 1972. National Engineering Handbook, Section IV, Hydrology https://directives.sc.egov.usda.gov/OpenNonWebContent.aspx?content=18393.wba.

Uribe, N. and Quintero, M. 2011. Aplicación del modelo hidrológico SWAT (Soil and Water Assessment Tool) a la cuenca del río Cañete. Primer Informe. Cali, Colombia.

Urrutia, R. and Vuille, M. 2009. Climate change projections for the tropical Andes using a regional climate model: Temperature and precipitation simulations for the end of the 21st century. J. Geoph. Res. Atmospheres 114(D2).

Valdivia, C., Dunn, E.G. and Jetté, C. 1996. Diversification as a risk management strategy in an Andean agropastoral community. Am. J. Agr. Econ. 78(5): 1329–1334.

Valdivia, C. 2004. Andean livelihood strategies and the livestock portfolio. Cult. Agr. 26(1-2): 69–79.

Valdivia, C., Seth, A., Gilles, J.L., García, M., Jiménez, E., Cusicanqui, J., Navia, F. and Yucra, E. 2010. Adapting to climate change in Andean ecosystems: Landscapes, capitals, and perceptions shaping rural livelihood strategies and linking knowledge systems. Ann. Assoc. Am. Geogr. 100(4): 818–834.

Valdivia, C., Thibeault, J., Gilles, J.L., García, M. and Seth, A. 2013. Climate trends and projections for the Andean Altiplano and strategies for adaptation. Adv. Geosci. 33: 69–77.

van Griensven, A., Meixner, T., Grunwald, S., Bishop, T., Diluzio, M. and Srinivasan, R. 2006. A global sensitivity analysis tool for the parameters of multi-variable catchment models. J. Hydrol. 324: 10–23.

Williams, J. and Berndt, H. 1977. Sediment yield prediction based on watershed hydrology. Trans. ASAE 20: 1100–1104.

Williams, J. 1980. SPNM, a model for predicting sediment, phosphorus, and nitrogen yields from agricultural basins. J. Am. Water. Resour. As. 16: 843–848.

Williams, J.R. 1969. Flood routing with variable travel time or variable storage coefficients. Trans. ASAE 2: 100–103.

Willmott, C.J. 1981. On the validation of models. Phys. Geogr. 2(2): 184–194.

CHAPTER 5

Mass Balance and Meteorological Conditions at Universidad Glacier, Central Chile

∞∞

Christophe Kinnard,[1,*] *Shelley MacDonell,*[2] *Michal Petlicki,*[3] *Carlos Mendoza Martinez,*[1,4] *Jakob Abermann*[5] *and Roberto Urrutia*[2,6]

Introduction

The Andes Cordillera is the water tower for several countries of South America. This is especially true in semi arid areas where water availability is limited due to low precipitation and where the bulk of the surface runoff and groundwater recharge is generated in the high mountains, far from the coastal population centers and cultivated lowland valleys (e.g., Viviroli et al., 2007). In the central region of Chile (32–37° S), runoff generation from mountains and its response to climate change have significant impacts on society. This region boasts the highest concentration (two thirds) of the country's population and includes the three largest metropolitan areas— Santiago, Valparaíso and Concepción. Its fertile central valley, also known as the 'intermediate depression', extends north-south between the Andes to

---

[1] Department of Environmental Sciences, Centre de Recherche sur les Interractions Bassins Versants - Écosystèmes Aquatiques, Université du Québec à Trois-Rivières, Trois-Rivières, Canada.
[2] Centro de Estudios Avanzados en Zonas Áridas, La Serena, Chile.
[3] Institute of Geophysics, Polish Academy of Sciences, Warsaw, Poland.
[4] Centro de Ciencias Ambientales EULA, Universidad de Concepción, Concepción, Chile.
[5] ASIAQ, Nuuk, Greenland.
[6] Centro de Recursos Hídricos para la Agricultura y Minería CRHIAM/CONICYT/FONDAP/15130015.
* Corresponding author: christophe.kinnard@uqtr.ca

the east and the coastal range to the west and is the agricultural heartland of the country. The climate of central Chile ranges from semi arid to the North, to temperate Mediterranean in the central and southern parts, being characterized on average by hot and dry summers and wet and cool winters The region's economy mainly lies on the exploitation of natural resources, by way of copper mining, logging, agriculture, wine production and fishing, as well as on the manufacturing sector, and is thus heavily dependent on the availability of water resources, especially during the summer growing season when precipitation is scarce and the demand for water is highest (Valdes-Pineda et al., 2014). A large portion of the annual precipitation falls as snow in the Andes during the winter, part of which accumulates on glaciers. Melting of snow and ice surfaces during the following spring and summer results in a major seasonal meltwater pulse which sustains river flow and groundwater recharge. This nivo-glacial Andean river regime is essential for agriculture—water accumulated as snow during the cold and humid winter is released by melting of snow and glacier ice during the warm, dry summer growing season when water demand is the highest. On the other hand, the amount and relative contribution of rainfall to streamflow increases southward and westward, following the decreasing elevation of the Andes. As a result, the main river catchments (e.g., Aconcagua, Maipo, Cachapoal, Tinguiririca) are mainly fed by a mix of rain and snow melt, to which glacier melt contributes during the summer period of low discharge. The total glaciated area in the Central Glaciological Zone (32–36° S), as defined by the Chilean Water Directorate (Dirección General de Aguas—DGA) is 855 km²[*MOP-DGA*, 2016]. Approximately 64% of this area corresponds to glaciers and 36% to rock and debris-covered glaciers (Janke et al., 2015). Hence despite the sometimes considerable glacier cover in headwater catchments, seasonal snow remains the principal source of meltwater in the central and northern Chilean Andes (Favier et al., 2009; Masiokas et al., 2006). Still, the hydrological contribution of glaciers can become significant during drought periods and very dry summers (Bown et al., 2008; Gascoin et al., 2011; Huss, 2011; Rabatel et al., 2011). As such, glaciers provide important ecosystem services related to water regulation and supply (Brauman et al., 2007) by releasing more water during warm and dry years, and storing water during colder and moister years (Fountain and Tangbor, 1985; Jansson et al., 2003). This is especially important in climates regulated by low-frequency ocean-atmosphere oscillations, such as the El Niño Southern Oscillation (ENSO) affecting central Chile (Garreaud, 2009; Garreaud et al., 2009). Currently increasing temperature along the western slope of the Andes (Falvey and Garreaud, 2009; Vuille et al., 2015), together with an observed decrease in precipitation over the previous decades, has resulted in an acute hydrological stress in central Chile (Boisier et al., 2016). Ongoing and projected climate change may strongly impact glacial and snow storage (Carrasco et al., 2005), thereby posing a long-term threat to future water resource availability in the cultivated lowlands downstream. As in other semi arid regions of the world, decreasing (Barnett et al., 2005) and reduced glacier storage (Kaser et al., 2010) may lower the capacity of catchments to buffer the impacts of climate variability, including extreme events, on the hydrological cycle. The prospect of long-term reduced glacier storage and associated increased inter annual variability in stream flow may exacerbate conflicts in central-northern Chile, where water demand is high (Aitken et al., 2016).

Knowledge of glacier processes and their interactions with climate is essential in order to assess the importance of glaciers within the regional hydrological cycle, and to diagnose their sensitivity to climate change. There have been various studies reporting changes in glacier length and area in Chile over the past two decades (Bown and Rivera, 2007; Bown et al., 2008; Malmros et al., 2016; Nicholson et al., 2010; Rabatel et al., 2011).

Recent summaries for Central Chile report a general decrease in glacier lengths and areas, however with large differences between glaciers, which highlights the importance of site specific factors (e.g., local topography, glacier geometry, microclimates) in controlling the response of glaciers to climate change (Malmros et al., 2016; Pellicciotti et al., 2014). Observations regarding changes in glacier length and area are more abundant due to the relative ease to map glaciers from remote-sensing sources. However while still valuable, they give an incomplete picture of the hydrological contribution and response of glaciers to climate variability, as the dynamical response of glaciers to climate can be slow and depend on their size and altitudinal distribution, among other factors (Cuffey and Paterson, 2010). Instead, the *glacier mass balance* is considered the best indicator of a glacier's 'health status'. It represents the direct link between a glacier and the atmosphere, because climate variations will result in direct changes to the glacier surface through variations in snow accumulation and snow and ice ablation rates. In contrast, the dynamic response of glaciers to climate variations—i.e., the advance or retreat of a glacier—occurs following a prolonged period of positive or negative mass balance, and thus represents a delayed response to climate change. The mass balance of mountain glaciers can be obtained (i) from direct field measurements using the glaciological method, (ii) via indirect observations using the geodetic method which derives mass balance from topographic changes over time, or (iii) via the hydrological method which infers mass balance as a residual term of the water balance equation (see Cuffey and Paterson, 2010 for a detailed review of respective methods). The glaciological method is recognized as the benchmark for measuring glacier mass balance. However it is field-intensive, costly, and sometimes logistically complicated and risky. For these reasons there are only few glaciers in the world with long mass balance records, but these data have been highly valuable for assessing worldwide trends in glacier mass balance (Medwedeff and Roe, 2016; Zemp et al., 2009; Zemp et al., 2015). Field mass balance data is particularly scarce in Chile given its extensive glacial cover. Mass balance measurements have been carried out since 1977 on Echaurren Norte Glacier in Central Chile (33.5° S, 0.23 km² in 2008); results indicate a positive net mass balance for the period 1977–1991 (Escobar et al., 1995) and an overall pronounced negative net mass balance afterward until 2008 (MOP-DGA, 2009). This is the only long-term glacier monitoring site in Chile. Other ongoing glacier monitoring programs have existed since 2002, for example, within the Pascua-Lama mining project in northern Chile (29° S) (Gascoin et al., 2011; Rabatel et al., 2011) and in the Lake District, further South (40° S) (Bown et al., 2007). Using published mass balance data along the whole Andes range Mernild et al. (2015) calculated mean glacier mass balance estimates for central-northern Chile of -650 ± 530 mm w.e. a^{-1} (mm of water equivalent per annum) for the period 1993–2002, and -770 ± 220 mm w.e. a^{-1} for the period 2003–2012. The large uncertainty points to the critical lack of data in the Chilean Andes. The creation in 2008 of a Glaciology Unit under the auspice of the Chilean Water Directorate DGA and a National Glacier Strategy is sparkling new and better coordinated monitoring and research efforts in glaciology, including new glacier inventories, glacier change detection studies, and field mass balance and hydrometeorological studies (MOP-DGS, 2009).

Due to the scarcity of glacier mass balance data, models have been developed and used extensively to assess glacier mass balance and its response to observed and projected climate conditions (e.g., Gabbi et al., 2014; Radic et al., 2014). Yet these models remain dependant on field mass balance observations and hydrometeorological data, however limited these may be, for calibration and validation. Hence temporary but intensive glacier monitoring efforts have high values for (i) characterizing the present mass balance status and driving meteorological conditions; (ii) providing calibration and validation data for glacier models, which can be further used to (iii) reconstruct or project glacier mass balance in response to climate change (e.g., Pellicciotti et al., 2008; Ragettli and Pelliciotti, 2012).

The present chapter presents results from such an intensive glacier monitoring program, conducted over a period (2012–2014) on Universidad Glacier (34°40' S, 70°20' W), the largest glacier in Chile outside Patagonia. The objectives of this chapter are twofold: (i) to outline the methods used to measure, calculate and interpret glacier mass and energy balance data, thereby serving as a guide for future field studies in the region; (ii) to present the first complete mass balance estimates for Universidad Glacier and discuss meteorological controls on glacier ablation over two complete hydrological years.

Study Area

Universidad Glacier (34°40′ S, 70°20′ W) is located above the commune of San Fernando in the O'Higgins Region of Chile, in the headwaters of the Tinguiririca River, itself a sub-catchment of the Rapel River basin (Fig. 5.15). It is part of the third glaciological zone that extends from 32–36° south (MOP-DGA, 2009). Morphologically speaking, it is a valley glacier that develops from the union of two contiguous sub-basins or cirques from which glacial tongues flow down, one toward the South and the other toward the West; both tongues coalesce together below icefalls. The main, lower glacier tongue displays ogives, with alternating bands of clear and debris-rich ice (Lliboutry, 1958) (Fig. 5.15). The glacier area in 2014 was 27.6 km^2 and its linear length 10 km, from its front to the upper part of the western sub-basin. The glacier elevation ranges from 2450 m a.s.l. to a maximum of 4550 m a.s.l. in its northern sub-basin. The climate in the region is Mediterranean semi arid-temperate with annual precipitation around 700 mm a^{-1} (MOP-DGA, 2010). The area is located in a climatic transition zone: precipitation at high altitudes (above 2500 m a.s.l.) fluctuates between 500 mm a^{-1} in the northern semi-arid part of the region (32° S), to up to 2500 mm a^{-1} at 36° S (Pellicciotti et al., 2014). Interannual variability in precipitation is largely influenced by ENSO and the Pacific Decadal Oscillation (Garreaud, 2009; Garreaud et al., 2009). The 0°C isotherm altitude decreases in the same latitudinal range, from about 4000 m a.s.l. at 32° S to 3000 m a.s.l. at 36° S (Carrasco et al., 2005). According to previous studies carried out by the Glaciology Unit of the (DGA), Universidad Glacier lost an area of 1.99 km^2 during the period 1945–2011, which amounts to an average of –0.03 km^2 per year and a 6% total areal loss from the initial surface in 1945. The front retreated 1430 m during the same period, which amounts to –22 m per year on average. Universidad Glacier is the source of the San Andrés River, which changes its name downstream to Azufre River and drains into the Tinguiririca River, and eventually into Rapel River.

Methods

Mass balance measurements

Glacier mass balance (*b*) is the amount of mass gained or lost at a given point and time, usually expressed in millimeters or meters of water equivalent (mm w.e. or m w.e.) (e.g., Cuffey and Paterson, 2010). In practical terms, it is the difference between the volume of water deposited as snow (the sum of precipitation, vapor deposition, wind transport and avalanches), and the volume of water lost through ablation (the sum of melt, sublimation, wind

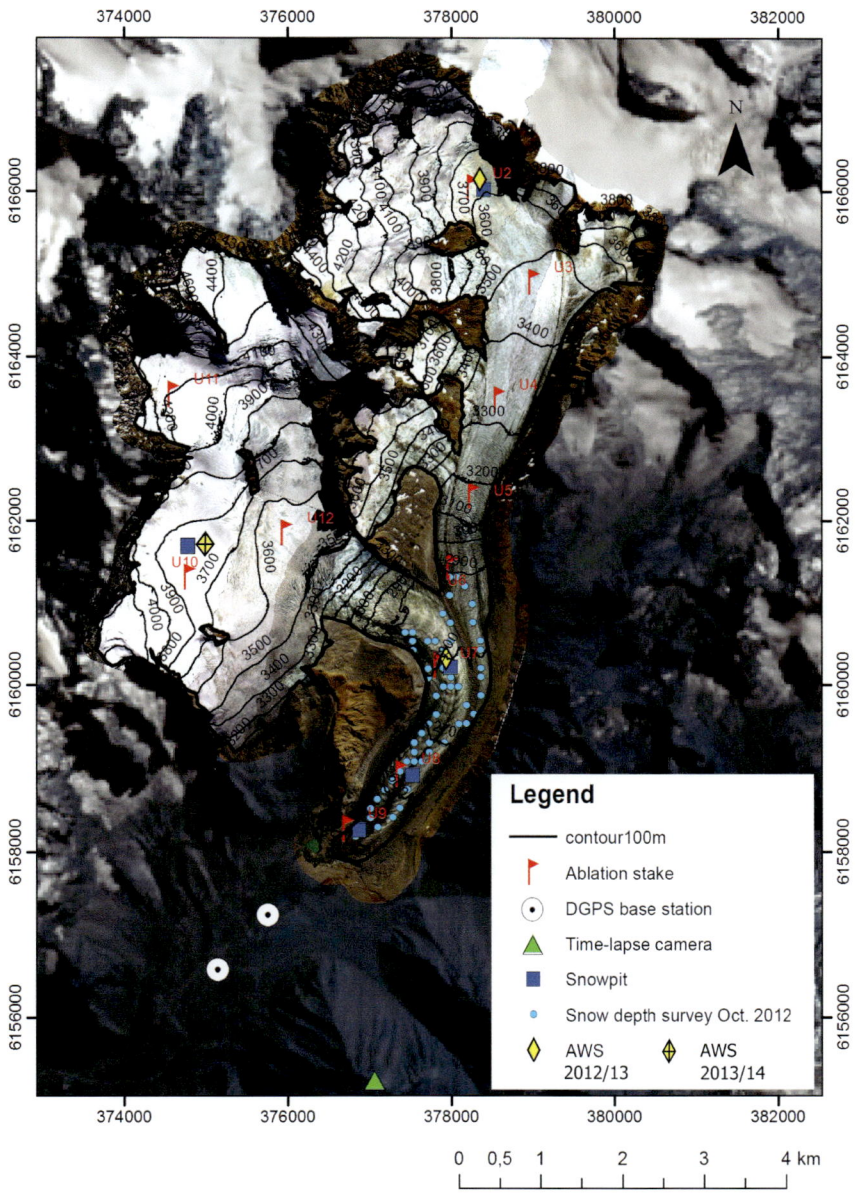

Fig. 5.15 Location of monitoring points on Universidad Glacier. In 2013 the upper AWS was moved from the eastern basin to the western basin. Background image: Landsat, 12 April 2013. Image of the glacier: hyperspectral image from 17–23 April 2013.

erosion and calving) on and from the glacier surface over a given period of time (*dt*). Mass changes arising from ice melt and refreezing of melt or rainwater at the glacier bed (basal mass balance) and within the glacier (internal mass balance), are typically of much smaller magnitude than the surface mass balance, and these terms are commonly ignored for mountain glaciers. Thus in this chapter the mass balance refers to the surface mass balance, i.e., mass changes resulting from the exchange of water and energy between the atmosphere and the glacier surface. The mass balance *b* at a point *x* on a glacier can be defined by:

$$b(x) = \int_{t_1}^{t_2} \dot{b}(x,t)\,dt \tag{1}$$

where $\dot{b}$ is the mass balance rate in time (Kaser et al., 2002). Typically, a glacier's *annual* mass balance is measured over the hydrologic year, i.e., from the beginning of winter until the end of summer. By convention, the hydrological year in the southern hemisphere is defined from 1 May to 30 April. In regions where the climate displays pronounced seasonality, such as in the central Chilean Andes, snow accumulation occurs mostly during winter and ablation predominantly in summer. Thus the glacier's annual mass balance can be expressed as:

$$b_a = c_a + a_a = b_w + b_s \tag{2}$$

where: b_a is the annual mass balance; c_a is the annual accumulation; a_a is the annual ablation; b_w is the winter balance; and b_s is the summer balance (Cogley et al., 2011). In reality, some ablation can occur during winter and some accumulation can take place in summer, which makes $b_w < c_t$ and $b_s < a_t$, but this does not affect the annual balance calculation over the hydrological year. Mass balance can be measured between fixed dates, such as at the beginning and end of the hydrological year (fixed date system) (Cogley et al., 2011; Østrem and Brugman, 1991). In the fixed date system corrections are applied to the measured balances if the measurement dates are not constant from one year to the other, so as to ensure that the mass balance is constantly referenced to the same period. The floating date system is more common, wherein the glacier is visited on variable dates near the beginning and end of the hydrological year, but without applying corrections. Another common method is the stratigraphic method, which defines mass balance in relation to the snow stratigraphy, i.e., a recognizable layer in the snowpack that marks the end of the previous summer surface, regardless of the date assigned to it. In practice however, most field studies use a system combining these last two methods (the combined system). The present study uses the combined system—the glacier was visited at about the same time each year, but the winter balance was measured with reference to the last summer layer identified in snowpits in the accumulation area or by probing to the ice surface in the ablation zone, while ablation was measured repeatedly at reference stakes. The glacier was visited once by skiing and by helicopter the remaining times, due to complicated access. Mass balance was measured over two hydrological years, 2012/13 and 2013/14, at 10–11 stakes on the glacier (Fig. 5.15). Stakes consisted of aluminium or polyvinyl chloride (PVC) tubes drilled into the firn or ice surface using a Kovacs ice drill, and plugged with PVC (in case of aluminium stakes) at the bottom end to restrict heat conduction and sinking of the stakes. Snow depth was measured by probing to the ice surface at nine points within a 3 × 3 m square centerd on each ablation stake, and averaging the results. Additionally, three (2012) and four (2013) snowpits were excavated, in which the snow density and temperature profiles were measured (Fig. 5.16). In 2012 an extensive snow depth survey was conducted on the lower glacier (Fig. 5.15). Snow depths (when present) and ablation were measured on a near-monthly basis between late October and late April/early May of each year. Not all measurement points could be surveyed at each visit, but all points were surveyed in late winter/early spring and at the end of summer.

Fig. 5.16 Field mass balance and meteorological measurements on Universidad Glacier. (A) snow stratigraphic measurements at point u2 in the accumulation zone, October 2012. (B) Measuring the position of the lowest stake (u9) in the ablation zone, April 2013. (C) AWS1 on 25 October 2012; (D) AWS2 on 25 October, 2012; (E) AWS1 on March 5, 2013; (F) AWS2 on 23 March, 2013. Refer to Fig. 5.15 for AWS positions.

Glacier-wide mass balance

A typical valley glacier can be separated into an accumulation zone where the annual mass balance is positive ($b_a > 0$), and an ablation zone where the annual mass balance is negative ($b_a < 0$), with the two zones being separated by the equilibrium line where the balance is nil ($b_a = 0$).

Mass balance on mountain glaciers typically displays a strong relationship with altitude. The *mass balance gradient* is the rate of change of mass balance with altitude (db/dz in mm w.e. m^{-1}); it drives the ice flow from the accumulation to the ablation zone. Its value near the Equilibrium Line Altitude (ELA) is called the *activity index* (Lliboutry, 1964) and is positively related with the ice turnover rate of the glacier. The mass balance gradient is typically greater in the ablation zone (10 to 20 mm m^{-1}) where it depends primarily on temperature-related ablation processes, but reduces in the glacier accumulation zone where it depends primarily on accumulation processes (Francou and Pouyaud, 2004).

The glacier-wide, specific mass balance (B_a, in mm w.e. a^{-1}) is obtained by integrating the point balance measurements (b_a) to the whole glacier surface (A) (Cuffey and Paterson, 2010):

$$B_a = \frac{1}{A} \int_A b_a dA \qquad (3)$$

The area of Universidad Glacier is 27.44 km^2, as mapped manually on a mosaic of hyperspectral images acquired in April 2013 and orthorectified with a Digital Elevation Model (DEM) acquired from an airborne lidar survey on the same date. The traditional method to derive B_a is to interpolate the point mass balance measurements to the whole surface using manual contouring or automatic spatial interpolation methods (Østrem and Burman, 1991). This requires that observations be sufficient and well distributed on the glacier surface. This is often the case for glaciers with a smooth surface and easy access. However extensive work is required to install and maintain such an observational network. Conversely, the strong relationship between mass balance and elevation means that a more limited number of observations can be reliably interpolated to whole surface, using the vertical balance gradient (Foutain and Vecchia, 1999). This method also allows calculating a standard error on the glacier-wide mass balance. This method was used in this study, given the difficult access to the glacier and the resulting sparse observation network (Fig. 5.15). The seasonal mass balance-altitude relationship, $b(z)$, was estimated by fitting a second-order polynomial to the winter (b_w) and summer (b_s) balance data for each year, using least-square regression. The polynomial coefficients were then summed to provide the annual mass balance-altitude function, $b_a(z)$:

$$b_w(z) = p1_w z^2 + p2_w z + p3_w \qquad \text{(4a)}$$

$$b_s(z) = p1_s z^2 + p2_s z + p3_s \qquad \text{(4b)}$$

$$b_a(z) = (p1_w + p1_s)z^2 + (p2_w + p2_s)z + (p3_w + p3_s) \qquad \text{(4c)}$$

The glacier-wide mass balance (B), omitting the subscripts s, w, and a, is then:

$$B = p1\overline{z^2} + p2\overline{z} + p3 \qquad \text{(5)}$$

where: $\overline{z^2}$ is the glacier average square elevation; and $\overline{z}$ is the glacier's average elevation (Fountain and Vecchia, 1999). The standard error on B is then:

$$SE(B) = \left(\frac{\sigma^2}{A} + x^T V x\right)^{1/2} \qquad \text{(6)}$$

where: σ^2 is the regression standard error using the least squares method (root mean squared error of the residuals in Equation 5); A is the glacier area, x is a column vector $(1, \overline{z^2}, \overline{z})$; T is the matrix transpose; and V the variance-covariance matrix of the regression coefficients (Fountain and Vecchia, 1999).

Meteorological measurements

Two Automated Weather Stations (AWS) were operated on the glacier during the study period (Fig. 5.15). The upper station ('AWS1') was installed at 3629 meters above sea level (m a.s.l.) in the eastern basin and operated continuously between 25 November 2012 and 15 May 2013. The station was moved later to the western basin (3724 m a.s.l.) where it operated between 29 October 2013 and 29 April 2014. The lower station ('AWS2') was installed on the glacier tongue at 2790 m a.s.l. and operated continuously between 25 November 2012 and 24 April 2014. Meteorological sensors (Table 5.8), data loggers and batteries were mounted on tripods resting on the glacier surface, while an ultrasonic gauge measured surface height changes relative to a fixed aluminium frame inserted into glacier ice/firn (Fig. 5.16). Both stations sampled air temperature, humidity, wind speed and direction, incoming and outgoing short- and longwave radiation, and surface height changes every 10 seconds. Air pressure and liquid precipitation were also measured at AWS2. Hourly averages were stored in Campbell Scientific CR1000 dataloggers and transmitted via a satellite link. A cloudiness factor was computed as one minus the ratio of measured incoming solar radiation to the theoretical clear sky solar radiation. The stations were visited as frequently as possible during the monthly visits to the glacier. Sensor heights were measured each time and the radiometers were checked and leveled when necessary.

Table 5.8 AWS sensors and meteorological variables used for energy balance calculations.

Variable	Instrument	Measurement range	Nominal accuracy
Air temperature	Vaisala HMP155	–80 to +60°C	0.226–0.0028 × airT (°C) (–80 to 20°C)
Relative humidity	Vaisala HMP155	0.8 to 100%	1.0 + 0.008 × RH (%)
Wind speed	Young 05103	1 to 60 m s^{-1}	0.3 m s^{-1} or 3%
Solar radiation	Kipp and Zonen CNR4	0 to 2000 W m^{-2}	< 5% for daily sums
Longwave radiation	Kipp and Zonen CNR4	–250 to +250 W m^{-2}	< 10% for daily sums
Surface height	Campbell SR50A	0.5 to 10 m	0.01 m or 0.4%
Air pressure	CS106	500–1100 mbar	1 mbar (–20 to 45°C)

Point Energy and Mass Balance Calculations

The surface energy balance was calculated at each meteorological station, using model formulations by Mölg et al. (2008). The model is described in detail by Mölg et al. (2008). MacDonell et al. (2013b) successfully applied the model to Guanaco glacier in the northern Andes of Chile. The surface energy balance is calculated as the sum of all energy fluxes entering and leaving the glacier surface by unit time interval (1 hour):

$$R + QS + QL + QP + QG = F \tag{7}$$

where R is the net radiation flux; QS is the turbulent sensible heat flux; QL is the turbulent latent heat flux; QP is the heat flux supplied by precipitation; and QG is the subsurface heat flux (Mölg et al., 2008). QG can be further separated into the conductive heat flux in the subsurface (QC), and the energy flux from shortwave radiation penetrating through the subsurface (QPS). F is the resultant energy flux at the surface. If the surface temperature is at the melting point (0°C) and F is positive on the right-hand side of Equation 7, then F represents the energy available for melt (QM), otherwise surface cooling occurs. In this study, QP is ignored because all precipitation falls as snow and precipitation intensity is low, which means that heat addition due to precipitation is likely to be negligible.

The net radiation (R) is calculated as the sum of incoming ($SWin$) and reflected ($SWout$) solar radiation and incoming ($LWin$) and emitted ($LWout$) long wave radiation as measured by the radiometers on the AWSs. $SWin$ is corrected for local slope at the AWS locations. $LWout$ is used to calculate surface temperature using the Stefan law, assuming a surface emissivity of 1. The turbulent sensible and latent heat fluxes (QH, QL) are calculated using the 'bulk' method outlined in Mölg et al. (2008), in which measurements of air and surface temperature, relative humidity and wind speed are used to estimate conditions at the glacier surface according to neutral logarithmic gradient profiles, along with surface roughness lengths for momentum (z_{0m}) and heat/vapor (z_{0h}). Representative values were selected from the literature (Cuffey and Paterson, 2010), and varied according to surface type (ice, fresh and old snow). Values of 4 mm, 2 mm and 1 mm were used as representative roughness lengths for momentum for old snow, ice and fresh snow, respectively. Values for the roughness lengths for heat/vapor were an order of magnitude smaller in each case. A correction term accounts for the stability of the boundary layer based on the Richardson number. The englacial temperature profile is calculated in the model using the thermodynamic equation, and includes energy release from penetrating solar radiation (QPS) and englacial refreezing in the snowpack (Mölg et al., 2009; Mölg et al., 2008). QPS was estimated to be 20% of the net solar radiation for ice, and 10% for snow (Bintanja and Van Den Broeke, 1995). The subsurface model in this study uses 14 vertical layers with thicknesses (dz) increasing progressively from the surface (dz = 0.09 m) down to 15 m depth (dz = 3 m). Conduction (QC) is then calculated from the temperature profile.

The mass balance at the AWS locations is modelled As:

$$b = c_{sp} + \frac{QM}{L_M} + \frac{QL}{L_S} + c_{cn} \tag{8}$$

where: c_{sp} is the sum of solid precipitation; L_M and L_S are the latent heat of melting and sublimation, respectively; and c_{en} is englacial accumulation by refreezing of meltwater in snow or at the interface between ice and snow. c_{sp} is calculated from surface change measurements measured by the sonic ranger on the AWS and using a fresh density of 60 kg m^{-3} (Cuffey and Paterson, 2010). L_M and L_S are known values, and QM, QL and c_{en} are calculated during the energy balance modelling (Equation 7). Mass or energy loss from (gain to) the surface is defined as negative (positive). The model was validated at each AWS using surface height changes measured by the sonic rangers.

Results and Discussion

Mass balance measurements

The glacier-wide mass balance ($B \pm$ standard error) of Universidad Glacier calculated by the glaciological method was –0.32 ± 0.40 m w.e. in 2012/13 and –2.53 ± 0.57 m w.e. in 2013/14 (Table 5.9). While measured accumulation at the end of winter (October) is similar between both years (2012/13: 1.43 ± 0.06 m w.e., 2013/14: 1.48 ± 0.15 m w.e.), ablation was much larger in summer 2013 (–4.01 ± 0.56 m w.e.) than in 2012 (–1.75 ± 0.38 m w.e.). The mass balance shows strong relationships with altitude, with r^2 coefficients of 0.79 to 0.92 in winter and 0.90 to 0.97 in summer (Fig. 5.17; Table 5.9). Despite the relatively strong elevation dependence of the seasonal mass balance components, scattering of individual observation is evident; these arise from combined measurements errors and spatial variability in accumulation and ablation processes unrelated to elevation (Zemp et al., 2013), and are responsible for the standard errors around the mass balance estimates (Table 5.9). The gradient decreases near the equilibrium line, a common phenomenon on valley glaciers, which occurs mainly because the snow surface above the equilibrium line has a greater albedo than the ice below the equilibrium line and thus the same added energy leads to lower melt above than below the ELA. The mean mass balance gradient was calculated as:

$$\frac{db}{dz} = 2p1 + p2\bar{z} \tag{9}$$

where: p1 and p2 are the first two polynomial coefficients from Equations 4 and 5; and $\bar{z}$ is the mean glacier altitude (3664 m a.s.l.). The winter balance gradient is rather small, with accumulation (b_w) varying between 0.5 m w.e. a^{-1} in the lower part to ~ 2 m w.e. a^{-1} in the higher part, with mean gradients of 0.01 and 0.13 m w.e. per 100 m for 2012/13 and 2013/14, respectively. The summer balance gradient is steeper, with b_s values varying from ~ 0 m w.e. in the higher section of the glacier, to –11 m w.e. in the lower section (–2 to –14 m w.e. in 2013/14), with similar mean gradients of –0.37 and –0.38 m w.e. per 100 m for 2012/13 and 2013/14, respectively. The annual mean mass balance gradients were –0.36 and –0.52 m w.e. per 100 m for 2012/13 and 2013/14, respectively. Hence ablation was more pronounced in 2013/14 compared to 2012/13 but the mass balance gradient was also steeper, mainly due to the steeper winter balance gradient measured in 2013/14. The Equilibrium Line Altitude (ELA), estimated from the mass balance-elevation relationships, was 755 m higher in 2013/14 (ELA = 4233 m) than in 2012/13 (ELA = 3478 m). The Accumulation Area Ratio (AAR), defined as the ratio between the accumulation area and the total glacier area, was 0.71 in 2012/13 but decreased to 0.11 in 2013/14. Hence in 2013/14 only 11% of the surface of Universidad Glacier accumulated mass, compared to 71% in 2012/13. Glaciers tend to have a mass balance in equilibrium ($B_a = 0$) when their AAR is between 0.4 and 0.8, with an average value of 0.57 ± 0.09 (Dyugerov et al., 2009).

Taken together, these results illustrate two contrasting years with respect to mass balance conditions on Universidad Glacier, with strong ablation driving a marked decrease in annual mass balance in 2013/14 compared to the preceding year 2012/13. We next analyze the meteorological and energy

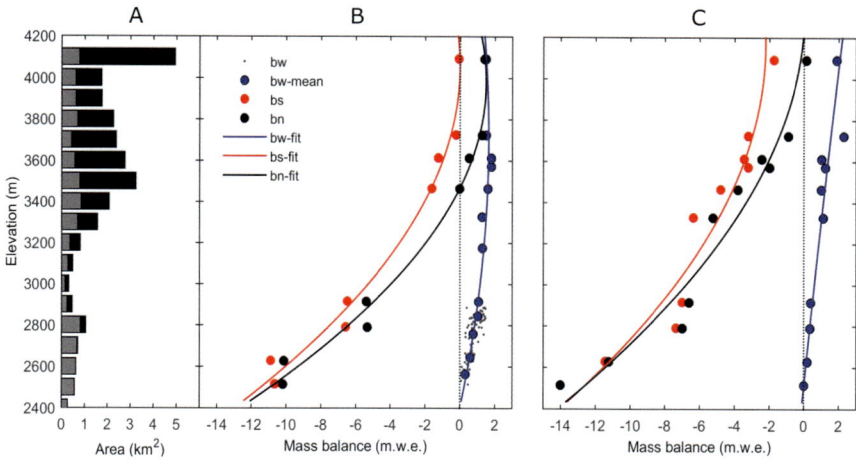

Fig. 5.17 Measured mass balance versus elevation. (A) Glacier hypsometry derived from the 2011 lidar survey. Gray (black) bars represent the eastern (western) basins of the glacier. (B) Mass balance, hydrological year 2012–2013; (C) Mass balance, hydrological year 2013–2014. Small gray points are manual snow measurements between stakes (2012 only). All available snow depths were averaged per 100 m elevation bin. Continuous lines are 2nd-order polynomial functions fitted on observations (blue: winter, red: summer, black: annual).

Table 5.9 Mass balance results for Universidad Glacier. S.E. = standard error on the mean mass balance (Fountain and Vecchia, 1999); r^2 = percent variance explained by the polynomial regression of mass balance against altitude; ELA: Equilibrium Line Altitude. AAR: Accumulation Area Ratio.

Year	Parameter	B_w (m w.e.)	B_s (m w.e.)	B_a (m w.e.)	ELA (m)	AAR [ ]
2012–2013	B ± S.E.	1.43 ± 0.06	−1.75 ± 0.38	−0.32 ± 0.40	3478	0.71
	r^2	0.92	0.97	0.97		
2013–2014	B ± S.E.	1.48 ± 0.15	−4.01 ± 0.56	−2.53 ± 0.57	4233	0.11
	r^2	0.79	0.90	0.92		

balance data collected on the glacier during the study period in order to identify the main ablation drivers and their differences in both years.

Point Mass and Energy Balance Calculation

Model validation

Total mass changes at AWS1 and AWS2 calculated by the mass balance model are presented in Table 5.10 for summer (Nov–Apr) 2012/13 and 2013/14. Surface height changes simulated by the mass balance model agree reasonably well with those measured by the sonic range sensors

Table 5.10 Calculated mass changes at AWS1 and AWS2 in summer (November–April) for hydrological years 2012–2013 and 2013–2014.

	2012–2013		2013–2014	
	AWS1	AWS2	AWS1	AWS2
Precipitation	44	28	26	50
Melt (mm)	1541	5889	3280	6920
Sublimation (mm)	106	223	117	238
Refreezing (mm)	603	44	450	3
Ablation (mm)	1043	6068	2947	7155
Sublimation ratio (%)	6	4	3	3

installed near the AWSs (Fig. 5.18). At AWS1 the Nash-Sutcliffe coefficient of efficiency (NSE), which measures the model performance relative to the mean measured height change (NSE = 1 being a perfect fit) was 0.93 and the root mean square error (RMSE) was 2.3 cm in 2012/13. Slightly lower values are calculated in 2013/14, with NSE = 0.90 and RMSE = 5.4 cm. The model slightly underestimates ablation toward the end of the 2013/14 period, which could be due to the appearance of penitentes (ice pinnacles) on the surface, which were not present at the AWS1 location during the first year, and whose effect is not taken into account in the energy balance model. The agreement between observed and simulated changes is still considered reasonable. At the lower station, AWS2, the fit between simulated and observed height changes is very good, with an NSE value of 0.98 and RMSE of 6.5 cm. The main differences between observed and simulated changes occur in winter during and after precipitation events, and at the end of September 2013 when larger ablation was calculated by the model in response to above-freezing temperatures over a few days.

Mass and Energy Fluxes at AWS1 and AWS2

Mean monthly energy fluxes and total monthly mass changes were calculated at both stations (Fig. 5.19, Fig. 5.20). At both stations and during both years the net solar radiation flux (SW^*) largely dominates the energy input to the glacier surface in summer, with the sensible heat flux (QS) provides a lesser energy contribution. The dominance of the net solar radiation flux is characteristic of the Mediterranean climate and latitude (36° S) of Universidad Glacier: high solar angles and a dry atmosphere in summer causes intense solar radiation on the glacier surface. At AWS1 the energy gained is mainly dissipated by melt (QM) and radiative cooling, with the net longwave heat flux (LW^*) being negative on average. The latent heat (QL) and ground heat fluxes (QG) are the smallest energy loss terms.

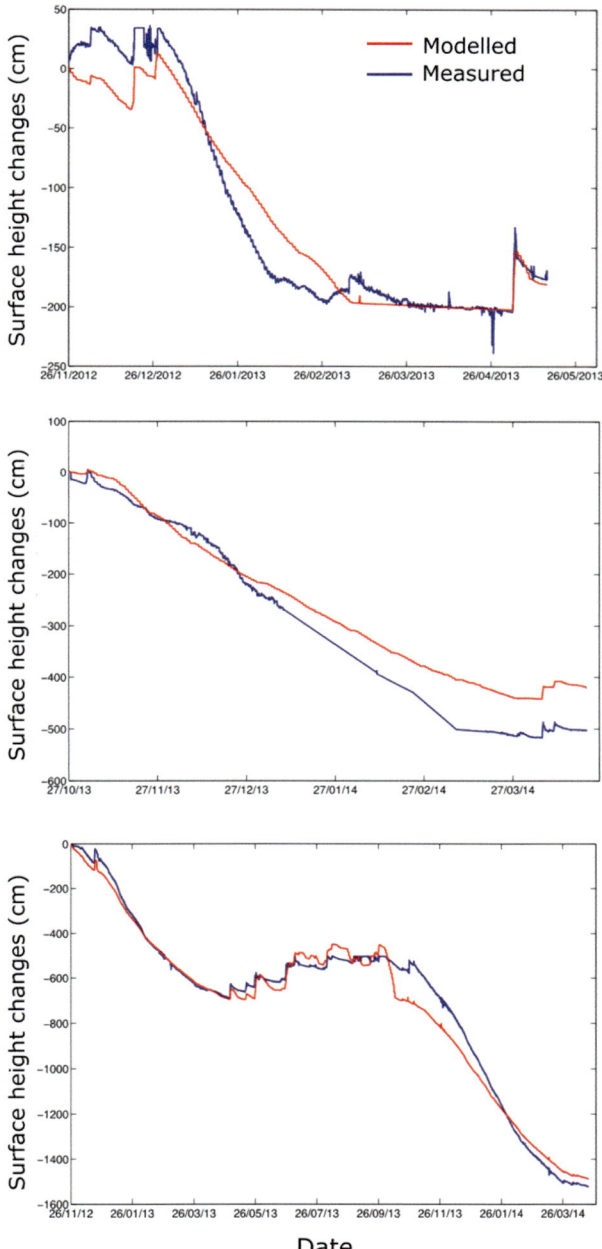

Fig. 5.18 Comparison between simulated (red) and measured surface height changes (blue). (A) AWS1, eastern basin: 2012–2013; (B) AWS1, western basin: 2013–2014; (C) AWS2 lower glacier: 2012–2014.

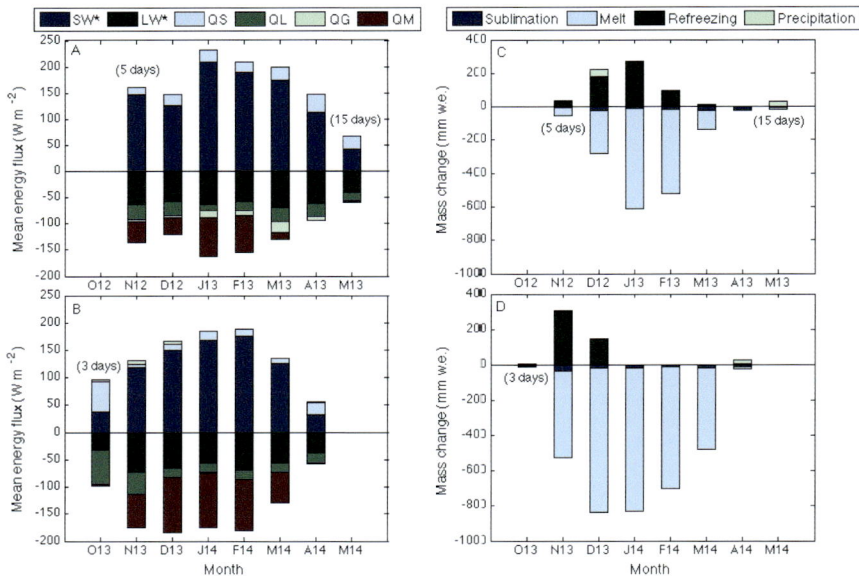

Fig. 5.19 Calculated monthly energy and mass balance at upper weather station AWS1. (A) energy balance, 2012–2013; (B) energy balance, 2013–2014; (C) mass balance, 2012–2013; (D) mass balance, 2013–2014. See Fig. 5.15 for location cf AWS1. The number of available days is labeled for months with incomplete data. Note the different x and y axes scaling.

Hence ablation occurred mainly by melting, with a sublimation ratio (the proportion of ablation occurring by sublimation) of only 6% in 2012/13 and 3% in 2013/14 (Table 5.10), but with significant refreezing of melt water within snow/firn: 39% of the melt water in 2012/13 and 14% in 2013/14. As AWS1 was moved from to the western basin in 2013/14 conditions cannot be directly compared. But despite its slightly higher altitude in 2013/14 (3724 m) compared to 2012/13 (3629 m), melt at AWS1 was more intense in 2013/14 than in 2012/13. Also, snowfall events occurred in December 2012, which increased albedo and decreased the net solar radiation flux (SW^*), and allowed refreezing of meltwater in the snowpack (Fig. 5.19A,C).

A continuous record is available at AWS2 on the lower glacier between November 2012 and April 2014 (Fig. 5.20). Net shortwave radiation (SW^*) also dominates the energy gains at AWS2 during summer (October–April), with a lesser contribution from the sensible heat flux (QS). Energy losses occur mainly through melt (QM), followed by sub surface heat flux (QG) and latent heat flux (QL). Hence the relative contribution of the different heat fluxes is similar to the upper glacier (AWS1), but melt is more intense and the sub surface heat flow is more pronounced due to increased thermal conductivity and light transmissivity of the ice surface, which favor heat conduction and transmission of solar radiation, respectively. During winter

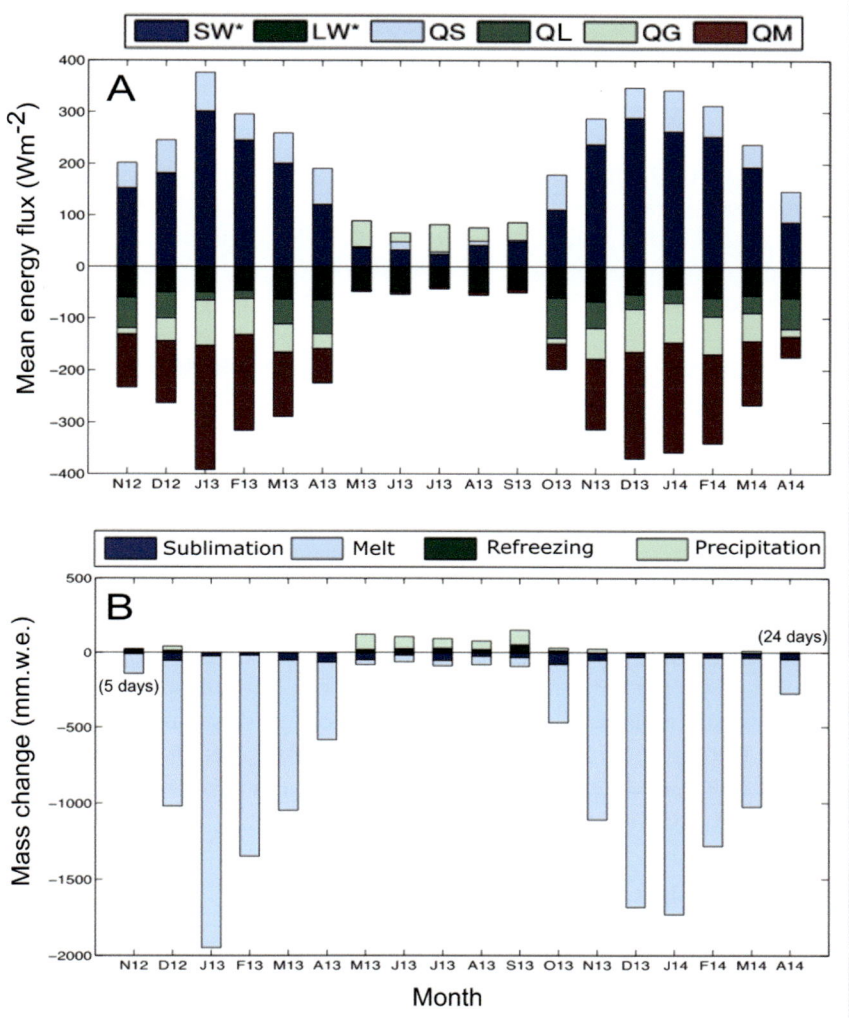

Fig. 5.20 Calculated monthly energy (A) and mass (B) balance at lower weather station AWS2 between November 2012 and April 2014. The number of available days is labeled for months with incomplete data.

(May–September) the net shortwave radiation flux is reduced due to lower sun angles and the higher snow albedo, and the glacier surface is warmed by upward heat conduction (QG) from the subsurface. Almost all the energy is dissipated radiatively (LW^*), with only occasional melt. Like at AWS1, sublimation represents a small portion of total ablation (3–4%), and refreezing of meltwater is much reduced compared to AWS1, due to the thinner and more transient snowpack covering the impervious ice surface

(Table 5.10, Fig. 5.20). Comparison between the summer periods of both years with complete data (December–April) shows that melt in December 2013 was more than twice that in 2012, and refreezing was less due to a thinner snow cover in 2013 (Fig. 5.20 C–D). Like at AWS1, the increased SW^* flux in December appears to be the main cause for increased melt in early summer 2013/14 compared to 2012/13. These differences are investigated next in more detail.

Calculated daily sublimation rates at AWS2 varied between 0.6 mm d^{-1} and 2.6 mm d^{-1}, with an average rate of 1.4 mm d^{-1} for the two-year period. Comparable sublimation rates were found on glaciers in the semi arid Andes further north: MacDonell et al. (2013b) modeled a mean daily sublimation rate of 1.3 mm d^{-1} on Guanaco glacier, a high-altitude (~ 5000 m a.s.l.) cold glacier at 29.3° S, while Gascoin et al. (2011) reported a mean sublimation rate of 2.5 mm d^{-1} from a suite of lysimeter experiments in the same area. On Universidad Glacier, such sublimation rates represented only a small fraction of ablation (3–6%), compared to the high sublimation ratio reported for the semi arid Andes (70–90%). This is because warmer air temperatures and higher humidity on Universidad Glacier increase long wave radiation from the atmosphere, which partly offset the long wave radiation lost from the glacier surface. On high-altitude, semi arid glaciers ~ 50% of the energy gained in summer (Nov–Apr) from net solar radiation is removed by the net long wave radiation loss; this amount decreases to ~ 30% on Universidad Glacier. In addition, cold glaciers have a larger cold content to be removed at the end of winter compared to the temperate ice on Universidad Glacier, so that much of the net positive energy flux is initially used to warm the glacier surface.

Comparison Between 2012/13 and 2013/14

The continuous meteorological record at the lower weather station AWS2 enables closer investigation of the differences in meteorological and energy balance conditions that led to greater ablation and reduced annual mass balance in 2013/14 compared to 2012/13. The difference between monthly mass and energy fluxes as well as for key meteorological conditions are presented in Table 5.11, with the color scale highlighting the relative magnitude of the differences between both years. Differences in mass fluxes over the period with available data in both years (December to April) show increased melt in December but moderately less melt in January and April for 2013/14 compared to 2012/13. Differences in sublimation and melt water refreezing rates are comparatively small between both years. In terms of energy fluxes, the net surface energy balance (F) was larger in December but smaller or similar in all other months. This increase in excess energy explains the larger melt rate in 2013/14 compared to 2012/13.

Breaking down the respective energy contribution to F shows that a large increase in the net shortwave radiation (SW^*) flux occurred in December 2013/14, as discussed qualitatively earlier, while moderate decreases in SW^* occurred later, in January and April. About 40% of the December 2013 increase in SW^* is due to higher incoming solar radiation ($SWin$), which itself correlates with decreased cloudiness during that month, as captured by the cloudiness factor (Table 5.11). A much decreased albedo (–0.23) explains the remaining December 2013 increase in SW^* compared to 2012 (Table 5.11). The persistence of winter snow as well as snowfall events in mid- to late December 2012 increased the albedo significantly during that month compared to the previous year (Fig. 5.21). Although the meteorological

Table 5.11 Monthly and summer (Nov–April) mean/total differences in simulated mass and energy fluxes and measured meteorological variables at AWS2. Differences are computed as 2013/14 minus 2012/13. The color scale highlights the most negative (blue) to most positive (red) differences (0 = white). The color formatting is applied separately for energy fluxes (Wm^{-2}), mass fluxes (mm w.e.), and meteorological variables. For the latter, the mean standardized difference is used in order to account for the different measurement scales, but absolute difference values are shown in the table with their respective units.

	Dec	Jan	Feb	Mar	Apr	Mean/total
Ablation (mm)						
melt	684	–229	–82	–5	–286	82
Sublimation	–22	9	16	–15	–21	–33
Refreezing	–15	0	0	0	1	–14
Energy fluxes (Wm^{-2})*						
F	81	–28	–20	1	–21	3
SW*	107	–39	7	–7	–34	7
SWin	42	–43	9	–17	–15	–5
LW*	–4	7	–14	8	4	0
QS	–5	5	10	–14	–10	–3
QL	22	–12	–20	14	6	2
QG	–39	11	–3	0	13	–4
Meteorological variables						
AirT (°C)	2.6	0.2	0.3	–0.3	–1.0	0.3
RH (%)	–3.9	–0.9	–10.5	7.9	0.3	–1.4
Windspeed (ms^{-1})	–0.5	0.0	0.5	–0.3	–0.1	–0.1
Albedo []	–0.23	0.02	0.00	–0.01	0.15	–0.02
Cloudiness []	–0.09	0.13	–0.04	0.06	0.09	0.03

* Positive fluxes are toward the surface

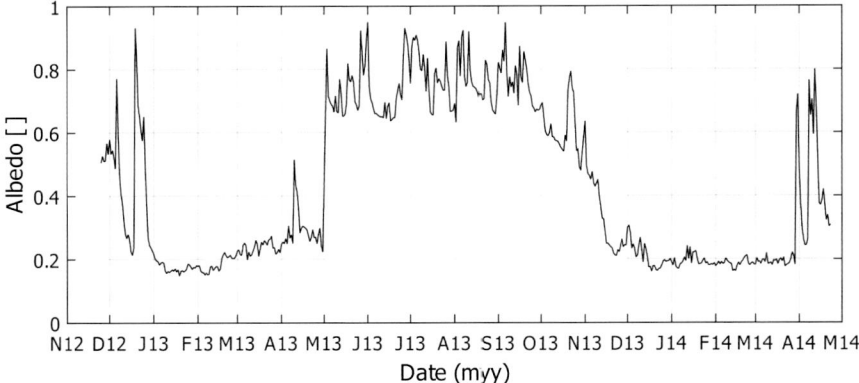

Fig. 5.21 Daily mean albedo at AWS2 between 25 November 2012 and 23 April 2014. Daily averages were calculated after excluding data with a solar zenith angle larger than 65°. The shaded areas highlight the summer period (Dec–Apr) discussed in section 4.2.

record only begins in late November 2012, the albedo prior to that must have been even higher since winter snow was still present, and melt largely suppressed compared to the year, when the snow cover disappeared in mid-November and the exposed ice lowered the albedo to ~ 0.20 (Fig. 5.21). Despite warmer air temperatures in December 2013 compared to 2012, no increase in the net longwave (LW^*) or sensible (QS) heat fluxes are noted. The moderate decrease in SW^* in January 2014 compared to 2013 was mainly due to increased cloudiness, while that in April 2014 was caused by a combination of increased cloudiness and a higher albedo resulting from earlier snowfall.

Conclusions

This chapter has outlined the methodological framework and initial results of a two-year monitoring effort on Universidad Glacier, a large valley glacier in the central Chilean Andes. Given the remoteness of this glacier, only a sparse network of mass balance stakes could be maintained during the period, and calculation of the glacier-wide mass balance was based on mass balance gradients to extrapolate point observations. The glacier-wide mass balance was slightly negative in 2012/13 ($B_a = -0.32 \pm 0.40$ m w.e. a^{-1}) but much lower in 2013/14 ($B_a = -2.53 \pm 0.57$ m w.e. a^{-1}), due to increased summer ablation. Analysis of the meteorological records collected at two on-glacier automatic weather stations revealed that the summer (November–April) energy balance is largely dominated by the net shortwave solar radiation flux (SW^*), and that this flux was significantly larger in early summer 2013/14 compared to 2012/13. Hence our results indicate that significant inter annual differences in ablation can occur in response to inter annual variations in net solar radiation. These were caused by (i) variations is snowfall rate and timing, which changes in albedo, and (ii) varying cloudiness. The persistent snow cover in November 2012 combined with fresh snowfalls in December 2012, a period when ablation is expected to increase rapidly, have had a very strong effect on the annual mass balance by drastically reducing ablation in the early ablation period. Hence, even in the absence of significant differences in air temperature between both years, large differences in ablation still occurred through albedo and cloudiness feedbacks on the net solar radiation flux. The current drought conditions affecting the extratropical region of Chile could not only reduce mass balance through decreasing snow accumulation, but also through increased ablation resulting from an earlier exposure of the ice surface in the ablation zone and a corresponding decrease in albedo. Continued monitoring or mass balance reconstruction using geodetic methods and/or modeling will be needed in order to interpret these results within a longer-term perspective.

Acknowledgements

Funding for this project was awarded from the Fondo de Innovación para la Competitividad del Ministerio de Economía, Fomento y Turismo, Gobierno de Chile (R. Urrutia, C. Kinnard), the Fondo Nacional de Desarrollo Científico y Tecnológico (CONICYT-FONDECYT 11130484-S. MacDonell) and Project CRHIAM/CONICYT/FONDAP/15130015.

References

Aitken, D., Rivera, D., Godoy-Faúndez, A. and Holzapfel, E. 2016. Water scarcity and the impact of the mining and agricultural sectors in Chile. Sustainability 8(2): 128.

Barnett, T.P., Adam, J.C. and Lettenmaier, D.P. 2005. Potential impacts of a warming climate on water availability in snow-dominated regions. Nature 438(7066): 303–309.

Bintanja, R. and Van Den Broeke, M.R. 1995. The surface energy balance of Antarctic snow and blue ice. J. Appl. Meteorol. 34(4): 902–926.

Boisier, J.P., Rondanelli, R., Garreaud, R.D. and Muñoz, F. 2016. Anthropogenic and natural contributions to the Southeast Pacific precipitation decline and recent megadrought in central Chile. Geophys. Res. Lett. 43(1): 413–421.

Bown, F. and Rivera, A. 2007. Climate changes and recent glacier behaviour in the Chilean Lake District. Global Planet Change 59(1): 79–86.

Bown, F., Rivera, A., Acuna, C. and Casassa, G. 2007. Recent glacier mass balance calculations at Volcán Mocho-Choshuenco (40 S), Chilean Lake District. IAHS-AISH Publication 143–152.

Bown, F., Rivera, A. and Acuña, C. 2008. Recent glacier variations at the Aconcagua basin, central Chilean Andes. Ann. Glaciol. 48(1): 43–48.

Brauman, K.A., Daily, G.C., Duarte, T.K. and Mooney, H.A. 2007. The nature and value of ecosystem services: An overview highlighting hydrologic services. Annu. Rev. Environ. Resour. 32: 67–98.

Carrasco, J.F., Casassa, G. and Quintana, J. 2005. Changes of the 0 C isotherm and the equilibrium line altitude in central Chile during the last quarter of the 20th century/ Changements de l'isotherme 0 C et de la ligne d'équilibre des neiges dans le Chili central durant le dernier quart du 20ème siècle. Hydrol. Sci. J. 50(6).

Cogley, J., Hock, R., Rasmussen, L., Arendt, A., Bauder, A., Kaser, G., Möller, M. and Nicholson, L. 2011. Glossary of glacier mass balance and related terms, UNESCO-IHP, Paris.

Cuffey, K.M. and Paterson, W.S.B. 2010. The Physics of Glaciers.

Dyurgerov, M., Meier, M.F. and Bahr, D.B. 2009. A new index of glacier area change: A tool for glacier monitoring. Journal of Glaciology 55(192): 710–716.

Escobar, F., Casassa, G. and Pozo, V. 1995. Variaciones de un glaciar de montana en los Andes de Chile central en las últimas dos décadas. Bull. Inst. fr. études andines 24(3): 683–695.

Falvey, M. and Garreaud, R.D. 2009. Regional cooling in a warming world: Recent temperature trends in the southeast Pacific and along the west coast of subtropical South America (1979–2006). Journal of Geophysical Research: Atmospheres 114(D4).

Favier, V., Falvey, M., Rabatel, A., Praderio, E. and López, D. 2009. Interpreting discrepancies between discharge and precipitation in high-altitude area of Chile's Norte Chico region (26–32° S). Water Resour. Res. 45(2).

Fountain, A.G. and Tangborn, W.V. 1985. The effect of glaciers on streamflow variations. Water Resour. Res. 21(4): 579–586.

Fountain, A.G. and Vecchia, A. 1999. How many stakes are required to measure the mass balance of a glacier? Geografiska Annaler: Series A, Physical Geography 81(4): 563–573.

Francou, B. and Pouyaud, B. 2004. Método de observación de glaciares en los Andes tropicales. Mediciones de terreno y procesamiento de datos. Documento GREAT ICE (IRD) en versión CD.

Gabbi, J., Carenzo, M., Pellicciotti, F., Bauder, A. and Funk, M. 2014. A comparison of empirical and physically based glacier surface melt models for long-term simulations of glacier response. Journal of Glaciology 60(224): 1140–1154.

Garreaud, R. 2009. The Andes climate and weather. Advances in Geosciences 22: 3–11.

Garreaud, R.D., Vuille, M., Compagnucci, R. and Marengo, J. 2009. Present-day south american climate. Palaeogeography, Palaeoclimatology, Palaeoecology 281(3): 180–195.

Gascoin, S., Kinnard, C., Ponce, R., Macdonell, S., Lhermitte, S. and Rabatel, A. 2011. Glacier contribution to streamflow in two headwaters of the Huasco River, Dry Andes of Chile. The Cryosphere (5): 1099–1113.

Gascoin, S., Lhermitte, S., Kinnard, C., Bortels, K. and Liston, G.E. 2013. Wind effects on snow cover in Pascua-Lama, Dry Andes of Chile. Advances in Water Resources 55: 25–39.

Ginot, P., Kull, C., Schotterer, U., Schwikowski, M. and Gäggeler, H. 2006. Glacier mass balance reconstruction by sublimation induced enrichment of chemical species on Cerro Tapado (Chilean Andes). Climate of the Past 2(1): 21–30.

Huss, M. 2011. Present and future contribution of glacier storage change to runoff from macroscale drainage basins in Europe. Water Resour. Res. 47(7): W07511.

Janke, J.R., Bellisario, A.C. and Ferrando, F.A. 2015. Classification of debris-covered glaciers and rock glaciers in the Andes of central Chile. Geomorphology 241: 98–121.

Jansson, P., Hock, R. and Schneider, T. 2003. The concept of glacier storage: A review. J. Hydrol. 282(1): 116–129.

Kaser, G., Fountain, A. and Jansson, P. 2002. A manual for monitoring the mass balance of mountain glaciers.

Kaser, G., Großhauser, M. and Marzeion, B. 2010. Contribution potential of glaciers to water availability in different climate regimes. Proceedings of the National Academy of Sciences 107(47): 20223–20227.

Lhermitte, S., Abermann, J. and Kinnard, C. 2014. Albedo over rough snow and ice surfaces. The Cryosphere 8(3): 1069–1086.

Lliboutry, L. 1958. Studies of the shrinkage after a sudden advance, blue bands and wave ogives on Glaciar Universidad (central Chilean Andes). J. Glaciol. 3: 261–270.

Lliboutry, L. 1964. Traité de glaciologie, Masson.

MacDonell, S., Kinnard, C., Mölg, T., Nicholson, L. and Abermann, J. 2013b. Meteorological drivers of ablation processes on a cold glacier in the semi-arid Andes of Chile. The Cryosphere 7(5): 1513–1526.

Malmros, J.K., Mernild, S.H., Wilson, R., Yde, J.C. and Fensholt, R. 2016. Glacier area changes in the central Chilean and Argentinean Andes 1955–2013/14. J. Glaciol. 62(232): 391–401.

Masiokas, M.H., Villalba, R., Luckman, B.H., Le Quesne, C. and Aravena, J.C. 2006. Snowpack variations in the central Andes of Argentina and Chile, 1951–2005: Large-scale atmospheric influences and implications for water resources in the region. J. Clim. 19(24): 6334–6352.

Medwedeff, W.G. and Roe, G.H. 2016. Trends and variability in the global dataset of glacier mass balance. Clim. Dyn. 1–13.

Mernild, S.H., Beckerman, A.P., Yde, J.C., Hanna, E., Malmros, J.K., Wilson, R. and Zemp, M. 2015. Mass loss and imbalance of glaciers along the Andes Cordillera to the sub-Antarctic islands. Global Planet Change 133: 109–119.

Meza, F.J. 2013. Recent trends and ENSO influence on droughts in Northern Chile: An application of the standardized precipitation evapotranspiration index. Weather and Climate Extremes 1: 51–58.

Minetti, J.L., Poblete, A.G., Vargas, W.M. and Ovejero, D.P. 2014. Trends of the drought indices in Southern hemisphere subtropical regions. Journal of Earth Science Research 2(2): 36.

Mölg, T., Cullen, N.J., Hardy, D.R., Kaser, G. and Klok, L. 2008. Mass balance of a slope glacier on Kilimanjaro and its sensitivity to climate. Int. J. Climatol. 28(7): 881–892.

Mölg, T., Cullen, N.J. and Kaser, G. 2009. Solar radiation, cloudiness and longwave radiation over low-latitude glaciers: implications for mass-balance modelling. J. Glaciol. 55(190): 292–302.

Ministerio de Obras Publicas Dirección General de Aguas de Chile. 2009. Estrategia Nacional de Glaciares S.I.T. N° 205. Realizado por el Centro de estudios Científicos. http://documentos.dga.cl/GLA5194v1.pdf.

Ministerio de Obras Publicas Dirección General de Aguas de Chile. 2010. Análisis de la composición físico química de los sedimentos fluviales y su relación con la disponibilidad de metales en agua: cuenca del río Cachapoal. Realizado por CENMA (Chile). http://documentos.dga.cl/CQA5191v5.pdf.

Ministerio de Obras Publicas Dirección General de Aguas de Chile. 2016. Atlas del Agua, Chile 2016. http://www.dga.cl/atlasdelagua/.

Nicholson, L., Marín, J., Lopez, D., Rabatel, A., Bown, F. and Rivera, A. 2010. Glacier inventory of the upper Huasco valley, Norte Chico, Chile: glacier characteristics, glacier change and comparison with central Chile. Ann. Glaciol. 50(53): 111–118.

Østrem, G. and Brugman, M. 1991. Mass balance measurement techniques. A manual for field and office work. National Hydrology Research Institute (NHRI) Science Report 4.

Pellicciotti, F., Helbing, J., Rivera, A., Favier, V., Corripio, J., Araos, J., Sicart, J.E. and Carenzo, M. 2008. A study of the energy balance and melt regime on Juncal Norte Glacier, semi-arid Andes of central Chile, using melt models of different complexity. Hydrol. Processes 22(19): 3980–3997.

Pellicciotti, F., Ragettli, S., Carenzo, M. and McPhee, J. 2014. Changes of glaciers in the Andes of Chile and priorities for future work. Sci. Tot. Environ. 493: 1197–1210.

Rabatel, A., Castebrunet, H., Favier, V., Nicholson, L. and Kinnard, C. 2011. Glacier changes in the Pascua-Lama region, Chilean Andes (29 S): recent mass balance and 50 yr surface area variations. The Cryosphere 5(4): 1029–1041.

Radić, V., Bliss, A., Beedlow, A.C., Hock, R., Miles, E. and Cogley, J.G. 2014. Regional and global projections of twenty-first century glacier mass changes in response to climate scenarios from global climate models. Clim. Dyn. 42(1-2): 37–58.

Ragettli, S. and Pellicciotti, F. 2012. Calibration of a physically based, spatially distributed hydrological model in a glacierized basin: On the use of knowledge from glaciometeorological processes to constrain model parameters. Water Resour. Res. 48(3).

Rivera, A., Casassa, G., Acuna, C. and Lange, H. 2000. Variaciones recientes de glaciares en Chile. Investigaciones Geográficas (34): 29–60.

Rivera, A., Acuña, C., Casassa, G. and Bown, F. 2002. Use of remotely sensed and field data to estimate the contribution of Chilean glaciers to eustatic sea-level rise. Ann. Glaciol. 34(1): 367–372.

Rivera, A., Bown, F., Casassa, G., Acuña, C. and Clavero, J. 2005. Glacier shrinkage and negative mass balance in the Chilean Lake District (40° S)/Rétrécissement glaciaire et bilan massique négatif dans la Région des Lacs du Chili (40° S). Hydrol. Sci. J. 50(6).

Valdés-Pineda, R., Pizarro, R., García-Chevesich, P., Valdés, J.B., Olivares, C., Vera, M., Balocchi, F., Pérez, F., Vallejos, C. and Fuentes, R. 2014. Water governance in Chile: Availability, management and climate change. J. Hydrol. 519: 2538–2567.

Viviroli, D., Dürr, H.H., Messerli, B., Meybeck, M. and Weingartner, R. 2007. Mountains of the world, water towers for humanity: Typology, mapping, and global significance. Water Resour. Res. 43(7): W07447.

Vuille, M., Franquist, E., Garreaud, R., Casimiro, L., Sven, W. and Cáceres, B. 2015. Impact of the global warming hiatus on Andean temperature. Journal of Geophysical Research: Atmospheres 120(9): 3745–3757.

Zemp, M., Hoelzle, M. and Haeberli, W. 2009. Six decades of glacier mass-balance observations: A review of the worldwide monitoring network. Ann. Glaciol. 50(50): 101–111.

Zemp, M., Thibert, E., Huss, M., Stumm, D., Denby, C.R., Nuth, C., Nussbaumer, S., Moholdt, G., Mercer, A. and Mayer, C. 2013. Reanalysing glacier mass balance measurement series. The Cryosphere 7(4): 1227–1245.

Zemp, M., Frey, H., Gärtner-Roer, I., Nussbaumer, S.U., Hoelzle, M., Paul, F., Haeberli, W., Denzinger, F., Ahlstrøm, A.P. and Anderson, B. 2015. Historically unprecedented global glacier decline in the early 21st century. J. Glaciol. 61(228): 745–762.

CHAPTER 6

Participatory Monitoring of the Impact of Watershed Interventions in the Tropical Andes

Boris F. Ochoa-Tocachi,[1,]* Wouter Buytaert[2] and Bert De Bièvre[3]

Introduction

This chapter documents the efforts of building the Regional Initiative for Hydrological Monitoring of Andean Ecosystems (iMHEA). First, we explain the background and motivations that led to the formation of a diverse consortium of institutions with a joint interest. Then we present the methodological approach that the monitoring network has adopted. Lastly, we discuss in brief the main results, the most relevant milestones and breakthroughs, and the major challenges remaining and perspectives in the

[1] Imperial College London, Department of Civil and Environmental Engineering & Grantham Institute – Climate Change and the Environment, South Kensington Campus, London, United Kingdom SW72AZ.
   Consorcio para el Desarrollo Sostenible de la Ecorregión Andina (CONDESAN), Área de Cuencas Andinas, Lima, Peru 15024.
   Regional Initiative for Hydrological Monitoring of Andean Ecosystems (iMHEA), Lima, Peru.
[2] Imperial College London, Department of Civil and Environmental Engineering & Grantham Institute – Climate Change and the Environment, South Kensington Campus, London, United Kingdom SW72AZ.
   Regional Initiative for Hydrological Monitoring of Andean Ecosystems (iMHEA), Lima, Peru.
   Email: w.buytaert@imperial.ac.uk
[3] Fondo para la Protección del Agua (FONAG), Technical Secretary, Quito, EC 170137.
   Regional Initiative for Hydrological Monitoring of Andean Ecosystems (iMHEA), Lima, Peru.
   Email: bert.debievre@fonag.org.ec
* Corresponding author: boris.ochoa13@imperial.ac.uk

scientific, technological and social domains. We argue that the correct use of the generated knowledge, from community level to national governance entities, proves crucial to increase catchment intervention efficiency and improve water resources management.

The tropical Andes are hotspots for ecosystem services provision and environmental change. The naturally high diversity of geographical and climatic characteristics results in equally variable and non-stationary hydrometeorological features (Vuille et al., 2000; Manz et al., 2016; 2017; Ulloa et al., 2017). These characteristics are closely related to the large portfolio of ecosystem services, especially in terms of the scale and variability of the discharge of rivers that support up- and downstream livelihoods. However, this region suffers from extensive data scarcity and an acute lack of understanding on how to leverage ecosystem services to support human development (Célleri and Feyen, 2009; Balvanera et al., 2012). One of the reasons for this is that the official national monitoring networks do not have an ecosystem assessment focus, and tend not to cover remote headwater areas (Célleri et al., 2010; Buytaert et al., 2016; Martínez et al., 2017), and therefore are not ideal to study the nature, distribution and evolution of hydrological ecosystem services. These issues, in combination with rapid changes in land use and climate, as well as increasing population and water demand, put severe pressure on water resources (Buytaert and De Bièvre, 2012).

Changes in land use and land cover, which are driven by anthropogenic pressure from local users and by increasingly intensive watershed management, have a large impact on hydrology and ecosystem services. In the last decades, local research has delivered relevant knowledge about the natural hydrological regime of Andean catchments and the impacts of several human activities that are commonly detrimental to water yield and hydrological regulation, e.g., Luteyn (1992), Inbar and Llerena (2000; 2004), Díaz and Paz (2002), Hofstede (2002), Bruijnzeel (2004), Farley et al. (2004), Buytaert et al. (2002; 2004; 2005; 2006a; 2006b; 2007), Célleri et al. (2007), Favier et al. (2008), Quichimbo (2008), Tobón (2009), Crespo et al. (2010; 2011; 2012), Carlos et al. (2014), Córdova et al. (2015), Mosquera et al. (2015), Padrón et al. (2015), Ochoa-Tocachi et al. (2016a; 2016b). However, the effectiveness of different interventions in the region, be it in a context of ecosystem management, payment for ecosystem services, adaptation to climate change or investment in green infrastructure in watersheds, is far off from being thoroughly assessed or even fully understood. For instance, many catchment conservation strategies and common restoration efforts, such as re- or afforestation, have not been evaluated properly for their hydrological benefits and are often based on a very limited local evidence base. Additionally, those which consider the necessity of generating new knowledge about the natural environment, such as several climate change adaptation initiatives in mountain areas, are responding to information gaps by putting significant efforts in glacier monitoring. But the information gap for an effective watershed management in high-elevation and headwater areas persist, as such monitoring sites cannot capture the impact of interventions.

Emerging from a local awareness about the need for better information on ecosystem services, a partnership of academic and non-governmental institutions triggered the use of participatory hydrological monitoring to address this gap. In 2009, they formed the Regional Initiative for Hydrological Monitoring of Andean Ecosystems (iMHEA), which leverages the growing availability of inexpensive and robust sensor technology (Buytaert et al., 2014). The iMHEA network is a consortium of institutions interested in generating and strengthening the hydrological knowledge of Andean ecosystems to improve decision making on water resources management in this region. Increasing the knowledge of hydrology and meteorology in Andean catchments involves the implementation of new monitoring, in a way that ensures optimal complementarity with existing monitoring networks (Célleri et al., 2010). The iMHEA recognizes the role of water and environment authorities and of the offices of hydrology and meteorology as the rectors of water resources management in each country, and as a way to complement their efforts of data generation, they propose a bottom-up approach in which civil institutions can contribute with local scale and headwater monitoring.

The iMHEA network uses a design based on a 'trading-space-for-time approach' (Célleri et al., 2010; Buytaert et al., 2014; Ochoa-Tocachi et al., 2016b) illustrated in Fig. 6.22. This concept relies on strengthening the statistical significance of an intervention signal by monitoring several catchments in a regional setting (Buytaert and Beven, 2009; 2011; Oudin et al., 2010; Singh et al., 2011; Sivapalan et al., 2011; Wagener and Montanari, 2011). The increased number of monitoring sites also allows for a robust regionalization of the results by covering different ecosystems with diverse physiographic characteristics and contrasting land uses and degrees of conservation/alteration (Ochoa-Tocachi et al., 2016b). The proposal recommends the use of paired catchments, which allows comparisons on the short-term; however, single catchments can be monitored to analyze long-term changes using a baseline of several years to capture inter annual climatic variations and analyze non-stationarity.

In this way, the establishment of a 'minimum' hydrological monitoring in several sites is of a higher priority than implementing detailed monitoring in few locations. Such an indispensable monitoring consists of the measurement of rainfall and stream flow at high temporal resolution and at micro-catchment scale. To maximize the usefulness of these data for national hydrometeorological offices and local partners, the selected sites are commonly located in areas with low density of stations. The generated data will potentially have different characteristics and quality, which is

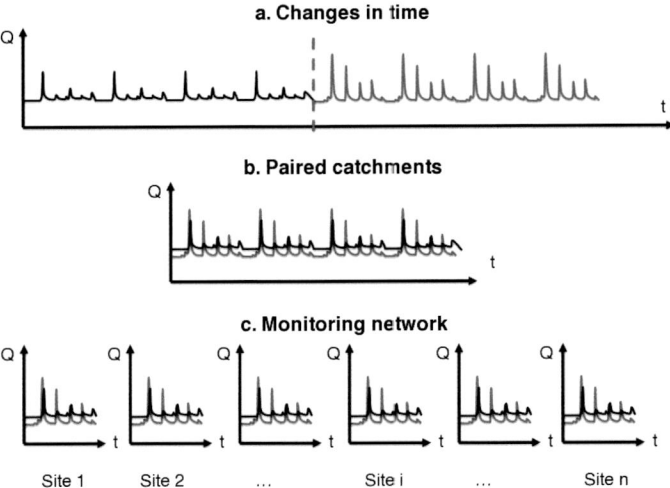

Fig. 6.22 Concept of space-for-time substitution. (a) Long-term monitoring allows to identify trends and changes in time when a perturbation has occurred. (b) Paired catchments reduce the monitoring length by adding a spatial variable to the analysis and comparing watersheds under contrasting conditions. (c) The replication of the perturbation in several sites or the consideration of different conditions in a monitoring network provides a robust framework in which to analyze impacts and facilitate the regionalization of results.

determined by the specific purposes of the local partner; however, the high-spatial and short-temporal resolution of these data is highly compatible to the long-term and low-spatial density of national networks (Buytaert et al., 2016).

iMHEA aims to tackle data-scarcity in the Andean region by generating information using a participatory environmental monitoring framework. As such, iMHEA is not a formal network but is aimed mostly at promoting exchange of information and experience, and involves the movement of resources from different partners and funding sources. The good quality of the data is achieved through a partnership with research institutions (Célleri et al., 2010). Therefore, the entry threshold for local partners to the network is relatively low and accessible. The monitoring system includes an institutional agreement between local communities and users of water and land, local governments and institutions of development, research groups and universities, and monitoring networks at national and regional scales (Fig. 6.23). The engagement and experience of local users are critical for the success of monitoring activities, and the interaction between the several different stakeholders is necessary to achieve long-term sustainability. The involvement of local universities and research groups working in the areas of influence guarantees scientific rigor and robustness in the generated data. This interaction between the general public and traditional scientists is often

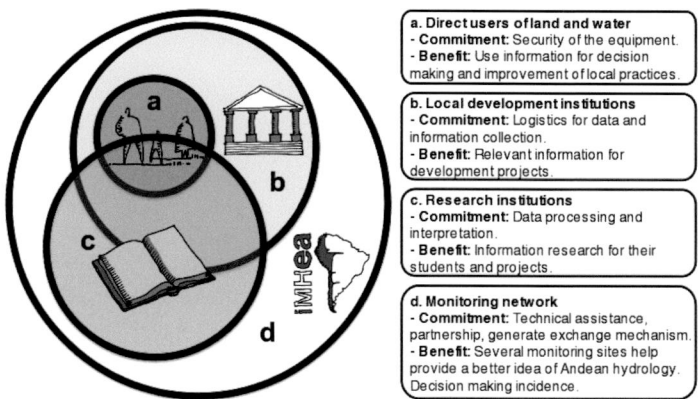

a. Direct users of land and water
- **Commitment**: Security of the equipment.
- **Benefit**: Use information for decision making and improvement of local practices.

b. Local development institutions
- **Commitment**: Logistics for data and information collection.
- **Benefit**: Relevant information for development projects.

c. Research institutions
- **Commitment**: Data processing and interpretation.
- **Benefit**: Information research for their students and projects.

d. Monitoring network
- **Commitment**: Technical assistance, partnership, generate exchange mechanism.
- **Benefit**: Several monitoring sites help provide a better idea of Andean hydrology. Decision making incidence.

Fig. 6.23 Institutional agreement in a participatory monitoring framework. The diverse partners in the network commit to different activities depending on their capacities and resources and obtain benefits for their own objectives and purposes.

referred to as citizen science (Buytaert et al., 2014), and has a strong potential to tackle data-scarcity in remote regions, such as the tropical Andes.

Lastly, a key element of the philosophy of iMHEA is that "information that is not shared is useless". Therefore, as a fundamental principle, the generated data and the derived knowledge is shared in common standards and at different levels. The information produced at a local scale can be used to draw regional conclusions about the hydrology of the Andes. This monitoring system aims to generate data at mid- and long-term to allow analyses of hydrological changes in time and consistent support of decision making. The iMHEA holds annually an Assembly of stakeholders to provide a space for reflection and action, as well as organizes international courses on Andean hydrology and hydrological monitoring for diverse audiences. Consequently, the newly generated information and knowledge can be useful and usable to achieve the objectives of water and land users, local partners, research groups and national institutions.

In this chapter, we present the methodological guidelines for a participatory hydrological monitoring of Andean ecosystems. The objectives of this hydrological monitoring system are to ensure that the observatories in each watershed will generate standardized data that can be used in future studies and assessments of ecosystem services. Such objectives are: (i) to increase the knowledge about hydrological ecosystem services in Andean watersheds, especially but not limited to water availability and hydrological regulation; and (b) to increase the knowledge about the impacts of watershed interventions (i.e., changes in land use and land cover, adaptation measures, deforestation, afforestation, restoration, etc.) (Ochoa-Tocachi et al., 2016a). Once good base information about hydrological processes and their relations exists, different models can be calibrated adequately with these data to support extrapolation and regional analyses (Ochoa-Tocachi et al., 2016b).

Hydrological Monitoring Setup

Monitoring methodology

In order to support the technical aspects of setting up the monitoring and to maximize the compatibility and consistency of data produced by each location, a core team within iMHEA designed a detailed monitoring protocol (Fig. 6.24).

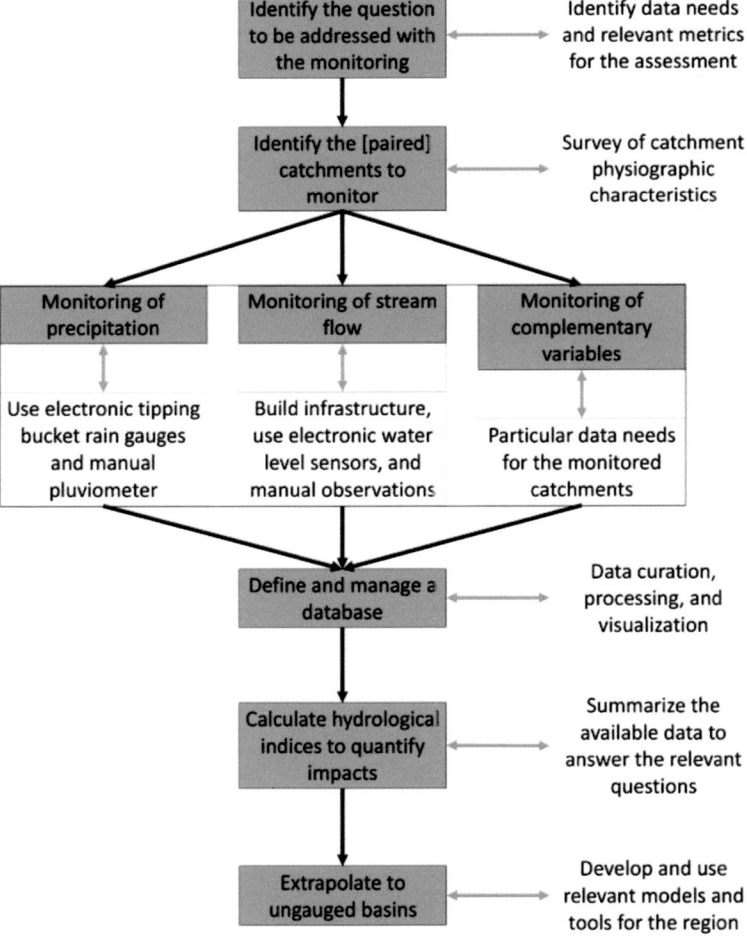

Fig. 6.24 Flowchart of procedure steps for the evaluation of watershed interventions. These steps are explained in detail in the monitoring protocol.

The minimum variables to be monitored are:

- Catchment physiographic characteristics.
- Rainfall inside the catchment.
- Stream flow at the outlet of the catchment.

With this information, a simple water balance equation can be used to estimate the total amount of water consumed by vegetation, evaporated from free water surfaces, infiltrated to the deeper soil strata and stored in the soils. The length of monitoring (at least one hydrological year) and the impermeable rocks generally found beneath the soils of Andean catchments (Buytaert et al., 2007), allows the assumption of equilibrium in the sub surface storage volumes. Therefore, the difference between rainfall and stream flow can be considered as a good approximation of the evapotranspiration in the catchment.

However, such assumption may not necessarily be true in all cases. Deep permeable soils may occur in some regions sustaining large aquifers (Buytaert et al., 2006b; Favier et al., 2008). Therefore, a careful selection of the catchments, ideally with the assessment of an experienced hydrologist, would help minimize potential issues due to the simplification of the water balance or to monitor those elements that influence greatly on such balance (Fig. 6.25(a)).

At the same time, this protocol recommends the monitoring of the hydrological variables using automatic electronic equipment supported by manual measurements. It has been observed that automatic measurements are more cost-effective than manual measurements (Célleri et al., 2010; Buytaert et al., 2014). Nevertheless, electronic equipment is also prone to failure or decalibration, in which cases the use of manual measurements guarantees a correct quality control and validation of the generated information.

Catchment Physiographic Characteristics

The main physical catchment attributes must be surveyed at the beginning of the monitoring project. The outlet of the catchment is defined at the point where the flow gauging station will be located. The key physiographic characteristics are: (i) those necessary to understand the functioning of the catchment (e.g., area, elevations, slopes, soils); and (ii) those necessary to compare with other catchments (e.g., land use and land cover, watershed interventions, special features). Generally, this characterization is done thoroughly once at the beginning of the project, due to the slow rate of change in such parameters, but it can be updated periodically if necessary (e.g., annually).

Fig. 6.25. Pictures depicting several steps in the iMHEA protocol: (a) Selection of monitoring catchments in Tambobamba, Peru, by a group of hydrologists, local partners and users of land and water. Photo: Boris Ochoa-Tocachi, 2011. (b) Example of a tipping bucket rain gauge with a resolution of 0.2 mm installed in Lloa, Ecuador. Photo: BOT, 2014. (c) Local partners of iMHEA installing a rain gauge in Piura, Peru. The rain gauge accuracy is revised using static and dynamic calibration methods on a regular basis. Photo: BOT, 2012. (d) A weir located at the outlet of an iMHEA catchment in Ica, Peru. The V-shaped section of the weir allows an accurate monitoring of low flows. Photo: Junior Gil-Rios, 2015. (e) A weir located at the outlet of an iMHEA catchment in Huaraz, Peru. The composite triangular-rectangular section allows the monitoring of high flows. Photo: JGR, 2014. (f) A weather station is located at a central location between the iMHEA catchments in Lloa, Ecuador. The monitoring of several climatic variables, in this case temperature, relative humidity, solar radiation, wind velocity and direction and rainfall, allows for a more accurate calculation of evapotranspiration. Photo: BOT, 2014. (g) Example of a tipping bucket rain gauge in Lloa, Ecuador, with some litter retained in the filter. Photo: BOT, 2014. (h) Hydrological monitoring of two catchments under different levels of conservation/alteration in Cochabamba, Bolivia. A local farmer contrasts visually (left) a highly degraded catchment produced by cultivation and overgrazing, against (right) a relatively conserved catchment situated closely to the former. In what we refer to as the 'Placebo iMHEA effect', local users recognize the relevance of monitoring impacts, even before seeing any data. Photos: Luis Acosta, 2012. (i) Highly seasonal catchment in Tambobamba, Peru. In puna, the landscape is markedly different during the (left) dry season and the (right) rain season. Photos: BOT, 2012, 2013. (j) An International Course on Andean Hydrology and Hydrological Monitoring organized by Nature and Culture International, CONDESAN, the University of Piura, Peru, and iMHEA in Piura in 2013. Photo: BOT, 2013. (k) A group of farmers in a meeting with local decision makers and hydrologists from iMHEA in Pacaipampa, Peru. Photo: BOT, 2012.

The main characteristics are:

- Catchment area and shape (indicating the calculation method, for instance, using a topographical map at 1:10,000 scale or GPS surveying).
- Elevation map, maximum and minimum points (location of the flow gauging station).
- Average catchment and riverbed slopes (indicating the calculation method).
- Initial characterization of land use and land cover (indicating the areal percentage of land cover, the land uses identified in the catchment, and the identification method).
- Soils and geology maps and characteristics (especially soil hydrophysics).
- Special features that can influence in the occurring hydrological processes (e.g., permeable bedrocks), or in other hydrological ecosystem services (e.g., important sediment transport and deposition due to the presence of roads).

This information can be organized using a standard iMHEA Inventory Form for each monitored catchment. A map of the catchment(s) at an appropriate scale must contain these elements, including the location of any equipment, monitoring points and operational routes.

Monitoring of Precipitation

Rainfall is the main component of precipitation in Andean catchments (Padrón et al., 2015), and is monitored using electronic tipping bucket rain gauges (Fig. 6.25(b)) and validated with manual measurements using a pluviometer. The rain gauges must have a resolution of at least 0.254 mm (0.1 in), but typically 0.2 mm and ideally 0.1 mm. Each rain gauge provides event rainfall data (time to tip), which is then aggregated at fixed time intervals depending on the posterior analysis. Such aggregation can be done simply by counting the number of tips during a time period or using linear interpolation (Ciach, 2003) or other techniques, for instance, a composite cubic spline interpolation on the cumulative rainfall curve (Sadler and Brusscher, 1989; Wang et al., 2008).

A correct measurement of precipitation, which implies regional representativeness, is conditioned by several factors such as reducing wind effects (World Meteorological Organization, 2012; Muñoz et al., 2016). The location of rain gauges must be chosen in such a way that wind velocity at the flume is as low as possible. When possible, both the rain gauge and the manual pluviometer should be protected against wind effects in all directions using natural or artificial elements but located at a distance of at least twice the height of such elements to avoid rainfall interception.

Additionally, the rain gauge must be leveled and installed at a standard height of 1.50 m above the ground (Fig. 6.25(c)).

The manual pluviometer must have a ruler with minimum graduations of 0.2 or 0.1 mm (i.e., the same as the automatic rain gauge resolution). To ensure accurate measurements, the maximum error of such graduations must not exceed ± 0.05%. In areas with snowfall occurrence, the equipment must be located above the maximum expected snow height accumulation in the ground. Additionally, an antifreeze substance must be used to melt ice and snow falling in the pluviometer without exceeding one third of its total capacity. On the other hand, to avoid evaporative losses from the pluviometer, the equipment must minimize heat absorption (e.g., using clear colors that reflect most on the incident sunlight), and could use a thin layer of oil ($\approx$ 8 mm) to prevent evaporation.

Monitoring of Stream Flow

Discharge is usually estimated as a function of water level at a gauging station, which has the function to facilitate the generation of continuous and systematic measurements to calibrate a stage-discharge curve (World Meteorological Organization, 2008). The water level is monitored using automatic electronic sensors, mainly pressure transducers and ultrasonic devices, and validated with manual observations. The gauging station must feature a control section (e.g., a weir or a stable river section) with known geometric characteristics unchanged over time, and able to contain the total discharge at the outlet of the catchment (Fig. 6.25(d)). The manual observations must be obtained using a ruler with graduations of 1 mm.

In mountain streams with moderate flows, such as those in Andean catchments, the discharge can be monitored using a sharp-crested weir equipped with automatic pressure transducers. The composite weir would feature a V-shaped section for low flows and a triangular-rectangular section for high flows (Fig. 6.25(e)). In lowland streams with sustained high flows and important sediment transport, such as those in Andean-Amazonian catchments, a weir is not recommended. After an appropriate hydrological assessment, the gauging station can be placed in a completely stable natural river section or building a canal structure such as a Parshall flume to avoid sediment deposition. The use of ultrasonic water level sensors is also recommended in this case.

The World Meteorological Organization (2008) recommends the following for the location of the flow gauging station:

- Identify a straight reach of approximately 100 m upstream and downstream to the monitoring station. This is sometimes impractical in Andean catchments, where reach lengths of approximately 10 m have been sufficient to provide stable values of discharge.

- The total flow must be confined in the control section for the entire range of water stage, avoiding the occurrence of subsurface flow.
- The control section must avoid erosion, sediment deposition and changes in geometry.
- The riverbanks must be stable and sufficiently tall to contain floods.
- In the case of using a weir, the water level must be monitored upstream of the control section at a distance of at least three times the maximum water level above the crest to avoid effects from the contraction of the water sheet.
- The velocity of water approaching the weir must be less than 0.15 m s^{-1}.
- Strong water level fluctuations in the stream surface must be avoided.
- The station should be sufficiently secure and accessible to allow taking measurements even when there is presence of ice, solids, sediments or high floods.

To calibrate the stage-discharge curve (World Meteorological Organization, 2012), simultaneous measurements of discharge and water level are needed along a range of flow rates. Although a control structure can be used in the gauging station, the theoretical equation needs to be calibrated under operational conditions to reduce errors in the estimation of discharge (USDI Bureau of Reclamation, 2001; Guallpa and Célleri, 2013). When flows are low, a volumetric method can be sufficient to estimate the discharge by dividing a measured volume over a period time. When flows are moderate and laminar, the discharge can be measured using the velocity-area method dividing the gauging section in several subsections and measuring velocity at different depths. When flows are high and turbulent, a dilution gauging method can be used, for instance, using table salt and measuring the electric conductivity that is related to substance concentration and discharge. In all cases, the discharge measurements must be repeated consecutively at least three times recording the water level, date and time of each observation.

Other monitored variables

Although not required, many iMHEA partners engage in monitoring of other variables, which provide added value for their particular management context. Examples include (i) monitoring of other meteorological variables such as temperature, humidity and wind speed/direction, (ii) soil properties, (iii) geological characterization and, (iv) tracer monitoring.

An approximation of evapotranspiration can be obtained using rainfall and stream flow on an annual basis, shorter-term time scales would require local meteorological data to improve the calculation, e.g., (Córdova et al., 2015). These data generally include air temperature, relative humidity, solar radiation, wind velocity and direction, and air pressure. A weather station can be located at a central location within a catchment or a group of catchments, generating data at least at an hourly basis (Fig. 6.25(f)).

At the same time, the effects of different watershed interventions on local hydrological ecosystem services are strongly linked to their impacts on soil properties. For example, soil compaction under overgrazing (Sarmiento, 2000; Díaz and Paz, 2002; Quichimbo, 2008) or enhanced soil infiltration in forested catchments (Bruijnzeel, 2004; Tobón, 2009; Beck et al., 2013a). Therefore, an extensive characterization of soil properties can improve the understanding of local hydrological processes. Similarly, an accurate identification of geologic influences (e.g., permeable rocks, aquifers, faults) can improve the posterior interpretation of results. This characterization can be complemented with tracer monitoring, for instance, using natural occurring isotopes (Mosquera et al., 2015).

Considerations for the Monitoring Design and Operation

Spatial scale, density and coverage of monitoring stations

It is important to obtain data at scales that are hydrologically representative of the ecosystems in the surrounding area. Commonly, the catchment of interest hosts various ecosystems or land uses (e.g., grasslands, forests, degraded areas, plantations), and thus the observed variables may not capture the hydrological signals of individual characteristics. In this case, it is difficult to attribute the stream flow response to a single land use or land cover that impacts on the hydrological functioning of an ecosystem.

Ideally, a catchment must host a homogeneous land use and land cover; however, this may be problematic in practice, especially for large catchments. For instance, Bosch and Hewlett (1982) found that impacts on stream flow due to changes in forest cover of less than 20% cannot be detected. Therefore, this protocol recommends that the monitored catchments must be smaller than 15 km^2 and have a single land use or land cover in at least 80% of their areas.

Similarly, other problems are present in small catchments (< 0.2 km^2). For example, in high Andean grasslands with significant wetland areas, sub surface water flow can become important. As only a fraction of the total discharge is effectively measured, the unmonitored circulation of water in the soil without draining to a common stream would invalidate the assumption considered to close the water balance. Another problem of small catchments is their regional representativeness, for instance, in terms of slopes, shape, storage capacity, or water yield, e.g., Mosquera et al. (2015), which could complicate the extrapolation of results to larger catchments (Ochoa-Tocachi et al., 2016b). In any case, the design must ensure that the total outflow can be monitored at the outlet of the catchment, avoiding zero-order basins, i.e., catchments without a permanent, defined discharge canal.

In terms of equipment, the number of stations necessary to measure rainfall depends on the catchment area and the expected spatial rainfall variability therein. For instance, Buytaert et al. (2006a) and Célleri et al. (2007) have found great differences in precipitation at small distances in mountain catchments of southern Ecuador. In these cases, a sub- or an over-estimation of total rainfall can lead to an erroneous application of the water balance equation, and thus to draw mistaken conclusions about the impacts of different watershed interventions. Furthermore, Padrón et al. (2015) argue that tipping bucket rain gauges may underestimate total precipitation by around 15% when rain falls mainly as low intensity events.

As a general rule, at least 2 rain gauges must be installed in each catchment even when these are small (< 1 km^2). Most of the iMHEA monitoring sites are equipped with 3 rain gauges at representative low, middle and high elevation points within the catchments (Ochoa-Tocachi et al., 2016a). This also allows to validate and correct the measurements or fill data gaps when one of the sensors fails, for example, due to battery or memory issues or inlet obstructions.

Frequency of Automatic Data Acquisition

Due to the rapid hydrological response of small catchments to precipitation events, flows can reach peak values in only a few minutes. Therefore, the frequency of data logging must be high, minimum at an interval of 15 minutes, but ideally 5 minutes. Most of the iMHEA's sites generate data at a high temporal resolution of 5 minutes, in order to identify impacts on the short-term hydrological regulation that can pass unnoticed on daily aggregated indices (Ochoa-Tocachi et al., 2016a).

Data stored by automatic sensors must be gathered approximately once a month. Although the most advanced sensors may have the capacity to store large amounts of data, the tough climatic conditions of Andean catchments increase the need of sensor maintenance. For example, some technicians have reported obstructions in the rain gauge flumes due to the presence of forest litter or small animals even in shorter periods (Fig. 6.25(g)). Another problem can be the limited capacity of batteries and data loggers to function for longer periods. Both issues may result in the loss of important data between the visits.

Monitoring Design for Impact Evaluation

Two general approaches are common to assess the impacts of land use and land cover change or other watershed interventions on local hydrology (McIntyre et al., 2014): long-term analysis and paired catchments.

However, in both cases quantifying these impacts is complicated by the difficulty of distinguishing the effects of changes in land use from those that are due to natural climatic variability or other confounding factors (Ashagrie et al., 2006; Bulygina et al., 2009). In long-term analysis, even though the same catchment is monitored before and after the change, the influence of natural climatic variability may be different during the two considered periods (Lørup et al., 1998). This is solved in the second approach by monitoring the paired catchments under the same climatic conditions and different land uses. However, all watersheds are unique and land use is not the only factor that affects their hydrological responses (Bosch and Hewlett, 1982; Thomas and Megahan, 1998; Beven, 2000; Brown et al., 2005; Adams et al., 2012; Biederman et al., 2015). In iMHEA, although some sites have robust combined before-after-intervention-reference setup (Fig. 6.26), impact of an intervention is evaluated based on spatial comparison rather than changes over time. On balance, the ability of paired catchments to deliver more rapid answers, makes it the preferred approach.

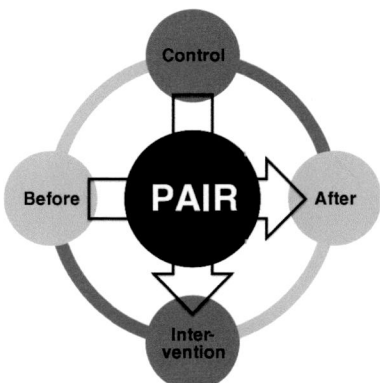

Fig. 6.26 The ideal design of the paired catchments involves the temporal (before-after) and spatial (control-intervention) dimensions. Although the definition of a baseline is more robust under this design, a simpler setup of 'control-intervention' has proven useful in many cases.

Paired-Catchment Monitoring

To understand the impacts of contrasting watershed interventions (e.g., different land use types), this protocol recommends the use of paired catchments. This design is based on the comparison of the hydrological responses of two catchments, where one acts as a reference of a particular state and the other represents the intervention to be assessed. For example, natural vs cultivated catchments, degraded vs restored catchments or forested vs grassland catchments. Each catchment must be equipped following the guidelines previously indicated.

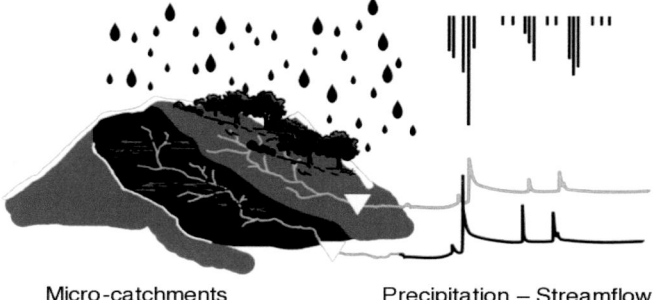

Micro-catchments   Precipitation – Streamflow

Fig. 6.27 Conceptual representation of the paired catchment setup. Under this design, catchments are commonly collocated to minimize climatic differences. The resulting hydrographs that characterize their hydrological responses are compared to identify the impacts of different watershed interventions.

Under this design, catchments are selected in such a way that their size, topography, soils, climate and other factors are as similar as possible, while the watershed intervention to be evaluated becomes their only significant difference. In this way, the discrepancies found (if present) in discharge values or in the water balance can be attributed to such differences (Fig. 6.27). To limit the variability in climatic and soil conditions, the catchments must be located as close as possible to each other. This also reduces the efforts in maintenance and operation works.

A major advantage of the paired catchment design is that important differences in their hydrological responses can be identified in relatively short time periods (e.g., 1 year), feeding rapidly into the decision-making process. A careful analysis and interpretation of the results is necessary to avoid an erroneous attribution of hydrological response impacts that are due to other confounding factors.

Figure 6.28 shows an example of a paired catchment design in northern Ecuador. The two selected catchments, Quebrada del Volcán (221 ha) and Quebrada Kachiyaku (179 ha), are equipped with 1 weir with pressure transducers and 2 rain gauges each, and 1 meteorological station that has been placed in a central location (Fig. 6.25(f)). The monitoring is projected to be maintained on the long-term to understand the ongoing hydrological processes of these ecosystems, identify differences due to land use and land cover, and to characterize subsurface water exchange mechanisms between the catchments.

Long-Term Monitoring

To understand changes in the hydrological response of a catchment over time (i.e., non-stationarity), each catchment monitored with the recommended instrumentation can be analyzed individually. Although

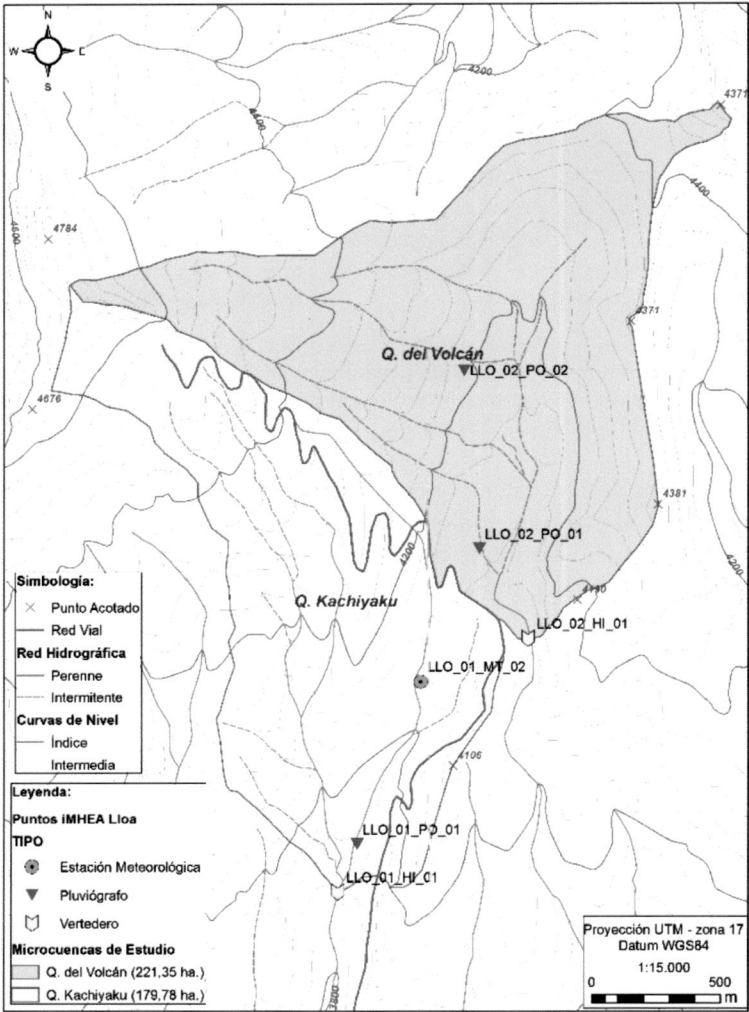

Fig. 6.28 Example of paired catchments monitored by local iMHEA partners in Lloa, Ecuador (in Spanish). This site aims to understand the hydrological processes occurring in the humid páramo of northern Ecuador, including the impacts of grazing exclusion, restoration efforts and deep water infiltration in the soils.

the principle of this protocol is to draw significant conclusions in the short term using space-for-time substitution, the long-term monitoring of the individual catchments can reveal important information. For example, many restoration efforts will show results after several years, which can be contrasted to an initial baseline. Under this design, the sustainability of the monitoring sites is extremely important, and surely benefits from a participatory approach and a strong institutional agreement.

It is important to differentiate between controlled interventions (e.g., restoration efforts) and uncontrolled changes (e.g., climate change), and between immediate impacts (e.g., infrastructure construction) and sustained changes (e.g., land-use change). Moreover, the monitoring should be set up ideally before the change to be assessed in order to generate a robust baseline that allows for a meaningful comparison of long term effects. A sensible combination of the paired catchment setup in the long term may reveal significant results of the evaluated interventions (Fig. 6.22).

Data Processing and analysis

Defining and managing a database

A quality control of the generated data must be done immediately after the data has been downloaded at every visit in order to identify errors (e.g., data outside the expected measurement range, such as negative water level values) or suspicious measurements (e.g., extreme rainfall intensities). Data must be curated thoroughly before their use in any posterior analysis. These data should feed a structured and organized database that could potentially have different purposes and users (Buytaert et al., 2012).

The high-resolution time series aggregated at different time intervals (5, 15, 60 minutes, or 1 day) may represent a large amount of information in the long term. However, it is imperative to store the finest resolution data to avoid losing important information that can be used in other studies not necessarily thought at the beginning of the project. For instance, a local partner may need daily values that can be obtained by aggregating the high-resolution data; however, disaggregating information from a larger to a shorter interval is not impossible without very restrictive assumptions that could invalidate the results or input large uncertainties. Furthermore, the high-resolution data can capture rapid hydrological processes and changes therein that can be omitted in aggregated indices (Ochoa-Tocachi et al., 2016a).

Hydrological Indices to Quantify Impacts

Hydrological indices are commonly used to summarize the hydrological response of a catchment, its state of conservation, and the impact of alterations. A large set of indices (also referred to as 'signatures') exist in the scientific literature, e.g., Walsh and Lawler (1981), Hughes and James (1989), Poff and Ward (1989), Richards (1989; 1990), Poff (1996), Poff et al. (1997), Richter et al. (1996; 1997; 1998), Clausen and Biggs (1997; 2000), Wood et al. (2000), Gippel et al. (2001), Baker et al. (2004), Mathews and Richter (2007), Beck et al. (2013a). Poff and Ward (1989) and Richter et al. (1996) defined five main components of the flow regime: (i) magnitude, (ii) frequency, (iii) duration, (iv) timing, and (v) rate of change in flow events. Olden and Poff (2003) have extended this classification with sub categories for average, low and high flows, and analyzed index redundancy and multicollinearity.

In the Andes, the great spatial variability of climatic conditions and the only recent development of research on Andean hydrology have delayed the development of indices to evaluate the hydrological response of these catchments. However, the emergent generation of large amounts of data requires the synthesis of information in a reduced set of hydrological indices that can reflect clearly the impacts of human interventions. Moreover, decision makers and local stakeholders require the presentation and visualization of results and the quantification of ecosystem services, including the associated uncertainty, in a common and understandable language. For instance, the water yield can be characterized by the runoff ratio, whereas the hydrological regulation is usually quantified in means of the base flow index and the slope of the flow duration curve (Sawicz et al., 2011; Ochoa-Tocachi et al., 2016b; Visessri and McIntyre, 2016). The calculation and use of hydrological indices to evaluate ecosystem functioning are simpler than other types of indicators (Pyrce, 2004), such as hydraulic evaluations, habitat assessment or holistic approaches.

To optimize the available information, indices must be selected in such a way that they are relatively independent (Sefton and Howarth, 1998; Bulygina et al., 2009), well-defined to minimize ambiguity (Olden and Poff, 2003), and susceptible to represent the impact of changes in the catchment (Archer et al., 2010). The use of indices to characterize the impacts of land-use change on the hydrological response of Andean catchments can be seen in Ochoa-Tocachi et al. (2016a).

Data Processing

Different methods exist to interpolate the tipping bucket rainfall data, but a commonly used method uses a composite cubic spline interpolation applied to the cumulative rainfall curve and then aggregated at different time steps. This allows a smoother estimation of rainfall intensities than simply counting the number of tips in a determined interval of time, which is deemed more realistic (Fig. 6.29). A 1-minute or 5-minutes moving window can be used to extend the calculation of rainfall intensities for different event durations under consideration. The normalized seasonality index (Walsh and Lawler, 1981) ranging from 0 (non-seasonal) to 1 (extremely seasonal) can be used to compare rainfall regimes between several catchments.

The discharge data must be normalized prior to any comparison between catchments, for instance, dividing by the catchment area (units of mm or $l\,s^{-1}\,km^{-2}$). For the calculation of some indices (e.g., runoff ratio) or to apply the water balance equation, flow units are transformed to match the rainfall units (mm). Daily flows can be used to calculate Flow Duration

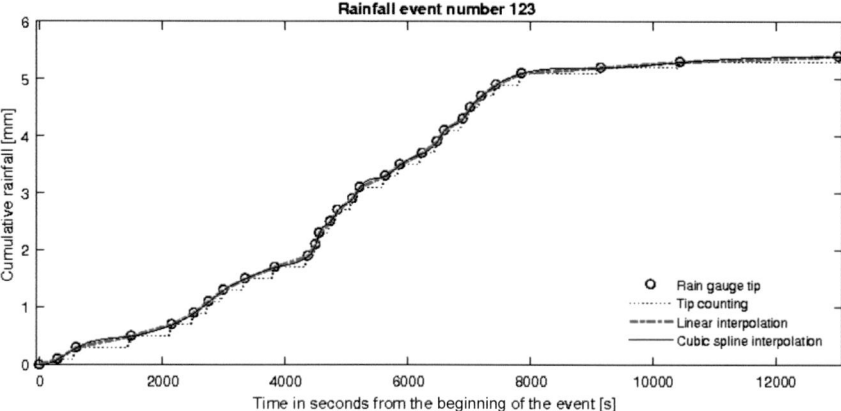

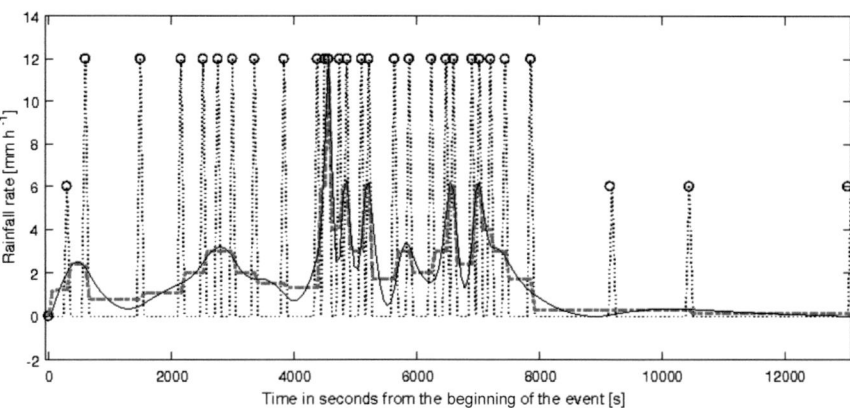

Fig. 6.29 Example of the calculation of 1-minute rainfall intensities based the tipping bucket rain gauge data. The simple tip counting provides a good calculation of rainfall totals but fails to provide realistic intensity estimations. Linear and polynomial interpolation techniques can be used to obtain a continuous time series of rainfall intensities that are considered a more accurate representation of real conditions.

Curves (FDC) and corresponding percentiles. Three indices are commonly considered in several hydrological studies and for watershed management:

- The Runoff Ratio (RR), calculated as the relation between average annual discharge and average annual rainfall, and represents an indication of water yield.
- The Base Flow Index (BFI), calculated as the ratio between base flow to total flow, e.g., Chapman (1999), Kabubi et al. (2005), Willems (2014), and can be associated to the short-term hydrological regulation at event scale.

- The slope of the FDC, calculated between the 33rd and 66th flow percentiles in logarithmic scale, commonly used as an indicator of hydrological regulation (Olden and Poff, 2003; Ochoa-Tocachi et al., 2016b). For example, a steep slope may indicate high flashiness in the hydrological response to precipitation events, whereas a flatter curve may represent a buffered behavior and larger storage capacity (Buytaert et al., 2007; Yadav et al., 2007).

Extrapolation to Ungauged Areas

Although several stakeholders are interested in data for specific watersheds, it is impractical to monitor every single catchment due to several constraints, such as the limited amount of resources. However, it is possible to regionalize the results obtained from the monitoring of individual catchments to a broader context based on their hydrological similarity (Sawicz et al., 2011) in contrast to their physical similarity, see also Muñoz et al. (2016). The objective of such relationships is not only to reduce the monitoring efforts (Correa et al., 2016), but to estimate and predict the hydrological response of ungauged catchments and the effects of different watershed interventions before their implementation. This allows for approximating several impacts by relating the information generated in a group of representative sites, including a quantification of the uncertainty associated with these estimations.

Regional analyses are used to generalize results from data gathered in the few monitored sites to ungauged basins and to account for differences due to catchment uniqueness and spatial variability (Sivapalan, 2003; Beck et al., 2013b). Two examples of regional models that are widely used in other regions are the BFIHOST in the UK (Boorman et al., 1995) and the Curve Number in the US (USDA, 1986). Predictions in ungauged basins can be done through different methods (Parajka et al., 2013; Visessri and McIntyre, 2016), from which two regionalization approaches are widely used (Bulygina et al., 2009; 2011; 2012):

- relationships between catchment attributes and model parameters, e.g., Lamb and Kay (2004), McIntyre et al. (2005), Parajka et al. (2005), Lee et al. (2006), Wagener and Wheater (2006), Young (2006), Beven (2007), Wagener (2007), Buytaert and Beven (2009); and,
- relationships between catchment attributes and hydrological indices, e.g., Berger and Entekhabi (2001), Brandes et al. (2005), Mazvimavi et al. (2005), Shamir et al. (2005a; 2005b), Bárdossy (2007), Yadav et al. (2007), Longobardi and Villani (2008), Oudin et al. (2010), Zhang et al. (2008), Peña-Arancibia et al. (2010), van Dijk (2010), Krakauer and Temimi (2011), Ahiablame et al. (2013), Visessri and McIntyre (2016).

The second approach has been implemented to the iMHEA network of paired catchments, where significant relationships were found between variables that describe land use and land cover and most of the tested hydrological indices (Ochoa-Tocachi et al., 2016b). Combining the paired catchment setup with the common regionalization approach has proven useful to improve both approaches (Fig. 6.30). Therefore, this design is seen as a useful strategy to optimize data collection to support and improve watershed interventions in data-scarce regions, such as the tropical Andes.

Current State of the iMHEA Network and Preliminary Results

Using a design based on a trading-space-for-time approach, over 30 catchments are currently being monitored for precipitation and stream flow by 18 local stakeholders in 15 sites located in Bolivia, Peru, Ecuador and Venezuela (Fig. 6.31). Where possible, one catchment is chosen such that it is representative for a reference state, while a second catchment represents the practice to be evaluated. The reference state may refer to either a conserved condition in contrast to a human intervention, or a degraded catchment without restoration activities or watershed management (Fig. 6.25(h)). The network focuses on small, homogeneous headwater catchments (between 0.2 km² and 10 km²) with a single land use or intervention in at least 80% of its area. As observed in Buytaert et al. (2016), the data generated by iMHEA is highly complementary to that of hydrometeorological stations. Figure 6.32 shows rainfall data in the Piura Basin from SENAMHI, the National Service of Meteorology and Hydrology of Peru, with stations located from 0 to 3000 masl and from iMHEA with stations between 2500 and 3800 masl. It is clear that the estimated rainfall trend (linear regression) as fitted on the SENAMHI stations underestimates precipitation in the highlands, and that the definition of climatic zones in official documents, e.g., Autoridad Nacional del Agua (2012), is far from realistic without the inclusion of the newly generated data.

As expected, the analyzed data clearly reflect the dominant regional climate patterns and the extraordinary wide spectrum of hydrological response behavior of the tropical Andes. Ochoa-Tocachi et al. (2016a) have summarized this range contrasting the perennially humid, highly buffered stream flow behavior of wet páramos with the strongly seasonal, flashy stream flow response of the drier puna (Table 6.12; Fig. 6.33). Rainfall seasonality is controlled by latitude, with minimum values of the seasonality index observed in the transition between the páramo and the puna biomes, especially in the region of the Andes where elevations are the lowest facilitating air fluxes. In other areas, the influence of the Amazonian warm and humid air masses contrasts to the cold and dry Pacific regime, generating notorious dissimilarities in both sides of the Cordillera

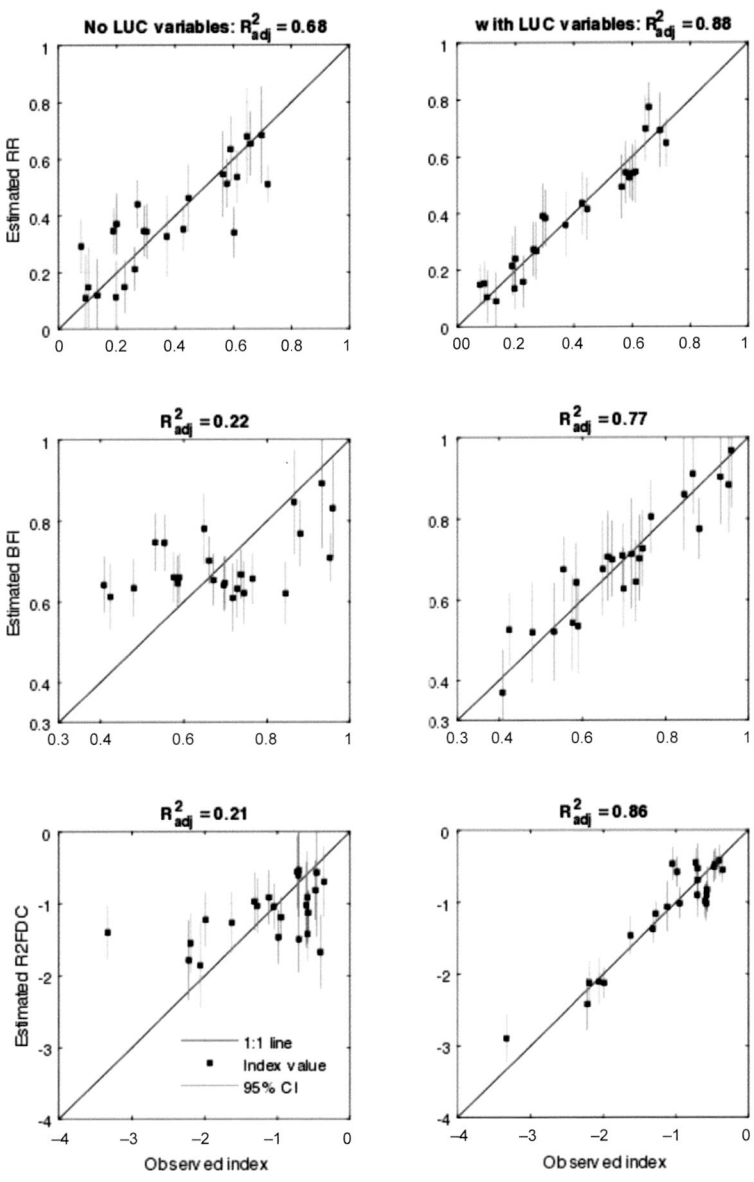

Fig. 6.30 Effects of including land use and land cover variables in the regional regressions derived using the iMHEA catchments, adapted from Ochoa-Tocachi et al., 2016b. As shown for the runoff ratio (RR), the base flow index (BFI) and the slope of the flow duration curve (R2FDC), the inclusion of land use and land cover variables improve the regression performance (R^2_{adj}) and reduce the associated uncertainty significantly (95% CI).

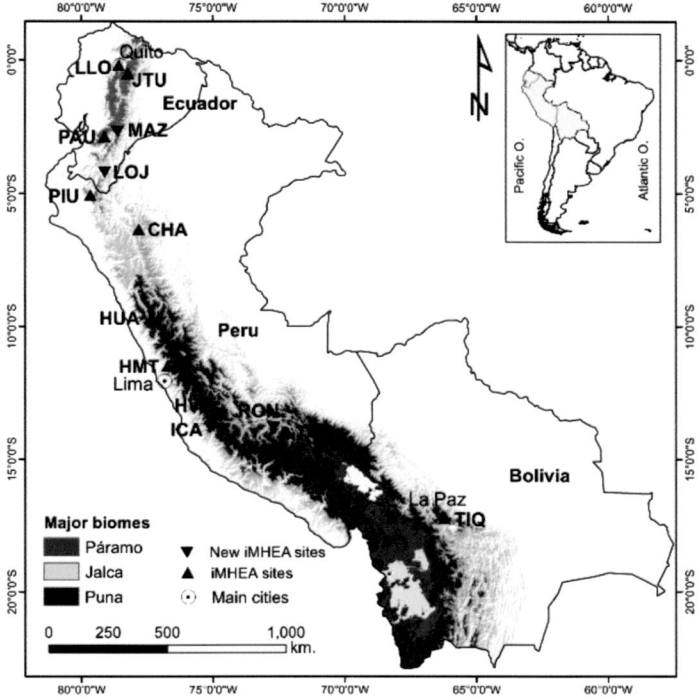

Fig. 6.31 Map of the tropical Andes and location of the iMHEA monitoring sites. Those represented a black up-pointing triangle where included in an assessment of the impacts of land use on the hydrological response of Andean catchments (Ochoa-Tocachi et al., 2016a), and where used to test the usefulness of paired catchments to regionalize land-use signals on stream flow (Ochoa-Tocachi et al., 2016b).

(Fig. 6.25(i)). Despite these differences, natural Andean catchments are characterized by high runoff ratios, indicating large water yields (Buytaert et al., 2007; Mosquera et al., 2015), and smaller slopes of the FDC, which are associated with a good hydrological regulation capacity and a base flow dominated response (Buytaert et al., 2006b; Crespo et al., 2011).

Similarly, the impacts of land use are found to be influenced by several factors, such as the catchments' physiographic characteristics, the original and replacement vegetation cover and soil properties and changes therein. Such impacts commonly result in more variable stream flows and in reduced water yields and worse hydrological regulation (Ochoa-Tocachi et al., 2016a). Despite the hydrological properties of the original biome, the effects of common human activities in the different Andean catchments are consistent (Fig. 6.34). It has been observed that cultivation increases flashiness and reduces low flows in particular (Sarmiento, 2000; Buytaert et al., 2007). Grazing effects depend on animal density and may pass

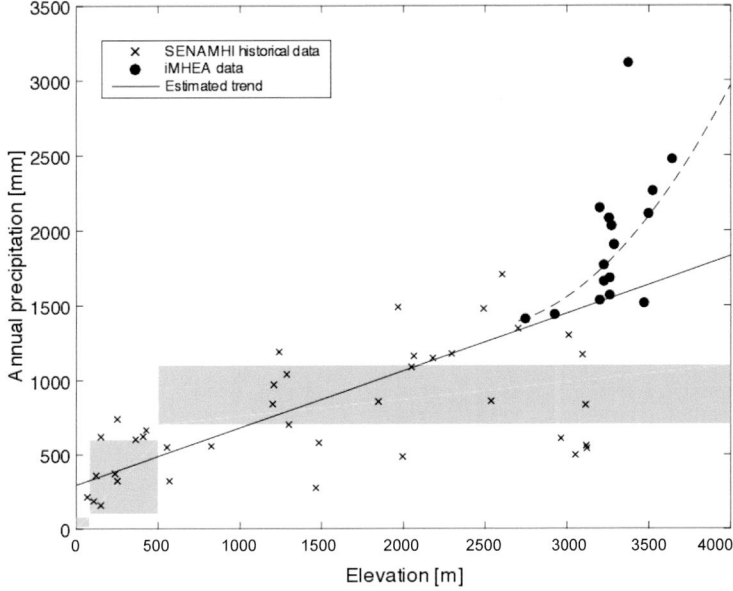

Fig. 6.32 Rainfall data from SENAMHI and iMHEA showing the complementarity of information generated by multiple sources in a polycentric approach (Buytaert et al., 2016).

unnoticeable in aggregated indices (Crespo et al., 2011), but they have the largest effect on the hydrological regulation. The afforestation of grasslands with exotic vegetation affects the entire range of discharges (Buytaert et al., 2007). Even though the specific magnitudes of such changes are variable, these trends are consistent in the different Andean biomes studied by the iMHEA network (Ochoa-Tocachi et al., 2016a).

An important case in point is afforestation with exotic species. When large extensions of Andean highlands were forested with pine, local authorities tried to replicate a successful experience to recover degraded lands occurred in Cajamarca, Peru. However, most frequently, pine plantations are introduced in non-degraded areas and, as seen in humid páramo studies (Buytaert et al., 2007), they change water and organic carbon retention and hydrological response features. We find similar trends in afforested jalcas and punas, including important reductions in water yield, mainly produced by higher evapotranspiration and canopy interception, and major impacts on discharge, especially over low flows (Ochoa-Tocachi et al., 2016a). Despite the negative evidenced impacts of afforestation on total water production and local biodiversity (Hofstede et al., 2002), this practice is still part of large regional efforts generally supported by the Ministries of Agriculture of Andean countries. In the last years, local awareness has increased recognizing such practices as productive interventions rather

Table 6.12 Summary of the iMHEA catchment characteristics for the selected sites in Fig. 6.30.

Code Units	Ecosystem [type]	Altitude [m]	Area [km²]	Land-use [type]	Monitoring period [dates]
LLO Lloa					
LLO_01	Páramo	3825–4700	1.79	Overgrazed	10/01/2013–27/01/2016
LLO_02	Páramo	4088–4680	2.21	Grazed	10/01/2013–27/01/2016
JTU Jatunhuaycu					
JTU_01	Páramo	4075–4225	0.65	Overgrazed	14/11/2013–15/02/2016
JTU_02	Páramo	4085–4322	2.42	Overgrazed	15/11/2013–15/02/2016
JTU_03	Páramo	4144–4500	2.25	Natural	13/11/2013–16/02/2016
PAU Paute					
PAU_01	Páramo	3665–4100	2.63	Natural	24/05/2001–16/08/2005
PAU_02	Páramo	2970–3810	1.00	Natural	29/02/2004–31/07/2007
PAU_03	Páramo	3245–3680	0.59	Afforested	29/05/2004–31/07/2007
PAU_04	Páramo	3560–3721	1.55	Cultivated	27/10/2001–14/10/2003
PIU Piura					
PIU_01	Páramo	3112–3900	6.60	Natural	05/07/2013–12/12/2015
PIU_02	Páramo	3245–3610	0.95	Grazed	06/07/2013–13/12/2015
PIU_03	Páramo	3425–3860	1.31	Overgrazed	11/04/2013–23/10/2015
PIU_04	Forest	2682–3408	2.32	Natural Forest	23/06/2013–14/01/2016
PIU_07	Dry puna	3110–3660	7.80	Overgrazed	11/07/2013–15/01/2015
CHA Chachapoyas					
CHA_01	Jalca	2940–3200	0.95	Afforested	18/08/2010–07/12/2015
CHA_02	Jalca	3000–3450	1.63	Natural	18/08/2010–07/12/2015
HUA Huaraz					
HUA_01	Humid puna	4280–4840	4.22	Natural	10/09/2012–20/06/2014
HUA_02	Humid puna	4235–4725	2.38	Grazed	10/09/2012–20/06/2014
HMT Huamantanga					
HMT_01	Dry puna	4025–4542	2.09	Overgrazed	28/06/2014–03/03/2016
HMT_02	Dry puna	3988–4532	1.69	Overgrazed	26/06/2014–03/03/2016
TAM Tambobamba					
TAM_01	Humid puna	3835–4026	0.82	Afforested	12/04/2012–02/01/2013
TAM_02	Humid puna	3650–4360	1.67	Natural	12/04/2012–16/04/2013
TIQ Tiquipaya					
TIQ_01	Humid puna	4140–4353	0.69	Cultivated	02/04/2013–25/01/2016
TIQ_02	Humid puna	4182–4489	1.73	Natural	18/02/2013–25/01/2016

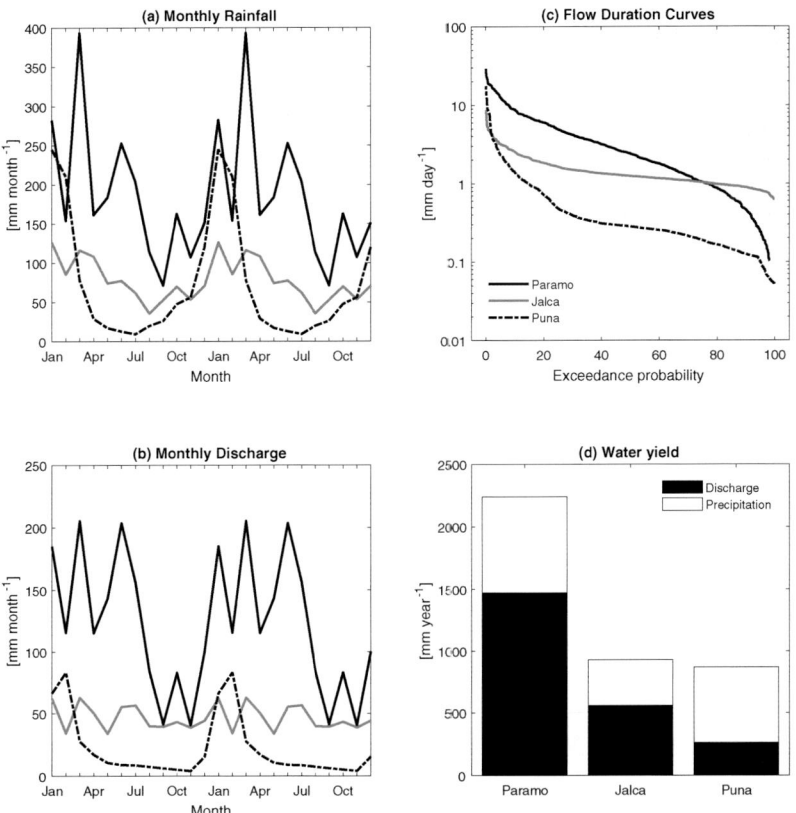

Fig. 6.33 Hydrological response of natural Andean catchments in different biomes. Monthly discharge (a) and monthly precipitation (b) evidence the higher seasonality of puna in contrast to the more sustained regimes in páramo and jalca catchments. Daily flows in jalca are more stable as seen in the flow duration curves (c), while water yield (d) is higher in parámo catchments and much lower in puna. See detailed data in Ochoa-Tocachi et al., 2016a.

than mistakenly camouflaging them as conservation efforts. This has made it possible to improve the identification of 'more suitable' areas of intervention and changing the approach towards the implementation of conservation agriculture in degraded lands by leveraging the enhancement in soil infiltration produced by tree roots to avoid or reduce erosive processes (Bruijnzeel, 2004; Tobón, 2009; Beck et al., 2013a). However, adequate watershed interventions using afforestation with exotic species in remote Andean catchments is still far from being solved.

In the context of growing investment in climate change adaptation under compensation schemes for ecosystem services in Peru,[1] iMHEA has started

---

[1] http://www.leyes.congreso.gob.pe/Documentos/Leyes/30215.pdf

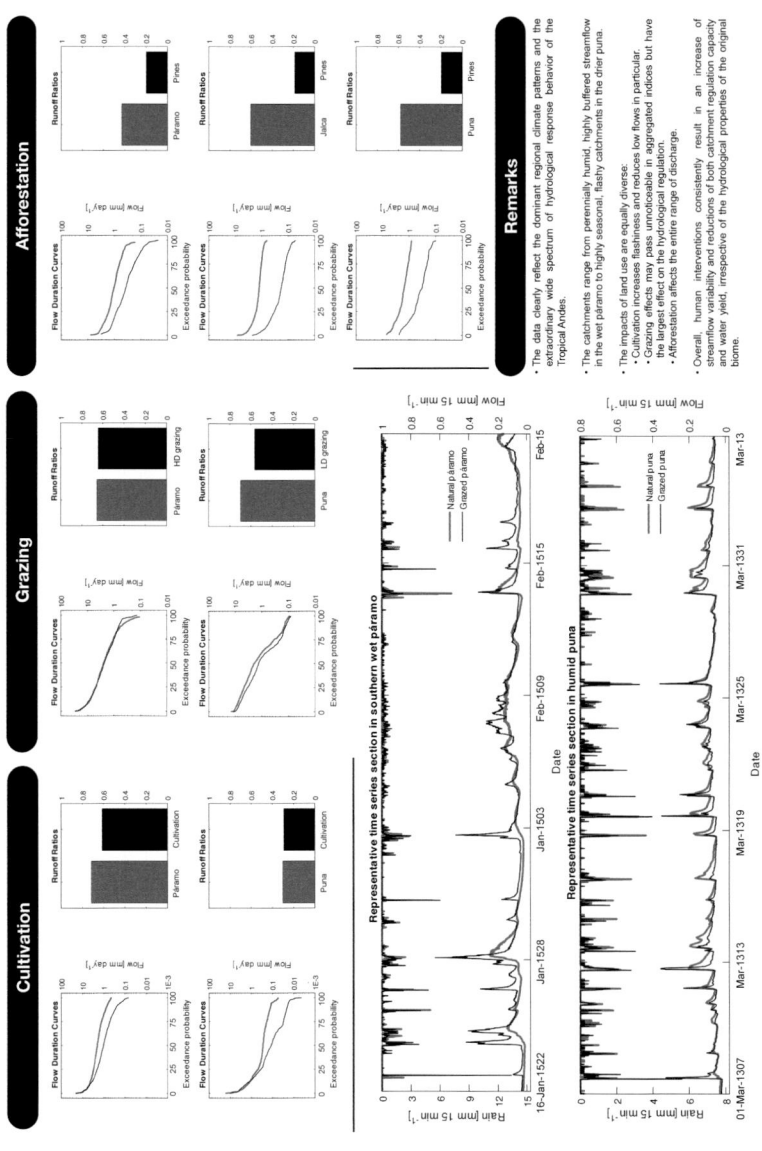

Fig. 6.34 Impacts of land use change showing consistent trends in the different Andean biomes analyzed, despite the differences in hydrological regime. In the case of livestock grazing, the impacts are noticeable in the high-resolution time series (bottom left) rather than in the aggregated indices (top middle).

to deliver useful information to multi-scale and multi-stakeholder decision-making activities, especially in previously sub-represented ecosystems. For example, the identification of livestock impacts on hydrological regulation in puna highlands provides also quantitative and complementary information on hydrological benefits of overgrazing elimination. Livestock grazing increases soil bulk density, which results in reduced hydraulic conductivity, increased overland flow and lower water yields (Díaz and Paz, 2002; Quichimbo, 2008; Crespo et al., 2010). In areas where seasonality and hydrological regulation are critical, water companies build expensive gray infrastructure to secure water for large cities downstream during long dry seasons. Recent changes in legislation now encourage companies to fight catchment degradation to improve natural hydrological regulation capacity. The sites provide a new generation of hydrological information that allows for economic analyses to study green infrastructure feasibility and cost-benefit comparisons between gray and green investments (Gammie and De Bièvre, 2015). For example, Fig. 6.35 shows a comparison of a set of gray and green infrastructure options for Lima, Peru.

The iMHEA initiative has now drawn the attention of major donors in the context of strengthening capacity for climate change adaptation, and green infrastructure investments in general, as they require evidence of the benefits of their investments. Other initiatives that are adopting the approach are the booming Water Funds in Latin America and the green infrastructure investments as they are promoted by recently approved

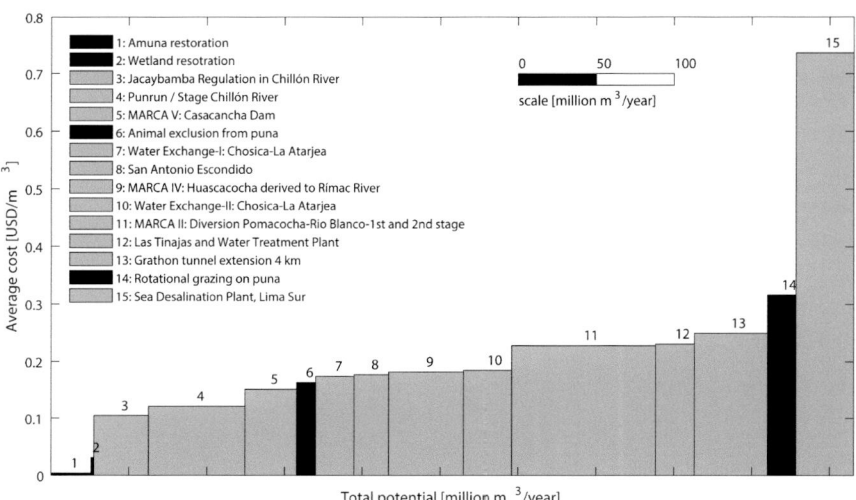

Fig. 6.35 Comparison of gray and green infrastructure in terms of Costs per Effectiveness (USD m^{-3}) from Gammie and De Bièvre (2015). The iMHEA has started to feed hydrological data into economic analyses to support large scale interventions, such as those to ensure water supply for Lima, Peru.

Peruvian legislation, being even compulsory in the case of water utilities of medium and large cities.

Conclusions

The research builds upon several years of extensive study, as part of the Regional Initiative for Hydrological Monitoring of Andean Ecosystems (iMHEA), to characterize the hydrology of tropical Andean catchments. By using data from a network of micro-catchments distributed from Ecuador to Bolivia, the iMHEA aims to tackle the scarcity of data and knowledge about the hydrology of high Andean catchments and the impacts of land use and cover change and degradation on their hydrological response and water yield, as well as those of many watershed interventions. By 'trading space-for-time', we find consistent impacts and trends in the different monitored biomes.

We recognize that a fixed common solution to the diversity of hydrological issues in the Andes does not exist, but the methodology proposed by the iMHEA has proven crucial to increase and strengthen the knowledge of Andean hydrology. Knowledge of hydrological change at this scale is limited but key, and this research output will significantly help decision makers inform policies related to development and conservation. Also, this network is an illustration of how information generated from participatory monitoring schemes, such as iMHEA, proves to be extremely relevant to overcome data-scarcity. The participatory nature of the network allows also for more rapidly feeding into decision-making processes and to promote mechanisms, opportunities and spaces to reflect, exchange experiences and provide feedback more easily (Fig. 6.25(j) and 6.25(k)). The results provided by iMHEA may be used to improve water resources management and the effectiveness of water conservation measures, and to support further research in the Andean region. Furthermore, the advent of new technologies and methods, as well as new questions that have been raised in the last years, involves a faster evolution of the network, but it certainly strengthens the route started a couple of years ago with promising and highly expected results.

As noted in Ochoa-Tocachi et al. (2016b), this methodology can be applied to evaluate human interventions both after their implementation (in the monitored catchments) and before (predicting responses in ungauged catchments). As the available database for human intervention assessment grows, including more catchments covering different ecosystems, characteristics and contrasting land use types and watershed interventions, more robust extrapolations can be expected with a better quantification of uncertainty. This approach is useful to generate information about the impact of human interventions on catchment hydrology, especially in data-scarce regions, but with potential application in other regions of the world.

Acknowledgments

The methodology presented on this chapter is based on the document "Guía Metodológica para el Monitoreo Hidrológico de Ecosistemas Andinos" by Rolando Célleri, Bert De Bièvre, Boris Ochoa-Tocachi, and Marcos Villacís (2013). The authors thank all partners of the Regional Initiative for Hydrological Monitoring of Andean Ecosystems (iMHEA, http://imhea. condesan.org) for their substantial collaboration and contributions. BOT was funded by the Imperial College PhD Scholarship Scheme and the UK Natural Environment Research Council (grant NE/L002515/1) "Science and Solutions for a Changing Planet DTP". WB acknowledges funding from the UK Natural Environment Research Council (grants NE/I004017/1 and NE-K010239-1). The fieldwork was funded by iMHEA through CONDESAN, Imperial College and the iMHEA partners. Special thanks to the people of the Andean communities and its authorities for providing important and constant consent and support to our fieldwork.

References and Bibliography

Adams, H., Luce, C., Breshears, D., Allen, C., Weiler, M., Hale, C., Smith, A. and Huxman, T. 2012. Ecohydrological consequences of drought- and infestation-triggered tree die-off: Insights and hypotheses. Ecohydrology 5: 145–159.

Adhikary, S., Yilmaz, A. and Muttil, N. 2015. Optimal design of rain gauge network in the Middle Yarra River catchment, Australia. Hydrological Processes 29(11): 2582–2599.

Ahiablame, L., Chaubey, I., Engel, B., Cherkauer, K. and Merwade, V. 2013. Estimation of annual baseflow at ungauged sites in Indiana USA. Journal of Hydrology 476: 13–27.

Archer, D., Climent-Soler, D. and Holman, I. 2010. Changes in discharge rise and fall rates applied to impact assessment of catchment land use. Hydrology Research 41(1): 13–26.

Ashagrie, A., Laat, P., Wit, M., Tu, M. and Uhlenbrook, S. 2006. Detecting the influence of land use changes on discharges and floods in the Meuse River Basin—the predictive power of a ninety-year rainfall-runoff relation? Hydrology and Earth System Sciences 10: 691–701.

Autoridad Nacional del Agua. 2012. Diagnóstico de la Gestión de los Recursos Hídricos de la Cuenca Chira-Piura, Piura: Ministerio de Agricultura, Autoridad Nacional del Agua, Peru.

Baker, D., Richards, R., Loftus, T. and Kramer, J. 2004. A new flashiness index: Characteristics and applications to midwestern rivers and streams. Journal of the American Water Resources Association (JAWRA) 40(2): 503–522.

Balvanera, P., Uriarte, M., Almeida-Leñero, L., Altesor, A., DeClerck, F., Gardner, T., Hall, J., Lara, A., Laterra, P., Peña-Claros, M., Silva Matos, D., Vogl, A., Romero-Duque, L., Arreola, L., Caro-Borrero, A., Gallego, F., Jain, M., Little, C., de Oliveira Xavier, R., Paruelo, J., Peinado, J., Poorter, L., Ascarrunz, N., Correa, F., Cunha-Santino, M., Hernández-Sánchez, A. and Vallejos M. 2012. Ecosystem services research in Latin America: The state of the art. Ecosystem Services 2: 56–70.

Bárdossy, A. 2007. Calibration of hydrological model parameters for ungauged catchments. Hydrology and Earth System Sciences 11(2): 703–710.

Beck, H., Bruijnzeel, L., van Dijk, A., McVicar, T., Scatena, F. and Schellekens, J. 2013a. The impact of forest regeneration on streamflow in 12 mesoscale humid tropical catchments. Hydrology and Earth System Sciences 17(7): 2613–2635.

Beck, H., van Dijk, A., Miralles, D., de Jeu, R., Bruijnzeel, L., McVicar, T. and Schellekens, J. 2013b. Global patterns in base flow index and recession based on streamflow observations from 3394 catchments. Water Resources Research 49(12): 7843–7863.

Bendix, J. 2000. Precipitation dynamics in Ecuador and northern Peru during the 1991/92 El Niño: a remote sensing perspective. International Journal of Remote Sensing 21: 533–548.

Bendix, J., Rollenbeck, R. and Reudenbach, C. 2006. Diurnal patterns of rainfall in a tropical Andean valley of southern Ecuador as seen by a vertically pointing K-band Doppler radar. International Journal of Climatology 26(6): 829–846.

Berger, K. and Entekhabi, D. 2001. Basin hydrologic response relations to distributed physiographic descriptors and climate. Journal of Hydrology 247(3-4): 169–182.

Beven, K. 2000. Uniqueness of place and process representations in hydrological modelling. Hydrology and Earth System Sciences 4(2): 203–213.

Beven, K. 2007. Towards integrated environmental models of everywhere: uncertainty, data and modelling as a learning process. Hydrology and Earth System Sciences 11: 460–467.

Biederman, J., Somor, A., Harpold, A., Gutmann, E., Breschears, D., Troch, P., Gochis, D., Scott, R., Meddens, A. and Brooks, P. 2015. Recent tree die-off has little effect on streamflow in contrast to expected increases from historical studies. Water Resources Research 51: 9775–9789.

Bond, B., Jones, J., Moore, G., Phillips, N., Post, D. and McDonnell, J. 2002. The zone of vegetation influence on baseflow revealed by diel patterns of streamflow and vegetation water use in a headwater basin. Hydrological Processes 16: 1671–1677.

Boorman, D., Hollist, J. and Lilly, A. 1995. Hydrology of soil types: a hydrologically based classification of the soils of the United Kingdom, IH Report 126, Wallingford: Institute of Hydrology.

Bosch, J. and Hewlett, J. 1982. A review of catchment experiments to determine the effect of vegetation changes on water yield and evapotranspiration. Journal of Hydrology 55: 3–23.

Bradley, R.S., Vuille, M., Diaz, H.F. and Vergara, W. 2006. Threats to water supplies in the tropical andes. Science 312: 1755–1756.

Bradshaw, C., Sodhi, N., Peh, K. and Brook, B. 2007. Global evidence that deforestation amplifies flood risk and severity in the developing world. Global Change Biology 13(11): 2379–2395.

Brandes, D., Hoffmann, J. and Mangarillo, J. 2005. Base flow recession rates, low flows, and hydrologic features of small watersheds in Pennsylvania, USA. Jorunal of the American Water Resources Association 41(5): 1177–1186.

Brauman, K., Daily, G., Duarte, T. and Mooney, H. 2007. The nature and value of ecosystem services: An overview highlighting hydrologic services. Annual Review of Environment and Resources 32: 67–98.

Brown, A., Zhang, L., McMahon, T., Western, A. and Vertessy, R. 2005. A review of paired catchment studies for determining changes in water yield resulting from alterations in vegetation. Journal of Hydrology 310: 28–61.

Bruhns, K.O. 1994. Ancient South America. New York: Cambridge University Press.

Bruijnzeel, L. 2004. Hydrological functions of tropical forests: not seeing the soil for the trees? Agriculture, Ecosystems and Environment 104: 185–228.

Bulygina, N., McIntyre, N. and Wheater, H. 2009. Conditioning rainfall-runoff model parameters for ungauged catchments and land management impacts analysis. Hydrology and Earth System Sciences 13: 893–904.

Bulygina, N., McIntyre, N. and Wheater, H. 2011. Bayesian conditioning of a rainfall-runoff model for predicting flows in ungauged catchments and under land use changes. Water Resources Research 47(2): W02503.

Bulygina, N., Ballard, C., McIntyre, N., O'Donnell, G. and Wheater, H. 2012. Integrating different types of information into hydrological model parameter estimation: Application to ungauged catchments and land use scenario analysis. Water Resources Research 48(6): W06519.

Buytaert, W., Deckers, J., Dercon, G., De Bièvre, B., Poesen, J. and Govers, G. 2002. Impact of land use changes on the hydrological properties of volcanic ash soils in South Ecuador. Soil Use and Management 18: 94–100.

Buytaert, W., De Bièvre, B., Wyseure, G. and Deckers, J. 2004. The use of the linear reservoir concept to quantify the impact of changes in land use on the hydrology of catchments in the Andes. Hydrology and Earth System Sciences 8(1): 108–114.

Buytaert, W., De Bièvre, B., Wyseure, G. and Deckers, J. 2005. The effect of land use changes on the hydrological behaviour of Histic Andosols in south Ecuador. Hydrological Processes 19: 3985–3997.

Buytaert, W., Célleri, R., Willems, P., De Bièvre, B. and Wyseure, G. 2006a. Spatial and temporal rainfall variability in mountainous areas: A case study from the south Ecuadorian Andes. Journal of Hydrology 329(3-4): 413–421.

Buytaert, W., Célleri, R., De Bièvre, R., Cisneros, F., Wyseure, G., Deckers, J. and Hofstede, R. 2006b. Human impact on the hydrology of the Andean páramos. Earth Science Reviews 79(1-2): 53–72.

Buytaert, W., Iñiguez, V. and De Bièvre, B. 2007. The effects of afforestation and cultivation on water yield in the Andean páramo. Forest Ecology and Management 251(1-2): 22–30.

Buytaert, W. and Beven, K. 2009. Regionalization as a learning process. Water Resources Research 45: W11419.

Buytaert, W., Vuille, M., Dewulf, A., Urrutia, R., Karmalkar, A. and Célleri, R. 2010. Uncertainties in climate change projections and regional downscaling in the tropical Andes: Implications for water resources management. Hydrology and Earth System Sciences 14: 1247–1258.

Buytaert, W. and Beven, K. 2011. Models as multiple working hypotheses: hydrological simulation of tropical alpine wetlands. Hydrological Processes 25(11): 1784–1799.

Buytaert, W. and De Bièvre, B. 2012. Water for cities: The impact of climate change and demographic growth in the tropical Andes. Water Resources Research 48: W08503.

Buytaert, W., Baez, S. and Bustamante, M. 2012. Web-based environmental simulation: Bridging the gap between scientific modeling and decision-making. Environmental Science & Technology 46: 1971–1976.

Buytaert, W., Zulkafli, Z., Grainger, S., Acosta, L., Alemie, T., Bastiaensen, J., De Bièvre, B., Bhusal, J., Clark, J., Dewulf, A., Foggin, M., Hannah, D., Hergarten C., Isaeva, A., Karpouzoglou, T., Pandeya, B., Paudel, D., Sharma, K., Steenhuis, T., Tilahun, S., Van Hecken, G. and Zhumanova, M. 2014. Citizen science in hydrology and water resources: opportunities for knowledge generation, ecosystem service management, and sustainable development. Frontiers in Earth Science 2(26): 1–21.

Buytaert, W., Dewulf, A., De Bièvre, B., Clark, J. and Hannah, D. 2016. Citizen science for water resources management: Toward polycentric monitoring and governance? Journal of Water Resources Planning and Management (ASCE) 142(4): 01816002.

Cain, S.F.I., Gregory, D.A., Loheide, S.P. and Butler, J.J.J. 2004. Noise in pressure transducers readings produced by variations in solar radiation. Ground Water 42(6): 939–944.

Carlos, G., Munive, R., Mallma, T. and Orihuela, C. 2014. Evaluation of the infiltration rate in farm, forestry and grazing land in the Shullcas River's basin. Apuntes de Ciencia & Sociedad 4(1): 32–43.

Célleri, R., Willems, P., Buytaert, W. and Feyen, J. 2007. Space-time rainfall variability in the Paute Basin, Ecuadorian Andes. Hydrological Processes 21: 3316–3327.

Célleri, R. and Feyen, J. 2009. The hydrology of tropical andean ecosystems: Importance, knowledge status, and perspectives. Mountain Research and Development (MRD) 29(4): 350–355.

Célleri, R. 2010. Estado del conocimiento técnico científico. pp. 24–45. In: Quintero, M. (ed.). Servicios Ambientales Hidrológicos en la RegiónAndina. Lima: CONDESAN & Instituo de Estudios Peruanos.

Célleri, R., Buytaert, W., De Bièvre, B., Tobón, C., Crespo, P., Molina, J. and Feyen, J. 2010. Understanding the hydrology of tropical Andean ecosystems through an Andean Network of Basins. Goslar-Hahnenklee, IAHS Publication, pp. 209–212.

Chapman, T. 1999. A comparison of algorithms for stream flow recession and base flow separation. Hydrological Processes 13: 701–714.

Ciach, G. 2003. Local random errors in tipping-bucket rain gauge measurements. Journal of Atmospheric and Oceanic Technology 20: 752–759.

Ciach, G. 2003. Local random errors in tipping-bucket rain gauge measurements. Journal of Atmospheric and Oceanic Technology 20: 752–759.

Clausen, B. and Biggs, B. 1997. Relationships between benthic biota and hydrological indices in New Zealand streams. Freshwater Biology 38(2): 327–342.

Clausen, B. and Biggs, B. 2000. Flow variables for ecological studies in temperate streams: Groupings based on covariance. Journal of Hydrology 237(3-4): 184–197.

Córdova, M., Carrillo-Rojas, G., Crespo, P., Wilcox, B. and Célleri, R. 2015. Evaluation of the Penman-Monteith (FAO 56 PM) method for calculating reference evapotranspiration using limited data. Mountain Research and Development 35(3): 230–239.

Correa, A., Windhorst, D., Crespo, P., Célleri, R., Feyen, J. and Breuer, L. 2016. Continuous versus event-based sampling: how many samples are required for deriving general hydrological understanding on Ecuador's páramo region? Hydrological Processes, 30(South American Hydrology): 4059–4073.

Crespo, P., Célleri, R., Buytaert, W., Feyen, J., Iñiguez, V., Borja, P. and De Bièvre, B. 2010. Land use change impacts on the hydrology of wet Andean páramo ecosystems. IAHS-AISH Publication 336: 71–76.

Crespo, P., Feyen, J., Buytaert, W., Bücker, A., Breuer, L., Frede, H. and Ramírez, M. 2011. Identifying controls of the rainfall-runoff response of small catchments in the tropical Andes (Ecuador). Journal of Hydrology 407(1-4): 164–174.

Crespo, P., Bücker, A., Feyen, J., Vaché, K., Frede, H. and Breuer, L. 2012. Preliminary evaluation of the runoff processes in a remote montane cloud forest basin using mixing model analysis and mean transit time. Hydrological Processes 26(25): 3896–3910.

Cuevas, J., Calvo, M., Little, C., Pino, M. and Dassori, P. 2010. Are diurnal fluctuations in streamflow real? Journal of Hydrology and Hydromechanics 58(3): 149–162.

Díaz, E. and Paz, L. 2002. Evaluación del regimen de humedad del suelo bajo diferentes usos, en los páramos Las Ánimas (Municipio de Silvia) y Piedra de León (Municipio de Sotará), Departamento del Cauca, Popayán: Fundación Universitaria de Popayán.

Etter, A. and van Wyngaarden, W. 2000. Patterns of landscape transformation in Colombia, with emphasis in the Andean region. Ambio 29: 432–439.

Farley, K., Kelly, E. and Hofstede, R. 2004. Soil organic carbon and water retention after conversion of grasslands to pine plantations in the ecuadorian andes. Ecosystems 7: 729–739.

Favier, V., Coudrain, A., Cadier, E., Francou, B., Ayabaca, E., Maisincho, L., Pradeiro, E., Villacís, M. and Wagnon, P. 2008. Evidence of groundwater flow on Antizana ice-covered volcano, Ecuador. Hydrological Sciences Journal 53(1): 278–291.

Fekete, B.M. and Vörösmarty, C.J. 2007. The current status of global river discharge monitoring and potential new technologies complementing traditional discharge measurements. Brasilia, IAHS Publication 309: 129–136.

Freeman, L., Carpenter, M., Rosenberry, D., Rousseau, J., Unger, R. and McLean, J. 2004. Use of submersible pressure transducers in water-resources investigations. Techniques of Water-Resources Investigations 8(A3).

Gammie, G. and De Bièvre, B. 2015. Assessing Green Interventions for the Water Supply of Lima, Peru. Cost-Effectiveness, Potential Impact, and Priority Research Areas. Washington, DC: Forest Trends Association.

Gippel, C. 2001. Hydrological analyses for environmental flow assessment. pp. 873–880. *In:* Ghasemi, F. and Whetton, P. (eds.). Proceedings MODSIM 2001, International Congress of Modelling and Simulation. Canberra: Modelling and Simulation Society of Australia & New Zealand, The Australian National University.

Gribovszki, Z., Szilágyi, J. and Kalicz, P. 2010. Diurnal fluctuations in shallow groundwater levels and streamflow rates and their interpretation—A review. Journal of Hydrology 385: 371–383.

Gribovszki, Z., Kalicz, P. and Szilágyi, J. 2013. Does the accuracy of fine-scale water level measurements by vented pressure transducers permit for diurnal evapotranspiration estimation? Journal of Hydrology 488: 166–169.

Groisman, P. and Legates, D. 1994. The accuracy of the United States precipitation data. Bulletin of the American Meteorological Society 75: 215–227.

Guallpa, M. and Célleri, R. 2013. Efecto de la estimación de la presión atmosférica sobre el cálculo de niveles de agua y caudales. Aqua-LAC 5(2): 56–68.

Hannah. 2011. Large-scale river flow archives: Importance, current status and future needs. Hydrological Processes 25(7): 1191–1200.

Harden, C.P. 2006. Human impacts on headwater fluvial systems in the northern and central Andes. Geomorphology 79: 249–263.

Hofstede, R., Groenendijk, J., Coppus, R., Fehse, J. and Sevink, J. 2002. Impact of pine plantations on soils and vegetation in the Ecuadorian High Andes. Mountain Research and Development 22: 159–167.

Hughes, J. and James, B. 1989. A hydrological regionalization of streams in Victoria, Australia, with implications for stream ecology. Australian Journal of Marine and Freshwater Research 40: 303–326.

Inbar, C. and Llerena, C. 2000. Erosion processes in high mountain agricultural terraces in Peru. Mountain Research and Development 20(1): 72–79.

Inbar, M. and Llerena, C. 2004. Procesos de erosión en andenes agrícolas andinos en la cuenca del río Santa Eulalia, Lima, Perú. pp. 141–148. *In*: Llerena, C., Inbar, M. and Benavides, M. (eds.). Conservación y Abandono de Andenes. Lima: Universidad Nacional Agraria La Molina, University of Haifa.

IUCN. 2002. High Andean Wetlands, s.l.: International Union for Conservation of Nature.

Kabubi, J., Mutua, F., Willems, P. and Mngodo, R. 2005. Low Flow Analysis using Filter Generated Series for Lake Victoria Basin, Cairo: FRIEND/NILE conference.

Körner, C. and Ohsawa, M. 2005. Mountain systems. pp. 681–716. *In*: Hassan, R., Scholes, R. and Ash, N. (eds.). Ecosystems and Human Well-being: Current State and Trends. s.l.: The Millenium Ecosystem Assessment.

Krakauer, N. and Temimi, M. 2011. Stream recession curves and storage variability in small watersheds. Hydrology and Earth System Sciences 15: 2377–2389.

Lamb, R. and Kay, A. 2004. Confidence intervals for a spatially generalized, continuous simulation flood frequency model for Great Britain. Water Resources Research 40: W07501.

Lee, H., Mcintyre, N. and Young, A. 2006. Predicting runoff in ungauged UK catchments. Proceedings of the Institution of Civil Engineers Water Management 159(2): 129–138.

Longobardi, A. and Villani, P. 2008. Baseflow index regionalization analysis in a mediterranean area and data scarcity context: Role of the catchment permeability index. Journal of Hydrology 355(1-4): 63–75.

Lørup, J., Refsgaard, J. and Mazvimavi, D. 1998. Assessing the effect of land use change on catchment runoff by combined use of statistical tests and hydrological modelling: Case studies from Zimbabwe. Journal of Hydrology 205(3-4): 147–163.

Luteyn, J.L. 1992. Páramos: why study them? pp. 1–14. *In*: Balslev, H. and Luteyn, J.L. (eds.). Páramo: An Andean Ecosystem Under Human Influence. London: Academic Press.

Lü, Y., Hu, J., Sun, F. and Zhang, L. 2015. Water retention and hydrological regulation: harmony but not the same in terrestrial hydrological ecosystem services (in Chinese). Acta Ecologica Sinica 35(15): 5191–5196.

Manz, B., Buytaert, W., Zulkafli, Z., Lavado, W., Willems, B., Robles, L. and Rodríguez-Sánchez, J. 2016. High-resolution satellite-gauge merged precipitation climatologies of the Tropical Andes. Journal of Geophysical Research: Atmospheres 121: 1190–1207.

Manz, B., Páez-Bimos, S., Horna, N., Buytaert, W., Ochoa-Tocachi, B., Lavado-Casimiro, W and Willems, B. 2017. Comparative Ground Validation of IMERG and TMPA at Variable Spatio-temporal Scales in the Tropical Andes. Journal of Hydrometeorology. doi:10.1175/ JHMD-16-0277.1, in press.

Martínez, E., Coello, C. and Feyen, J. 2017. Análisis comparativo del comportamiento de la escorrentía de tres microcuencas andinas con diferente régimen de precipitación y cobertura vegetal. Maskana 8(1): 129–144.

Mathews, R. and Richter, B. 2007. Application of the indicators of hydrologic alteration software in environmental flow setting. JAWRA Journal of the American Water Resources Association 43(6): 1400–1413.

Mazvimavi, D., Meijerink, A., Savenije, H. and Stein, A. 2005. Prediction of flow characteristics using multiple regression and neural networks: A case study in Zimbabwe. Physics and Chemistry of the Earth, Parts A/B/C 30(11-16): 639–647.

McIntyre, N., Lee, H., Wheater, H., Young, A. and Wagener, T. 2005. Ensemble predictions of runoff in ungauged catchments. Water Resources Research 41: W12434.

McIntyre, N., Ballard, C., Bruen, M., Bulygina, N., Buytaert, W., Cluckie, I., Dunn, S., Ehret, U., Ewen, J., Gelfan, A., Hess, T., Hughes, D., Jackson, B., Kjeldsen, T., Merz, R., Park, J., O'Connell, E., O'Donnell, G., Oudin, L., Todini, E., Wagener, T. and Wheater, H. 2014. Modelling the hydrological impacts of rural land use change. Hydrology Research 45(6): 737–754.

McLaughlin, D. and Cohen, M. 2011. Thermal artifacts in measurements of fine-scale water level variation. Water Resources Research 47: W09601.

Molina, A., Govers, G., Vanacker, V., Poesen, J., Zeelmaekers, E. and Cisneros, F. 2007. Runoff generation in a degraded Andean ecosystem: Interaction of vegetation cover and land use. Catena 71: 357–370.

Molina, A., Vanacker, V., Brisson, E., Mora, D. and Balthazar, V.a. 2015. Multidecadal change in streamflow associated with anthropogenic disturbances in the tropical Andes. Hydrology and Earth System Sciences 19: 4201–4213.

Mora, D. and Willems, P. 2012. Decadal oscillations in rainfall and air temperature in the Paute River Basin–Southern Andes of Ecuador. Theoretical and Applied Climatology 108(1): 267–282.

Mosquera, G., Lazo, P., Célleri, R., Wilcox, B. and Crespo, P. 2015. Runoff from tropical alpine grasslands increases with areal extent of wetlands. Catena 125: 120–128.

Muñoz, E., Arumi, J., Wagener, T., Oyarzún, R. and Parra, V. 2016. Unraveling complex hydrogeological processes in Andean basins in south-central Chile: An integrated assessment to understand hydrological dissimilarity. Hydrological Processes 30(South American Hydrology): 4934–4943.

Muñoz, P., Célleri, R. and Feyen, J. 2016. Effect of the Resolution of Tipping-Bucket Rain Gauge and Calculation Method on Rainfall Intensities in an Andean Mountain Gradient. Water 8(11): 534.

Ochoa-Tocachi, B., Buytaert, W., De Bièvre, B., Célleri, R., Crespo, P., Villacís, M., Llerena, C., Acosta, L., Villazón, M., Guallpa, M., Gil-Ríos, J., Fuentes, P., Olaya, D., Viñas, P., Rojas, G. and Arias, S. 2016a. Impacts of land use on the hydrological response of tropical Andean catchments. Hydrological Processes 30(South American Hydrology): 4074–4089.

Ochoa-Tocachi, B., Buytaert, W. and De Bievre, B. 2016b. Regionalization of land-use impacts on streamflow using a network of paired catchments. Water Resources Research 52: 6710–6729.

O'Connell, P., Beven, K., Carney, J., Clements, R., Ewen, J., Fowler, H., Harris, G., Hollis, J., Morris, J., O'Donnell, G., Packman, J., Parkin, A., Quinn, P., Shepherd, R. and Tellier, S. 2004. Review of Impacts of Rural Land Use and Management on Flood Generation, Impact Study Report, London: Defra/Environment Agency R&D Technical Report (FD2114).

Olden, J. and Poff, N. 2003. Redundancy and the choice of hydrologic indices for characterizing streamflow regimes. River Research and Applications 19(2): 101–121.

Oudin, L., Kay, A., Andréassian, V. and Perrin, C. 2010. Are seemingly physically similar catchments truly hydrologically similar? Water Resources Research 46: W11558.

Padrón, R., Wilcox, B., Crespo, P. and Célleri, R. 2015. Rainfall in the Andean Páramo: New insights from high-resolution monitoring in southern Ecuador. Journal of Hydrometeorology 16: 985–996.

Parajka, J., Merz, R. and Blöschl, G. 2005. A comparison of regionalisation methods for catchment model parameters. Hydrology and Earth System Sciences 9: 157–171.

Parajka, J., Andréassian, V., Archfield, S., Bárdossy, A., Blöschl, G., Chiew, F., Duan, Q., Gelfan, A., Hlavčová, K., Merz, R., McIntyre, N., Oudin, L., Perrin, C., Rogger, M., Salinas, J., Savenije, H., Skøien, J., Wagener, T., Zehe, E. and Zhang, Y. 2013. Prediction of runoff hydrographs in ungauged basins. pp. 53–69. *In*: Blöschl, G., Sivapalan, M., Wagener, T., Viglione, A. and Savenije, H. (eds.). Runoff Prediction in Ungauged Basins: Synthesis across Processes, Places and Scales. Cambridge,: Cambridge University Press.

Peña-Arancibia, J., van Dijk, A., Mulligan, M. and Bruijnzeel, L. 2010. The role of climatic and terrain attributes in estimating baseflow recession in tropical catchments. Hydrology and Earth System Sciences 14(11): 2193–2205.

Poff, N. and Ward, J. 1989. Implications of streamflow variability and predictability for lotic community structure: A regional analysis of streamflow patterns. Canadian Journal of Fisheries and Aquatic Sciences 46(10): 1805–1818.

Poff, N. 1996. A hydrogeography of unregulated streams in the United States and an examination of scale-dependance in some hydrological descriptors. Freshwater Biology 36: 71–91.

Poff, N., Allan, J., Bain, M., Karr, J., Prestegaard, K., Brian, D., Sparks, R. and Stromberg, J. 1997. The natural flow regime: A paradigm for river conservation and restoration. BioScience 47(11): 769–784.

Pyrce, R. 2004. Hydrological low flow indices and their uses. *In*: Watershed services Centre Report, 04-2004, Peterborough: Trent University, Symons Campus.

Quichimbo, P. 2008. Efecto de la forestación sobre la vegetación y el suelo, Cuenca: Universidad de Cuenca.

Richards, R. 1989. Measures of flow variability for great lakes tributaries. Environmental Monitoring and Assessment 13(2-3): 361–377.

Richards, R. 1990. Measures of flow variability and a new flow-based classification of great lakes tributaries. Journal of Great Lakes Research 16(1): 53–70.

Richter, B., Baumgartner, J., Powell, J. and Braun, D. 1996. A method for assessing hydrologic alteration within ecosystems. Conservation Biology 10(4): 1163–1174.

Richter, B., Baumgartner, Wigington and Braun, D. 1997. How much water does a river need? Freshwater Biology 37(1): 231–249.

Richter, B., Baumgartner, J., Braun, D. and Powell, J. 1998. A spatial assessment of hydrologic alteration within a river network. Regulated Rivers: Research & Management 14(4): 329–340.

Roa-García, C.E. and Brown, S. 2009. Assessing water use and quality through youth participatory research in a rural Andean watershed. Journal of Environmental Management 90: 3040–3047.

Roa-García, M.C., Brown, S., Schreier, H. and Lavkulich, L. 2011. The role of land use and soils in regulating water flow in small headwater catchments of the Andes. Water Resources Research 47: W05510.

Sadler, E. and Brusscher, W. 1989. High-intensity rainfall rate determination from tipping-bucket rain gauge data. Agronomy Journal 81: 930–934.

Sarmiento, L. 2000. Water balance and soil loss under long fallow agriculture in the Venezuelan Andes. Mountain Research and Development 20(3): 246–253.

Sawicz, K., Wagener, T., Sivapalan, M., Troch, P. and Garrillo, G. 2011. Catchment classification: empirical analysis of hydrologic similarity based on catchment function in the eastern USA. Hydrology and Earth System Sciences 15: 2895–2911.

Sefton, C. and Howarth, S. 1998. Relationships between dynamic response characteristics and physical descriptors of catchments in England and Wales. Journal of Hydrology 211(1-4): 1–16.

Shamir, E., Imam, B., Gupta, H. and Sorooshian, S. 2005a. Application of temporal streamflow descriptors in hydrologic model parameter estimation. Water Resources Research 41: W06021.

Shamir, E., Imam, B., Morin, E., Gupta, H. and Sorooshian, S. 2005b. The role of hydrograph indices in parameter estimation of rainfall–runoff models. Hydrological Processes 19(11): 2187–2207.

Singh, R., Wagener, T., van Werkhoven, K., Mann, M. and Crane, R. 2011. A trading-space-for-time approach to probabilistic continuous streamflow predictions in a changing climate—accounting for changing watershed behavior. Hydrology and Earth System Sciences 15(11): 3591–3603.

Sivapalan, M. 2003. Prediction in ungauged basins: A grand challenge for theoretical hydrology. Hydrological Processes 17: 3163–3170.

Sivapalan, M., Yaeger, M., Harman, C., Xu, X. and Troch, P. 2011. Functional model of water balance variability at the catchment scale: 1. Evidence of hydrologic similarity and space-time symmetry. Water Resources Research 47: W02522.

Thomas, R. and Megahan, W. 1998. Peak flow responses to clear-cutting and roads in small and large basins, Western Cascades, Oregon: A second opinion. Water Resources Research 34(12): 3393–3403.

Tobón, C. 2009. Los Bosques Andinos y el Agua. Serie investigación y sistematización #4. Quito: Programa Regional ECOBONA—INTERCOOPERATION & CONDESAN.

Ulloa, J., Ballari, D., Campozano, L. and Samaniego, E. 2017. Two-Step Downscaling of Trmm 3b43 V7 Precipitation in Contrasting Climatic Regions With Sparse Monitoring: The Case of Ecuador in Tropical South America. Remote Sensing 9(7): 758.

USDA. 1986. Urban Hydrology for Small Watersheds, Report TR-55, Washington DC: United States Department of Agriculture.

USDI Bureau of Reclamation. 2001. Water Measurement Manual., s.l.: Water Resources Research Laboratory, US Department of the Interior.

van Dijk, A. 2010. Climate and terrain factors explaining streamflow response and recession in Australian catchments. Hydrology and Earth System Sciences 14(1): 159–169.

Visessri, S. and McIntyre, N. 2016. Regionalisation of hydrological responses under land-use change and variable data quality. Hydrological Sciences Journal 61(2): 302–320.

Viviroli, D., Dürr, H., Messerli, B., Meybeck, M. and Weingartner, R. 2007. Mountains of the world, water towers for humanity: typology, mapping, and global significance. Water Resources Research 43: W07447.

Vuille, M., Bradley, R.S. and Keiming, F. 2000. Interannual climate variability in the Central Andes and its relation to tropical Pacific and Atlantic forcing. Journal of Geophysical Research 105(D10): 12447–12460.

Wagener, T. and Wheater, H. 2006. Parameter estimation and regionalization for continuous rainfall-runoff models including uncertainty. Journal of Hydrology 320: 132–154.

Wagener, T. 2007. Can we model the hydrological impacts of environmental change? Hydrological Processes 21: 3233–3236.

Wagener, T. and Montanari, A. 2011. Convergence of approaches toward reducing uncertainty in predictions in ungauged basins. Water Resources Research 47(6): W06301.

Walsh, R. and Lawler, D. 1981. Rainfall seasonality: description, spatial patterns and change through time. Weather 36(7): 201–208.

Wang, J., Fisher, B. and Wolff, D. 2008. Estimating rain rates from tipping-bucket rain gauge measurements. Journal of Atmospheric and Oceanic Technology 25: 43–56.

White, S. and Maldonado, F. 1991. The use and conservation of national resources in the Andes of southern Ecuador. Mountain Research and Development 11: 37–55.

White, W.N. 1932. A method of estimating ground-water supplies based on discharge by plants and evaporation from soil-results of investigation in Escalante Valley, Utah. Water Supply Paper 659(A): 1–105.

Willems, P. 2014. Parsimonious rainfall–runoff model construction supported by time series processing and validation of hydrological extremes—Part 1: step-wise model-structure identification and calibration approach. Journal of Hydrology 510: 578–590.

Wohl, E., Barros, A., Brunsell, N., Chappell, N., Coe, M., Giambelluca, T., Goldsmith, S., Harmon, R., Hendrickx, J., Juvik, J., McDonnell, J. and Ogden, F. 2012. The hydrology of the humid tropics. Nature Climate Change 2: 655–662.

Wondzell, S., Gooseff, M. and McGlynn, B. 2009. An analysis of alternative conceptual models relating hyporheic exchange flow to diel fluctuations in discharge during baseflow recession. Hydrological Processes 24(6): 686–694.

Wood, P., Agnew, M. and Petts, G. 2000. Flow variations and macroinvertebrate community responses in a small groundwater-dominated stream in south-east England. Hydrological Processes 14(16-17): 3133–3147.

World Meteorological Organization. 2008. Guide to Hydrological Practices, Volume I, Hydrology—From Measurement to Hydrological Information. Sixth edition, 2008 ed. Geneva: Chairperson, Publications Board, World Meteorological Organization (WMO).

World Meteorological Organization. 2012. Guide to Meteorological Instruments and Methods of Observation, WMO-No. 8. 2008 edition, updated in 2010 ed. Geneva: Chairperson, Publications Board, World Meteorological Organization (WMO).

Yadav, M., Wagener, T. and Gupta, H. 2007. Regionalization of constraints on expected watershed response behavior for improved predictions in ungauged basins. Advances in Water Resources 30: 1756–1774.

Young, A. 2006. Stream flow simulation within UK ungauged catchments using a daily rainfall-runoff model. Journal of Hydrology 320: 155–172.

Zhang, Z., Wagener, T., Reed, P. and Bhushan, R. 2008. Reducing uncertainty in predictions in ungauged basins by combining hydrologic indices regionalization and multiobjective optimization. Water Resources Research 44: W00B04.

Zulkafli, Z., Buytaert, W., Manz, B., Véliz Rosas, C., Willems, P., Lavado-Casimiro, W., Guyot, J. and Santini, W. 2016. Projected increases in the annual flood pulse of the Western Amazon. Environmental Research Letters 11: 014013.

CHAPTER 7

A Framework for Agricultural Water Management Support Following the 2010 Maule Earthquake

Rodriguez, Jenna,[1,*] *S.L. Ustin,*[1] *Sam Sandoval Solis,*[1]
Diego Rivera Salazar[2] *and Toby O'Geen*[1]

Introduction

Abrupt changes in the hydrologic system can severely alter environmental functionality, requiring immediate emergency responses to facilitate appropriate disaster management techniques. Disaster management has therefore become increasingly important for mitigation actions. However, emergency management of post-earthquake disasters largely focuses on infrastructural and humanitarian threats, which has left a significant gap in understanding of agricultural responses and potential for recovery. We address this problem to aid improved farm recovery from earthquakes at the food-water nexus, and ultimately, food security. Remote sensing—that is, imagery datasets through satellite or aerial platforms—can provide information about conditions that precede an abrupt disaster event and improve a grower's understanding of the impact, and consequently make better management decisions (Joyce et al., 2009a).

Significant gaps in knowledge of hydrologic vulnerability and resilience following earthquakes are best attributed to the unpredictability of the time and location of an extreme event which need data collection prior to

[1] University of California, Davis, One Shields Avenue, Davis, California, 95366, USA.
[2] University of Concepcion – Chillán. Avenue Vicente Mendez, Chillán, Chile.
* Corresponding author: jmmartin@ucdavis.edu

its occurrence (Geller, 1997; Hough, 2009). While comprehensive ground measurements cannot be established in advance for an unspecified time and place of an earthquake, remote sensing can supplement these gaps using current and archived data. Remote sensing enables us to gather information prior to agricultural damage, providing a technical opportunity to monitor disaster response and recovery across a variety of temporal, spatial and spectral resolutions (Joyce et al., 2009b). As a case study, we investigate agricultural recovery from hydrologic damage at the field scale following the 8.8 Maule earthquake by coupling ground and remote measurements. We specifically investigate orchard recovery management from seismically influenced waterlogging following the 2010 Maule, Chile earthquake.

Earthquake Hydrology

Earthquake-water dynamics are well studied, with observed connections to a variety of behaviors that can include changes to groundwater supply, surface water supply and water quality. These effects vary in timing and magnitude, dependent on earthquake magnitude, distance to epicenter and aquifer structure (Montgomery and Manga, 2003). Unexpected changes in local hydrology can be especially problematic for crop management, and hence threaten local food security. To improve our understanding of agricultural management following abrupt environmental changes, this study focuses on the effects of seismically related groundwater changes on agricultural land use. Table 7.13 displays observed groundwater level responses to a variety of earthquake magnitudes and locations. Varying behaviors of earthquakes observed show groundwater connections, positioning crop water distribution uniformity as vulnerable to such abrupt hydrologic shifts. As the need to explore crop recovery from natural disasters

Table 7.13 Recorded seismic effects on groundwater levels following earthquake events illustrate the frequency of earthquake impacts on groundwater behavior as well as the unpredictability of hydrologic responses.

Year	Location	Magnitude	Δ Depth (m)	Author
1989	Loma Prieta, California, USA	7.1	−21.0	Rojstaczer and Wolf, 1992
1993	Taiwan, China	7.3	1.0–11.1	Chia et al., 2001
1994	Parkfield, California, USA	4.7	−0.16 − + 0.34	Quilty and Roeloffs, 1997
1997	Tono, Japan	5.8	−0.29–1.8	King et al., 1999
2004	Japan	9.0	+/−5.0	Kitagawa et al., 2006
2010	Canterbury, New Zealand	7.1	5.0–20.0	Cox et al., 2012

is clear, we narrow our focus on local decision-making to facilitate crop recovery in the wake of earthquake-induced waterlogging.

Waterlogging of Agricultural Soils

Waterlogged agricultural soils can adversely affect crop health, dependent on factors such as time of flood, duration and crop affected. While waterlogging during cold, dormant months is known to cause minimal damage on dormant trees and crops, the same conditions during the growing season can eliminate entire crops or orchards (Kozlowski, 1984). Specifically, orchards and crops with poor drainage can suffer from hypoxic soils (Crawford, 1982), vegetative diseases (e.g., phytophthora) and salinity (Oster, 1994). It is thus important to better understand waterlogging effects upon orchards, specifically how orchard crops recovery from temporary waterlogging during the growing season. Capturing the spatial variation and progression of crop health over time to such stress can be especially useful for irrigation uniformity and responsible farm water management. Such time and space analysis can be optimally understood through remotely sensed imagery.

Remote Sensing in Agricultural Water Management

The usefulness and importance of satellite and airborne platforms for emergency management based on sensor capacities is widely acknowledged in the remote sensing community (Joyce et al., 2009a; 2009b). Joyce et al. (2010) reviews the use of remote sensing imagery in emergency management, identifying a large gap in understanding recovery to natural disasters and emphasizing the need for collaboration among various stakeholders to better understand disaster management in smaller niches such as agriculture. It is therefore clear that the substantial need to improve disaster response strategies in agricultural settings can be addressed in the utilization of remote sensing technologies. Remotely sensed imagery has been used to understand aspects of cropland recovery from disasters for land-use planning (Burby et al., 2000). In this chapter, we investigated the use of satellite imagery in a more narrow focus, specific to agricultural needs. We direct remote sensing to capture spatial heterogeneity of agricultural impacts following the earthquake event to pinpoint areas of orchard vulnerability (canopy stress) and resiliency (canopy vigor).

Remote Sensing of Orchard Stress

In this study, it is important to note that apple orchards can tolerate waterlogging during dormancy, but submergence during active growth—suggested for any length of time—has been known to cause root death

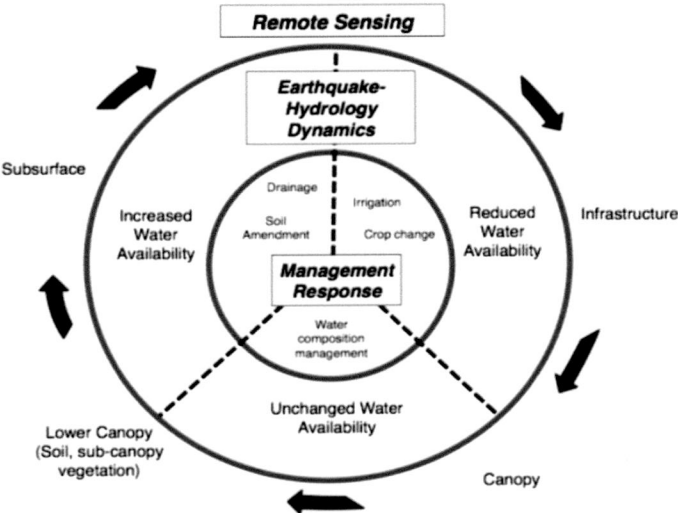

Fig. 7.36 A framework to support agricultural decision-making after earthquake events, specific to remote sensing of earthquake-water dynamics Rodriguez et al. (2016).

(Faust, 1978). Waterlogging observations associated with the Maule earthquake began in February 2010—Central Chile's growing season. To strategize remote sensing applications of this orchard following an abrupt hydrologic change, we follow the newly introduced framework by Rodriguez et al. (2016) for post-earthquake agrohydrologic remote sensing (PEARS, Fig. 7.36). This approach navigates through remote sensing of crop stress at the canopy, sub-canopy or subsurface level dependent upon the hydrologic behavior associated with the earthquake and what can be sensed from an aerial or orbital perspective—from which we will draw upon remote sensing exploration of the orchard canopy. While waterlogged soils are a sub surface impedance on tree health, summer orchard canopies are fully open and thus intercept sub-canopy and soil reflectance. Waterlogged soils can be visually identified with sufficient multispectral sensor data (Dwivedi et al., 2009), using specific wavelengths (Zarco-Tejada et al., 2012), or by calculating spectral ratios in vegetation indices (Glenn et al., 2008). It is important to note that ground data collection is crucial to eliminate unlikely stressors (water quality, over- or under-irrigation, abnormal weather, etc.) to narrow down the best-suited remote sensing approach for the context.

Applying a Framework

We apply a newly introduced conceptual model that dovetails current research of earthquake-water dynamics with applicable remote sensing of soil-plant-water relations. Figure 7.36 illustrates the PEARS framework to support post-earthquake agricultural management using remote sensing techniques specific to earthquake-water dynamics (Rodriguez et al., 2016). The framework of focus categorizes earthquake-water dynamics into three components: (1) changed surface water supplies, (2) changed groundwater supplies, and (3) water quality change. For this study, groundwater depths were coupled with remote sensing and *in situ* observations of orchard health to interpret crop responses to component 2—changed groundwater supplies—directed by the PEARS framework. We thus apply the framework to assess vulnerability of an apple orchard site to abruptly elevated water tables, monitoring canopy vigor. This approach allows us to assess the effectiveness of the grower's decision to facilitate drainage through trenching, as well as the resiliency of the orchard amidst mid-season waterlogging. We additionally set the stage for remote sensing guidelines to monitor and improve farm operations amidst agro-hydrologic disruptions that can be employed in similarly affected sites.

Objectives

This study characterizes field-scale responses of apple orchard operation to extreme waterlogging, in a case study of Coihueco, Chile (–37 latitude, –71.82 longitude). Specifically, we will improve understanding of agricultural land management at the field scale by monitoring local farm decisions implemented to sustain apple orchard production amidst abrupt waterlogging. Here, we investigate variables driving orchard stress (groundwater elevation) and characteristics that indicate stress (decline in vigor as indicated by decreasing canopy greenness). Assessment of rootstock and cultivar vigor responses to poor drainage conditions is critical for organic apple growers—the leading organic commodity in Chile. Varying rootstocks across the study site must be considered as apple trees have demonstrated varying abilities among rootstocks to conduct water to the scion (Olien, 1986). Furthermore, we employ remote sensing technologies in alignment with a consistent post-earthquake framework to support agricultural management.

Materials and Methods

Background & study site

An 8.8 magnitude earthquake event occurred February 27, 2010 off the coast of Concepcion, Chile that devastated the Maule and Bíobío Regions. Local observations recorded structural damages (Tang et al., 2010), liquefaction compaction (Verdugo, 2012) and increased streamflow (Mohr et al., 2012) following the earthquake. Groundwater supplies serve as an important source of irrigation water, especially during the growing season (December–

February) when seasonal mountain snowmelt is no longer available, thus changes in availability has secondary impacts. Abrupt hydrologic changes in this region can threaten local agricultural productivity, driving the motivation of our study.

The study site observed is located in Coihueco, Chile of the Cato River Watershed, approximately 500 kilometers south of the capital city of Santiago (Fig. 7.37a). Coihueco is in the Biobío Region (Region VIII, –37 latitude, –71 longitude) whose economy relies largely on agriculture, supporting 16% of the employment sector, behind personal services (28%) and commerce (18%, Dresdner et al., 2009). The Biobío Region traditionally experiences a Mediterranean climate of hot dry summers and cool wet winters, systematically bringing spring snowmelt to irrigate throughout the Cato River Watershed dominated by andisols, sands and fluvial deposits (OECD, 2009). Here, appropriate crop water deliveries are vulnerable to winter snowpack, irrigation infrastructure and good drainage. It is thus important to assess if there were changes in hydrologic conditions following the February 27, 2010 earthquake, how local crop management can adapt, and whether strategies were successful.

The specific site of study is a 20-hectare apple orchard (approx. 50 acres) located at –37 latitude, –71.82 longitude managed by Viva Tierra Organic

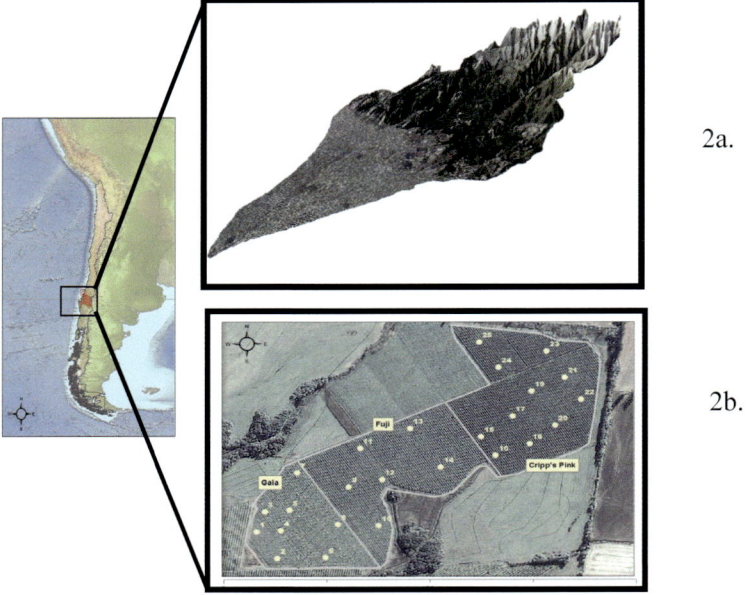

2a.

2b.

Fig. 7.37 (a) The Cato River Watershed in the Biobío Region (VIII) of Central Chile, and (b) The study site: a 20-hectare organic apple orchard (approx. 50 acres) of Gala, Fuji, and Cripp's Pink apples located at –37 latitude, –71.82 longitude, managed by Viva Tierra Organic within the Cato River Watershed.

Table 7.14 Apple cultivars planted with associated rootstocks and installed boreholes.

Cultivar	Rootstock	Boreholes
Gala	M106 + M9	1–8
Fuji	M106	9–14
Cripp's Pink	M106 + M9	15–26

(Fig. 7.37b). In 2007, three different varieties of apple cultivars—Gala, Fuji, and Cripp's Pink—were planted with various rootstocks (Table 7.14). The growing season for this crop runs approximately from green-up in November through harvest in early-mid March. Following the February 27, 2010 earthquake, elevated water tables were observed by local growers, provoking suspicion of waterlogging. To salvage the newly planted trees, emergency management decisions were made to trench along the entire southern border of the orchard to accelerate drainage, thus preventing hypoxia and onset of disease. The methodology conducted characterized field-scale apple orchard responses to earthquake-induced waterlogging, assessed the apple orchard resiliency to temporary waterlogging, and assessed effectiveness of local decision-making following a remote sensing framework. Results support framework development to identify actions and decision-making that promotes agricultural recovery following abrupt hydrologic disruption.

Borehole Observations

Twenty-six monitoring wells were distributed across the orchard in December 2009 using polyvinyl chloride (PVC) tubes 150 cm deep, providing records of pre-earthquake subsurface hydrology. While 26 wells were installed, only a fraction of the wells were selected to use for data purposes. Boreholes that did not experience any change in groundwater were not selected for data collection. Specifically, wells 5, 7, 8, 12, 15, 17, 18, 20, 21, 22 and 25 were the only wells showing elevation change after the earthquake event. These boreholes were selected to utilize data representative of the phenomena observed by the farm manager and owner. The borehole locations were geo-located using a Garmin Oregon Global Positioning System (GPS) to interpret spatial variation of water table fluctuations before and after the earthquake event. The depth to water table was measured consistently by the same irrigation manager using steel measuring tape from the top of the PVC tube as the reference point (Harter, 2008). The date, well depth and comments were logged and saved in an electronic spreadsheet file. Borehole observations were recorded from December 2009 until May 2014 when consistent records indicated 'good drainage', at 120 cm from observation surface or deeper. After good drainage was declared by the grower, data collection activities were reduced to bi-annual occurrences. A trench was excavated three days after the earthquake to mitigate waterlogging (March 2, 2010) along the south and southwest borders of the orchard to counteract elevated ground water levels. This data was also used for spatial interpolation to remotely monitor tree health variability throughout the orchard before and after the earthquake event. Water quality measurements were also sampled and provided by a contracted vendor following the earthquake event to ensure that salinity levels were not adversely elevated, thus narrowing our focus to monitor effects of groundwater elevation on orchard canopies using PEARS.

Precipitation Records

Agro-meteorological stations recorded the precipitation data used for this study, collected and managed by the Department of Hydrologic Resources at the University of Concepción–Chillán. The station is located on the university campus at –36.5667 latitude, –72.1 longitude, and 129.92 meters above mean sea level. Data was collected beginning January 1965 and is still recording weather data. Data was analyzed according to total sum precipitation recorded each month. This dataset provides data prior to the earthquake event to better interpret environmental conditions preceding the earthquake and associated groundwater changes, as well as precipitation following the event to confirm that abnormal groundwater levels were not influenced by an abnormally wet rainy season (Fig. 7.38).

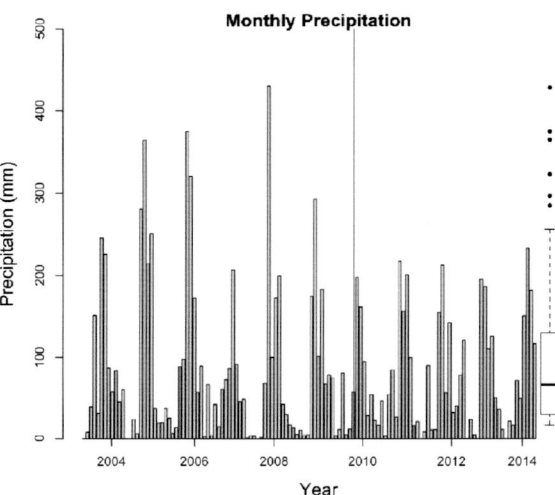

Fig. 7.38 Precipitation records from 2008 to 2014. Earthquake event is indicated by red vertical line to delineate precipitation events prior to and following the hydrologic disturbance. Mean distribution of precipitation measurements from 2004–2014 is displayed by boxplot at right.

Imagery

Imagery was collected from the Landsat Data Continuity Mission (LDCM), utilizing Landsat Thematic Mapper 5 and Landsat Enhanced Thematic Mapper (ETM+) 7. This sensor sufficiently covers the 20 hectare (20,000 m²) orchard study site, as Landsat 5 TM and Landsat 7 ETM+ provide coverage at 30×30 meter pixels, allowing spectral measurements across approximately 222 unique pixels. Atmospherically corrected imagery from LEDAPS was selected for Landsat 5 TM and Landsat 7 ETM+. Dates were selected prior to and following the earthquake with consideration of cloud cover, the apple growing season and irrigation scheduling. Satellite overpasses within 3 days of irrigation scheduling were not used. Irrigation events can cause artificially high spectral measurements, as sprinkler-irrigated water can cause abnormal blue band measurements from water, dampened reflectance from soil darkening and green or infrared measurements due to immediate leaf and cover crop green up the following day.

Vegetation Indices

While we know that the visible and near infrared are important in monitoring vegetative trends, there are many spectral ratios that have been demonstrated as useful throughout the literature. This study utilizes the Normalized Difference Vegetation Index (NDVI), calculated by the ratio:

$$\frac{p_{NIR} - p_{RED}}{p_{NIR} + p_{RED}} \tag{10}$$

NDVI has successfully monitored and identified vegetative trends with responses to groundwater elevation changes (Aguilar et al., 2012; Sun et al., 2008). NDVI is sufficient for this study, as the orchard study site provides near complete vegetative cover, removing the need for soil correction indexes (e.g., Soil Adjusted Vegetation Index) or overly dense tree canopies due to consistent pruning and younger trees (Enhanced Vegetation Index).

Results and Discussion

Groundwater responses

Knowledge of groundwater depths is important for farm management to facilitate proper drainage, maintain root zone aeration for crops and monitor subsurface irrigation supplies. Figure 7.39 displays groundwater stages before the earthquake, after the earthquake and longer-term water level tracking following emergency water management decision-making (i.e., trenching). Measurements prior to the earthquake from December 2009–February 2010 were categorized as 'Pre-Earthquake' observations,

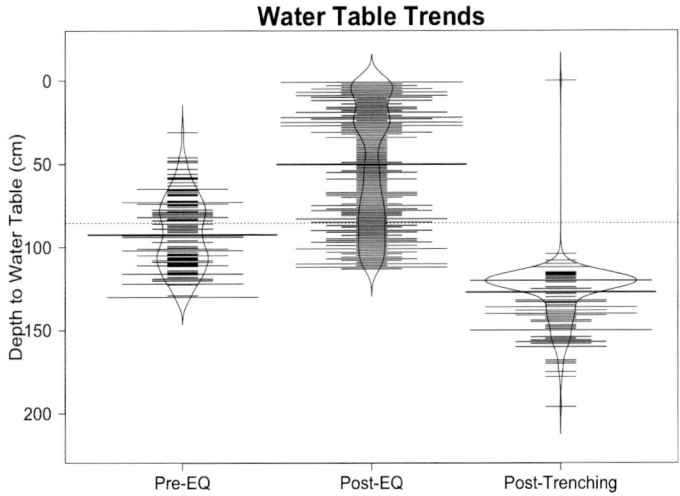

Fig. 7.39 Water table observation records suggesting elevated water table after the earthquake event (February 2010–2011) and longer term (October 2011–January 2014).

while records taken following the earthquake event were conducted from October 2010–December 2011 were labeled as 'Post-EQ' records. Records were later categorized as 'Post-Trenching' from 2012–2014, as these records revealed groundwater table levels returning to depths once again acceptable for apple production. Observation well records show an initial increase in average groundwater elevation after the earthquake, followed by a return to optimal depths (120–150 cm) likely due to management decisions (i.e., trenching) that facilitated groundwater drainage. Figure 7.39 shows initial groundwater table elevation at acceptable depths, with some borehole measurements indicating groundwater closer to the root zone than desired by the farmer—that is, above 120 cm impeding root aeration. These records also followed trenching decisions, yet were still influenced by waterlogging conditions. The lag time observed from the date of trenching to subsided groundwater levels is likely a combined product of soil properties and subterranean water sources. This can often be attributed to a sustained high groundwater table or underground spring released during the earthquake event, as can occur during earth movement (Montgomery and Manga, 2003).

Precipitation Trends

Precipitation records show that seasonal rainfall before and after the February 2010 earthquake, denoted by the red vertical line, did not exceed normal moisture regimes (Fig. 7.38). Figure 7.38 describes average historical precipitation to determine significantly higher or lower precipitation trends that could affect groundwater overdraft or recharge. Historical precipitation records for Chillán, Chile fall within 68.5–236.5 mm during the rainy season (May–October) and 14.9–26.2 mm during the dry season (November–April; Fontannaz, 2001). Average monthly precipitation records across the 2004–2014 period revealed that only 6 of 126 observations were significant outliers. Furthermore, the rise in subterranean water levels occurred during the growing season, at the trough of precipitation inputs and traditional peak of irrigation drawdown by local farmers. The station observations thus suggest that sustained groundwater level elevation was not influenced by above average rain events. We can therefore deduce that the shallow groundwater phenomena noticed abruptly after the earthquake event was likely driven by subterranean sources.

Orchard Spatial Responses to Groundwater Trends

Remote imagery used to calculate the Normalized Difference Vegetation Index (NDVI) across the orchard study site allowed identification of pre-earthquake orchard health and post disaster responses. It is clear that prior to the earthquake, canopy health was not unhealthy nor of concern to the grower, while moderate stress was detectable along the northeastern border of the Cripp's Pink block. Although orchard canopy health is still adversely affected one year after the earthquake event, yet shows improvement after the earthquake (2012–2014). Canopy health improvements are likely attributed to improved drainage facilitated by trenching, lowering a relatively high water table. Additionally, management decisions that preserved the life of the trees allowed younger trees planted prior to the earthquake to grow larger with age, consequently producing larger tree canopies. While struggling orchard health was evident immediately following the earthquake throughout the 2011 growing season, orchard health recovered fully and became more uniform with less hotspots of poor orchard health. Spatial mapping of orchard health before and after the earthquake indicated reduced orchard health in response to an abruptly elevated water table (0 to 120 cm) that gradually declined to optimal depths below 120 cm. Figure 7.40a–c displays NDVI maps that, over time, become more uniform, with less 'hotspots'—areas suffering from exceptionally low canopy health—after the earthquake event. This lag time to recovery is likely a function of the soil properties and groundwater table sources that sustained groundwater levels even after trenching. A time series analysis of canopy health using NDVI not only allows a time series analysis and identification of areas most adversely affected, but also allows correlation of remotely estimated canopy health to ground measurements (i.e., groundwater depths).

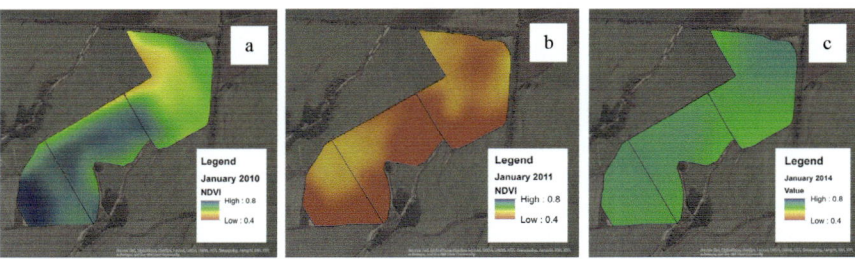

Fig. 7.40a–c Spatial visualization of orchard NDVI suggests variability of orchard health, reduced greenness after earthquake, and recovery after trenching.

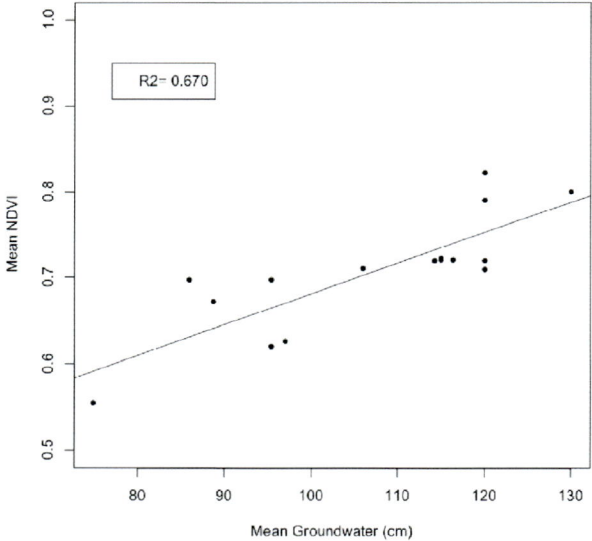

Fig. 7.41 Correlation of mean NDVI measurements to mean groundwater levels across 16 different observation dates, suggesting improved orchard health with a deeper groundwater table.

Groundwater Relationship to Orchard Behavior

In addition to spatial mapping and time series analysis, remotely estimated NDVI values were also utilized to identify a significant relationship to adverse waterlogging conditions. To do so, mean calculated groundwater levels recorded *in situ* across the orchard were correlated to corresponding mean NDVI estimations from atmospherically corrected Landsat imagery. The correlation revealed that mean groundwater table levels approaching the surface inhibited orchard greenness (NDVI), while groundwater levels that remained closer to target depth of 120–150 cm were tied to improved orchard canopy health. The correlation fit value of 0.67 suggests that groundwater depths influenced orchard health as represented by NDVI; this suggests that the rootstocks and apple cultivars selected were tolerant to short-term waterlogging during the growing season and likely recovered due to the receding groundwater table. The correlation detected also indicates that decisions to facilitate drainage by trenching did improve orchard canopy health, hence, prevent tree damage and death. These findings will be especially valuable if waterlogged (e.g., flooding) conditions persist in the Cato River watershed, or if extreme events forcing similar conditions recur in this region.

Conclusions

This study was largely limited by the data collection surrounding the earthquake. While it proved fortuitous to have pre-earthquake baseline information. Data collection stalled following the earthquake as efforts largely went to infrastructural, farm and personal recovery needs. Additionally, data collection followed varying protocols as water table levels were recorded largely at the convenience of farm operations, and by multiple people. These limitations are typical of data collection in a working farm site and largely contributes to the utility of remote sensing in agricultural practices.

As climate projections suggest a high likelihood of increasing extreme events, we must better prepare our vulnerable food supplies through coping strategies such as emergency management and disaster recovery. These efforts, though essential and urgently needed, are still immaturely explored. Further research is therefore needed, especially if we expect to create a more comprehensive and robust framework for efficient and rapid agricultural management following abrupt disruptions to agricultural water deliveries. Ample opportunities are present and contributing to this emerging field, with a more thorough understanding enabled by remote sensing applications. This research can support land-use management for recovery of irrigation supplies as well as aid in anthropogenic changes to local hydrology.

In the broad scope of food production and water resource management, groundwater supplies become increasingly depleted in arid and semi-arid climates. To better harness seasonal snowmelt that fuels seasonal irrigation, groundwater recharge projects are now a priority if not mandatory for sustainable population support. Target recharge locations must allow temporary inundation during snowmelt periods without damage to overlying land use. In many Mediterranean climates, valley farming practices are supported by spring snowmelt, requiring recharge locations to typically be rangeland or crops that can survive temporary waterlogging. Crops overlying unconfined aquifers are in prime target areas for groundwater recharge projects. While it is ideal to channel recharge areas while perennial are dormant, spring snowmelt can often times fall within the period of green up. Recharge during dormancy may become even less likely if changes in climatic trends begin to shorten. This study therefore provides a better understanding of crop responses to temporary water inundation during the growing season.

While natural causes of flooding are inevitable, it is also important to identify crops more tolerant to anthropogenic flooding. Our results suggest that temporary (3 days) flooding of the observed cultivar-rootstock combinations during the growing season did initially impose orchard stress, but permitted fruit production without tree death. The orchard of study was also found to reach full recovery after prolonged drainage. These results enable a better understanding for local crop management that can aid pre-hazard planning for local growers, as well as empower informed decision making for crop extension educators and agencies navigating projects to temporarily flood agricultural lands.

Acknowledgements

This study was funded by the National Science Foundation (Grant # DGE-1148897), the Henry A. Jastro research grant and the National Cattlemen's Foundation. We thank VivaTierra Organic for their time, resources and permission to publish results from this study. We are especially grateful to Luis Acuña and Nicholas Simian for access to the orchard site, data collection and support throughout the study. We also thank the collaborative power shared with University of Concepción–Chillan, especially that of Mario Lillo, Jose Luis Arumí and Carlos Cea.

References

Aguilar, C., Zinnert, J., Polo, M. and Young, D. 2012. NDVI as an indicator for changes in water availability to woody vegetation. Ecological Indicators 23: 290–300.

Burby, R., Deyle, R., Godschalk, D. and Olshansky, R. 2000. Creating hazard resilient communities through land-use planning. Natural Hazards Review 1: 99–106.

Chia, Y., Wang, Y.-S., Chiu, J.J. and Liu, C.-W. 2001. Changes of groundwater level due to the 1999 Chi-Chi earthquake in the Choshui River alluvial fan in Taiwan Bulletin of the Seismological Society of America 91: 1062–1068.

Cohrssen, J.J. and Covello, V.T. 1989. Risk analysis: A guide to principles and methods for analyzing health and environmental risks. Springfield, Virginia: Executive Office of the President of the U.S., Council on Environmental Quality.

Cox, S.C., Rutter, H.K., Sims, A., Manga, M., Weir, J.J., Ezzy, T. et al. 2012. Hydrological effects of the MW 7.1 Darfield (Canterbury) earthquake, 4 September 2010, New Zealand. New Zealand Journal of Geology and Geophysics 55: 231–247.

Crawford, R. 1982. Physiological responses to flooding. pp. 453–477. *In*: Lange, O.L., Nobel, P.S., Osmond, C.B. and Ziegler, H. (eds.). Physiological Plant Ecology II. Berlin Heidelberg: Springer.. pp. 453–477.

Dresdner, J., Acuña Duarte, A., Castro Ramirez, B., Suazo, M.Q., Cabrera, H.S., Oliva, A.U. et al. 2009. The Bio Bio Region, Chile: Self-Evaluation Report. Reviews of Higher Education in Regional and City Development. IMHE.

Dwivedi, R., Sreenivas, K. and Ramana, K. 1999. Inventory of salt-affected soils and waterlogged areas: a remote sensing approach. International Journal of Remote Sensing 20: 1589–1599.

Faust, M. 1978. Establishing and Managing Young Apple Orchards. Farmers' Bulletin. United States Department of Agriculture, Washington, D.C. pp. 1–26.

Gahalaut, K., Gahalaut, V.K. and Chadha, R.K. 2010. Analysis of coseismic water-level changes in the wells in the Koyna-Warna region, Western India. Bulletin of the Seismological Society of America 100: 1389–1394.

Geller, R. 1997. Earthquake prediction: a critical review. Geophysical Journal International 131: 425–450.

Glenn, E.P., Huete, A.R., Nagler, P.L. and Nelson, S.G. 2008. Relationship between remotely-sensed vegetation indices, canopy attributes and plant physiological processes: What vegetation indices can and cannot tell us about the landscape. Sensors 8: 2136–2160.

Grecksch, G., Roth, F. and Kümpel, H.-J. 1999. Coseismic well-level changes due to the 1992 Roermond earthquake compared to static deformation of half-space solutions. Geophysical Journal International 138: 470–478.

Harter, T. and Rollins, L. 2008. Watersheds, Groundwater and Drinking Water. United States of America: University of California Agriculture and Natural Resources.

Herrera, M. 2010. Chile Organic Products Report. United States Department of Agriculture, Santiago, Chile.

Hough, S. 2010. Predicting the unpredictable: The tumultuous science of earthquake prediction. Princeton, New Jersey: Princeton University Press.

Joyce, K.E., Belliss, S.E., Samsonov, S.V., McNeill, S.J. and Glassey, P.J. 2009a. A review of the status of satellite remote sensing and image processing techniques for mapping natural hazards and disasters. Progress in Physical Geography 33: 183–207.

Joyce, K.E., Wright, K.C., Samsonov, S.V. and Ambrosia, V.G. 2009b. Remote sensing and the disaster management cycle. INTECH Open Access Publisher.

King, C.Y., Azuma, S., Igarashi, G.M., Ohno, H.S. and Wakita, H. 1999. Earthquake-related water-level changes at 16 closely clustered wells in Tono, central Japan (1978–2012). Journal of Geophysical Research: Solid Earth 104: 13073–13082.

Kitagawa, Y., Koizumi, N., Takahashi, M., Matsumoto, N. and Sato, T. 2006. Changes in groundwater levels or pressures associated with the 2004 earthquake off the west coast of northern Sumatra (M9. 0). Earth Planets and Space 58: 173–179.

Kozlowski, T.T. 1984. Flooding and Plant Growth. San Diego, California: Academic Press, Inc.

Mohr, C. and Wang, C. 2011. Streamflow response to the 2010 M8. 8 Maule earthquake. AGU Fall: Meeting Abstracts 1: 1164.

Montgomery, D.R. and Manga, M. 2003. Streamflow and water well responses to earthquakes. Science 300: 2047–2049.

OECD/Bío Bío's Regional Steering Committee. 2009. The Bío Bío Region, Chile: Self-Evaluation Report, OECD Reviews of Higher Education in Regional and City Development, IMHE, www.oecd.org/edu/imhe/regionaldevelopment.

Olien, W. and Lakso, A. 1986. Effect of rootstock on apple (Malus domestica) tree water relations. Physiologia Plantarum 67: 421–430.

Oster, J. 1994. Irrigation with poor quality water. Agricultural Water Management 25: 271–297.

Quilty, E.G. and Roeloffs, E.A. 1997. Water-level changes in response to the 20 December 1994 earthquake near Parkfield, California. Bulletin of the Seismological Society of America 87: 310–317.

Rodriguez, J., Ustin, S., Sandoval-Solis, S. and O'Geen, A.T. 2016. Food, water, and fault lines: Remote sensing opportunities for earthquake-response management of agricultural water. Science of the Total Environment 565: 1020–7.

Roeloffs, E.A. 1988. Hydrologic precursors to earthquakes: A review. Pure and Applied Geophysics 126: 177–209.

Rojstaczer, S. and Wolf, S. 1992. Permeability changes associated with large earthquakes: An example from Loma Prieta, California. Geology 20: 211–214.

Salazar, D.R. 2012. Chile Environmental, Political and Social Issues. New York: Nova Science Publishers.

Singh, R., Mehdi, W., Gautam, R., Senthil Kumar, J., Zlotnicki, J. and Kafatos, M. 2010. Precursory signals using satellite and ground data associated with the Wenchuan Earthquake of 12 May 2008. International Journal of Remote Sensing 31: 3341–3354.

Sun, X., Jin, X. and Wan, L. 2008. Effect of groundwater on vegetation growth in Yinchuan plain. Geoscience 22: 321–324.

Tang, A. and John M. Eidinger. 2013. Chile Earthquake of 2010; Lifeline Performance.

Verdugo, R., Sitar, N., Frost, J.D., Bray, J.D., Candia, G., Eldridge, T. et al. 2012. Seismic performance of earth structures during the February 2010 Maule, Chile, Earthquake: Dams, levees, tailings dams, and retaining walls. Earthquake Spectra 28: S75–S96.

Wang, C. and Manga, M. 2009. Earthquakes and Water: Springer, New York, 2014. 1–38.

Zarco-Tejada, P.J., González-Dugo, V. and Berni, J.A. 2012. Fluorescence, temperature and narrow-band indices acquired from a UAV platform for water stress detection using a micro-hyperspectral imager and a thermal camera. Remote Sensing of Environment 117: 322–337.

CHAPTER 8

Hydrochemical and Tracer Monitoring to Assess Runoff Generation from Semi-arid Andean Headwater Catchments

Nauditt, A.,[1] *Rusman, A.,*[2] *Schüth, C.,*[2] *Ribbe, L.*[1] *and Álvarez, P.*[3,*]

Introduction

Mountainous catchments are often remote and difficult to monitor and are therefore rarely featured by long term, spatially distributed hydro-meteorological data. However, such headwater regions are of key importance for runoff generation as a result of higher precipitation as rain or snow providing valuable ecosystem services downstream (Price and Egan, 2014). Mountains as 'Water Towers' in semi arid and arid regions contribute up to 95% to total basin discharge (Messerli et al., 2004; Viviroli et al., 2007) supplying large parts of the world population with drinking water as well as with water resources for irrigation, hydropower and other industries and ecosystems. According to Barnett et al. (2005), more than one sixth of the world's population depends on montane snow covered glaciers and seasonal snow for water supplies which are at risk due to a warming climate. Water resources management and the equitable allocation of water for domestic, agricultural and industrial uses depend on reliable long term and seasonal discharge predictions. These are even more needed to address the additional challenges related to climate variability and change affecting

[1] Institute for Technology and Resources Management in the Tropics and Subtropics, Technical University Cologne.
[2] Institute for Applied Geosciences, University of Darmstadt.
[3] Department of Agricultural Engineering, Universidad de La Serena.
* Corresponding author: alexandra.nauditt@th-koeln.de

particularly mountain hydrology (Barnett et al., 2005; Immerzeel et al., 2010; Vuille et al., 2015). However, streamflow predictions from mountainous headwaters ideally require estimates of areal precipitation, snowpack extent, snow water equivalent and melt rates as well as groundwater recharge and storage dynamics. In addition, such predictions need to be contextualized with long term climate and discharge time series. At higher latitudes and elevations, ice and snow only allow access by helicopter during most of the year and monitoring equipment is exposed to extreme climatic conditions as strong winds, cold and ice. Therefore, for most mountainous headwater catchments worldwide, hydro-meteorological data or any other *in situ* information are not available and methodologies are needed to assess mountainous hydrological processes.

This chapter introduces to the features of the semi-arid Limarí Basin and illustrates the findings of synoptic seasonal tracer surveys in two remote mountainous headwater catchments with the aim to improve the understanding about hydrological processes in this data scarce and remote environment. During the sampling period (2012–2014) Central Chile was exposed to an extreme long term drought leading to severe water shortages for irrigation water supply (Boisier et al., 2016).

The Limari River Basin, Central Chile

The Limarí Basin forms part of the semi-arid Coquimbo region in northern-central Chile and covers an area of 11.696 km². It reaches from the Andes at 5.550 m of elevation to the east of the Pacific coast in the west and is a snowmelt-dominated hydrological system. Figure 8.42 shows the Limarí Basin with its hydrological network, topography and agricultural area and the la Paloma Irrigation System consisting of three reservoirs.

Water demand in the basin is dominated by irrigated agriculture and accounts for 724 million m³/year (DGA, 2005; Espinosa et al., 2011). The Limarí valley is the most relevant in terms of agricultural production and 70% of the regional export is produced there (Oyarzún, 2010), in particular cash crops as wine grapes and table grapes. The total cultivated area has increased by 70% since the construction of the Paloma irrigation system in 1973 alongside with a change in cultivation from annual crops as cereals to cash crops and perennial vineyards and fruit tree plantations representing 53% of the irrigated agricultural area (Álvarez, 2006; Oyarzún, 2010). The Paloma irrigation system is composed of three reservoirs with a total storage capacity of 1,000 million m³ and a complex channel network extending over more than 700 km.

The **climate** of the Limarí catchment is arid to semi-arid with marked Mediterranean seasonality (Peel et al., 2007). It largely varies along the

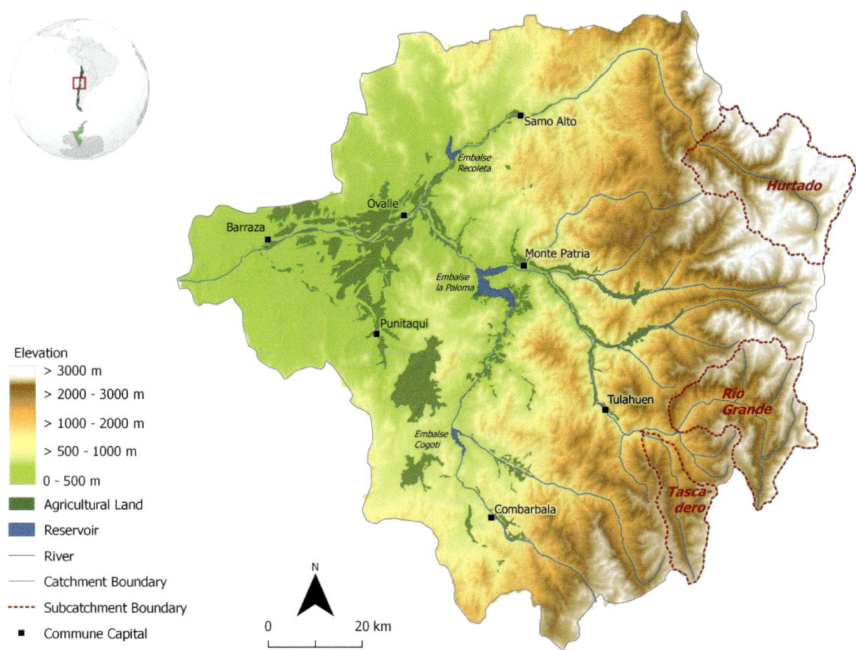

Fig. 8.42 The Limarí Basin with its hydrological network, topography and agricultural area and the la Paloma Irrigation System consisting of three reservoirs, studied headwater catchments are highlighted by red borders.

Fig. 8.43 Agricultural area next to the Recoleta dam in the Limarí Basin.

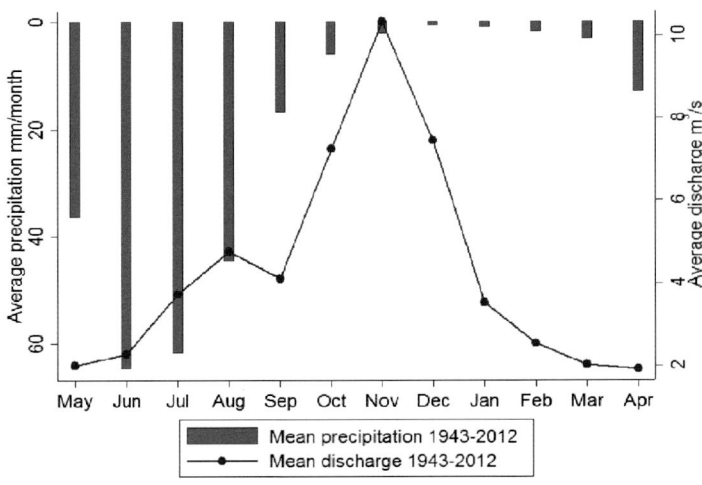

Fig. 8.44 Monthly mean precipitation-discharge distribution at Las Ramadas station.

140 km between the western Pacific coast and the eastern Andean Mountains due to the south-eastern pacific anticyclone, the cold Humboldt Current along the Pacific coast and the steep topography of the mountain range of the Andes (Kalthoff et al., 2006).

Precipitation ranges from 100 to 300 mm per year from the coastal area to the highest station in the Andes (recorded up to 1250 m of elevation) and also from north to south, with mean annual values of 70 mm in the north and 275 mm in the south at lower elevations (Oyarzún, 2010; Strauch et al., 2010; Favier et al., 2009). Inter-annual precipitation is characterized by high variability, years with high precipitation are typically linked with a High Oceanic Nino Index (ONI) representing so called 'El Niño years' while 'La Niña' years in this region are usually drier (Garreaud et al., 2003; Nuñez et al., 2013; Meza, 2015).

Figure 8.44 shows the mean monthly precipitation and discharge for the hydrological year at the Las Ramadas station at the outlet of the Rio Grande headwater catchment. The snowmelt dominated hydrograph does not immediately respond to the increased winter rainfall from May to August but has its peak during the melting period in summer between September and December.

However, the precipitation amounts strongly vary depending on the year. Figure 8.45 shows two examples of extreme precipitation and discharge behavior at Las Ramadas station. Two years have been selected: 2002/2003 as a hydrological year with more than average precipitation (total annual P = 652.9) and 1969/1970 with much less than average (total P = 52.8 mm).

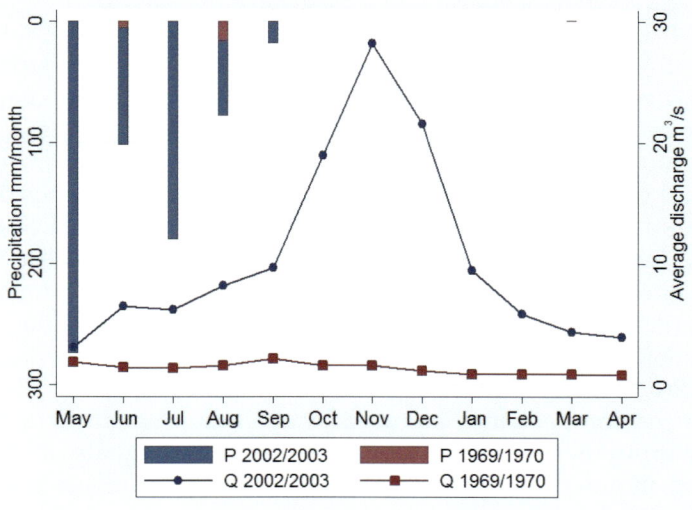

Fig. 8.45 Precipitation and discharge for two selected hydrological years of extreme climatic conditions at Las Ramadas station.

Fig. 8.46 Cushion peatland ('vegas') in the Rio Grande catchment.

Andean Headwater Catchments

Little geographical and climate information is available for the Andean Cordillera. The geology of the area is highly heterogeneous and fractured as evidenced by thermal springs. Plutonic and volcanic formations comprise the higher elevation areas, with fractured metamorphic and sedimentary formations at lower altitudes. Alluvial sediments occupy the main river valleys and a range of glacial deposits are also present. Soil development is limited, with shallow leptosols dominating. The semi-arid high elevation Andean environments are characterized by sparse shrub vegetation, bare rocks and cacti. Above 2000 masl, vegetation is exposed to exceptionally harsh environmental conditions, with little moisture, pronounced diurnal temperature variation with frequent frosts, high solar radiation, and low oxygen concentration. High elevation cushion peatland (called 'bofedales' or 'vegas') is a common feature of these valleys (Squeo et al., 2006; Schittek et al., 2014). Typical adaptive plant species can be found there such as tussock grasses, low stature woody shrubs, rosetted herbaceous perennials and cushion plants (Schittek et al., 2012). The main peat-accumulating species of these soligenous peatlands are the Juncaceae Distichia muscoides and Oxychloe andina (Schittek et al., 2012).

The cushion peatlands are used by the local human population as grazing grounds sending their animals, especially goats, to the high elevation mountains in the summertime as water and pastureland is available. The peatlands are artificially extended by the goat farmers who divert the water springs in order to accelerate the peatland development and grazing land for their animals. According to Schittek et al. (2012), the intense grazing by hoofed animals might degrade these sensitive ecosystems, resulting in erosion and destruction of the peat deposits, increased input of sediments and the expansion of alluvial fans upon the peat deposits. However, this hypothesis would require support by the assessment of a comparable totally pristine Andean catchment.

Recent studies show that there is a warming trend over the last 40 years with an increasing number of warm days especially for the high elevation Andean regions (Souvignet et al., 2012; Vuille et al., 2015). This will have serious implications for ecosystems and the regional hydrological cycle as earlier snow melt with higher discharges in springtime and decreasing discharge values in summer (Barnett et al., 2010; Vicuña et al., 2010; Souvignet et al., 2010).

Diurnal and inter-annual temperatures and precipitation vary extremely at the higher elevations (Nauditt et al., 2016). As there are no time series for temperature and precipitation available for higher altitudes above 1250 m, a climate station was installed in the Tascadero catchment at 3450 m.a.s.l. by ITT with research funding of the German Research Ministry (BMBF) (Figure 8.48). For the study period November 2012–Dec 2013, a mean temperature of 3.0°C with a minimum of −17.7°C in July and a maximum of 19.9°C in January was recorded. Figure 8.47 show recorded high elevation minimum air temperatures between 22.11.2012 and 23.12.2013 for the Tascadero station in the Cordillera compared to the minimum temperatures measured at the Las Ramadas station at 1250 m of elevation.

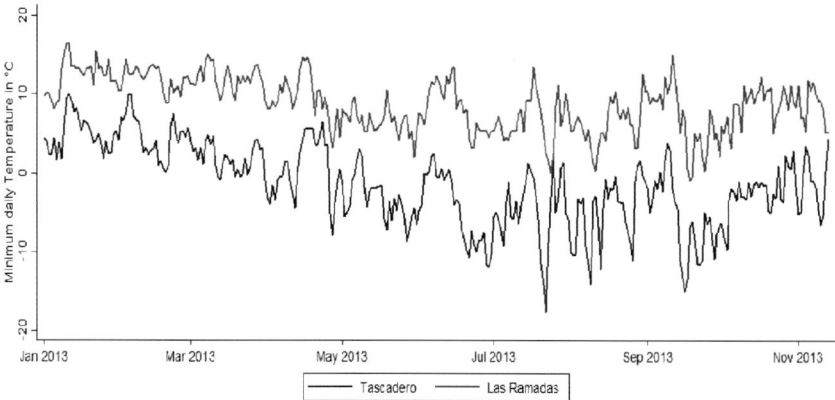

Fig. 8.47 High altitude minimum temperatures at 3450 m a.s.l. compared to minimum temperatures at Las Ramadas Station at 1250 m a.s.l.

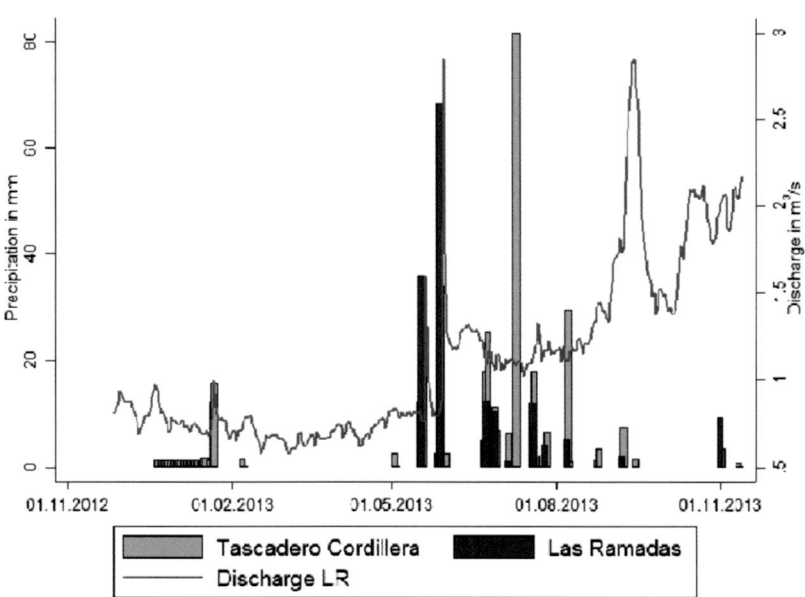

Fig. 8.48 Precipitation recorded by the climate station at the Tascadero headwaters at 3.450 m of elevation.

Fig. 8.49 Precipitation and discharge measured at Las Ramadas station (1250 m) for the study period from November 2012 to November 2013.

Synoptic Tracer Assessments

As described in the introduction, there are no hydro-meteorological time series for the high elevation Andean region and little information related to the bio-physical characteristics of the catchment. Besides monitoring climate and discharge, alternative methods need to be considered to improve the understanding of discharge generation especially in regard of a changing climate. Tracer surveys using stable isotopes and hydro-chemical data can provide valuable information to understand sources of runoff and seasonal system dynamics.

Stable isotope and geochemical tracer analyses have been widely used to improve the understanding of hydrological processes (Kendall and McDonnell, 1998; Mook et al., 2001; Leibundgut et al., 2009) by tracing the origin of surface waters, the mode of recharge of groundwater and to determine the age of the water (Craig, 1961). Also in headwater catchments, this method offers an opportunity to assess climate and discharge behaviour, residence times and the origin of headwaters (Soulsby et al., 2003; Soulsby et al., 2007; Hrachowitz et al., 2011).

Very few studies with stable isotopes have been carried out in the remote arid to semi-arid headwaters of Central and Northern Chile. Fritz et al. (1981) in the late 70s sampled and analyzed stable isotopes and ^{13}C to assess age and residence time of groundwater in the Pampa del Tamarugal, in the north of the Atacama Desert region suggesting that variation in isotopic composition stems from differing Amazon-Atlantic or Pacific moisture sources. According to their findings, groundwaters at this site have a residence time of > 5 years up to fossil ages. Aravena et al. (1999) found that strongly $\delta^{18}O$ depleted values in the high altitude altiplano are related to Atlantic-Amazonian-sourced moisture by the accumulation of the continental effect as well as the altitudinal and convective effect when precipitated in the Altiplano. Both, Fritz et al. (1981) and Aravena et al. (1999) highlight the strong altitudinal effect above of 3000 m of elevation compared to the altitudinal effect below 3000 m. Hence the relatively $\delta^{18}O$ enriched values observed at lower altitudes are associated with Pacific moisture sources. In both studies isotopic patterns in rain, spring and surface waters were similar (Aravena et al., 1999; Fritz et al., 1981). More recently, precipitation, river and soil water isotopic values were collected and analyzed for the Vth region of Chile and adjacent Argentinean Andean headwater catchments (latitude –32.5° to –35.5°) by Hoke et al. (2013) between 2008 and 2010. They suggested that precipitation on the Argentinean eastern slopes above 2 km is isotopically dominated by Pacific-sourced winter moisture and below 2 km by a mixture of Atlantic-Amazonian-sourced and westerly sources (Hoke et al., 2013). Ohlanders et al. (2013) evaluated the origin of glaciered headwaters for the same Aconcagua Basin (V. Region) regarding their provenance from glacial- or snowmelt. They detected a strong spatial and altitudinal gradient in stable isotopic composition of snowpack and its effect on the temporal evolution of streamflow isotopic composition during snowmelt. They calculated a glacier melt contribution to streamflow of 50–90% during the melting season of a dry hydrological year 2011/2012 indicating the strong acceleration of glacial retreat during dry years.

For the non-glaciered headwater catchments of the semiarid Limarí Basin so far, no tracer or geochemical data have been sampled or published. Oyarzún et al. (2014, 2016) assessed surface-groundwater relationships in the downstream area of the Rio Grande headwaters proving a strong connectivity. Surveys in the headwaters will provide valuable insights in the hydrological processes of the Andean headwaters to inform hydrological models and water management.

Sampling Strategy and Methods

Due to the difficulties to access the high Andean Cordillera only seasonal synoptic sampling was carried out. Precipitation events rarely occur and therefore only streamflow and groundwater springs could be sampled. Samples were taken at each accessible water source in stream or directly from groundwater springs or groundwater draining from the mountains as tributaries to the mainstream.

Sampling in Rio Grande was carried out in January and May 2013 as well as in September and October 2013. In November 2012 and May and September 2013, samples were collected in the Tascadero headwater catchment. Figure 8.50 shows the sampling locations in the Rio Grande catchment.

Samples for stable isotope analysis were collected in 5 ml bottles at each point during each sampling campaign from January 2013 to December 2014. Samples collected in January and May were analyzed for stable isotopes of water (^{2}H and ^{18}O) at the Northern Rivers Institute at the University of Aberdeen, with a Los Gatos Research (LGR) DLT-100 laser diode water isotope analyzer and transformed into the δ-notion (‰) according to Vienna Standard Mean Ocean Water (VSMOW) standards. Analytical precision is 1.0‰ for δ^{2}H and 0.2‰ for δ^{18}O. Samples collected in September and early December 2013 were analyzed to the same precision at the Technical University Darmstadt Laboratory with a Picarro L2130-i Cavity Rind Down Spectrometer (CRDS) isotope analyzer connected to a Picarro A0211 high precision vapourizer as described in Coplen (1996). Electric Conductivity (EC) of stream waters (µS/cm) was measured *in situ* with a Hach Lange HQ30D hand held device. A more detailed sampling methodology is

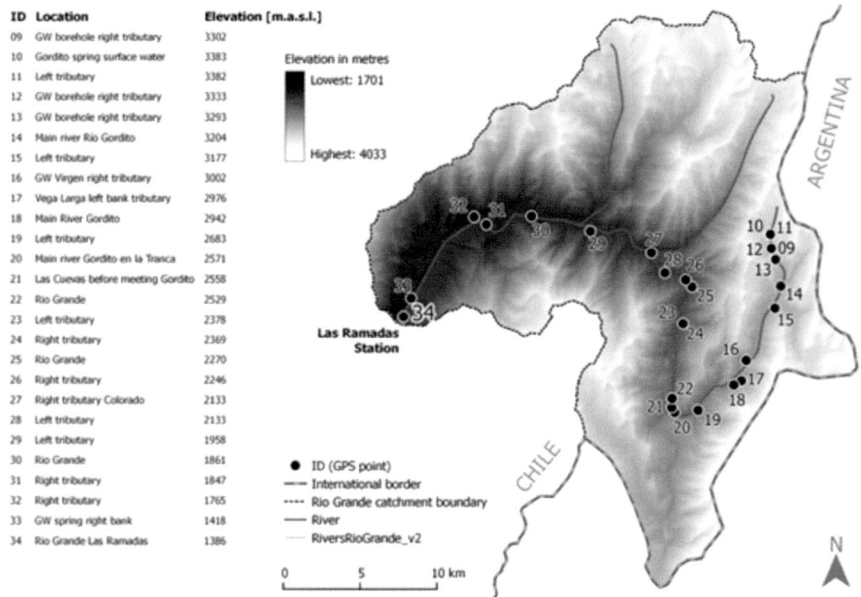

Fig. 8.50 Sampling locations in Rio Grande headwater catchments, simulated streams for water rich periods.

described in Nauditt et al., 2017. Major anions and cations were sampled in water scarce May of 2013 at the end of autumn using a Metrohm 882 Compact IC plus ion chromatograph with an IC Autosampler Plus 919 were used. Samples were analyzed at the laboratory of the Geosciences Faculty of Technical University of Darmstadt. The detailed methodology is described in Rusman (2014).

Stable Isotopes in the Rio Grande and Tascadero Headwater Catchments

We used stable isotope data from seasonal sampling campaigns to assess spatial variability in runoff sources in two Andean non-glaciered headwater catchments (Nauditt et al., 2017). Figure 8.51 shows the seasonal water lines for the example of the Rio Grande catchment for January, May September and December 2013. The isotopic composition shows a seasonal variation (Nauditt et al., 2017). The isotopic signal of surface waters is strongly affected by evaporation from late spring to late autumn, the period which is characterized by high temperatures and little precipitation leading to the enrichment in heavy isotopes—mainly $\delta^{18}O$ values—in both headwaters resulting in lower waterline gradients. δ^2H and $\delta^{18}O$ values are more depleted during wintertime. The early spring September samples hence show lighter values and a steeper gradient of the waterline due to more humid conditions, snow melt and colder climate during wintertime. From the beginning of the melting season during early spring, δ^2H and $\delta^{18}O$ decrease, which is characteristic for alpine snow melt driven rivers (Taylor et al., 2001).

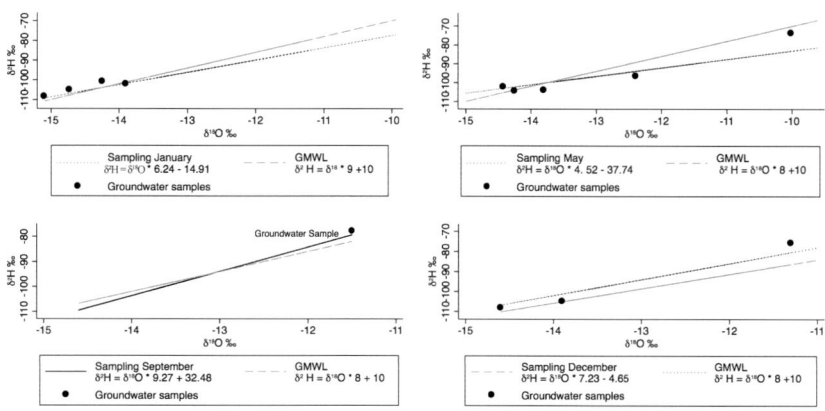

Fig. 8.51 Local seasonal water lines (surface water samples) for the Rio Grande samples from Jan, May, Sept and December compared to the Global Meteoric Water Line (GMWL) indicating groundwater samples.

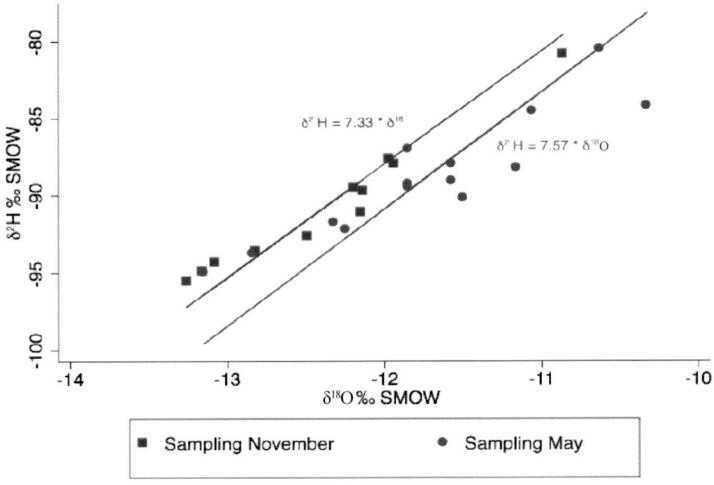

Fig. 8.52 Seasonal local water lines in the Tascadero for November 2012 and May 2013.

We calculated a local meteoric waterline (LMWL) for the Limari-Rio Grande headwaters with $\delta^2H = 7.8 * \delta^{18}O + 3.29$. Other LMWLs were established for the lower Limarí basin downstream of Rio Grande catchment: $\delta^2H = 8.5\ \delta^{18}O + 16$ (Oyarzún et al., 2016), the Chilean Meteoric Water Line $\delta^2H = 8.3\delta^{18}O + 9.8$ (IAEA, 2005). For the latitude of 21° the LMWL was defined as $\delta^2H = 7.8\ \delta^{18}O + 9.7$ (Aravena et al., 1999) and for the Salares de Atacama at 23° as $\delta^2H = 7.8\ \delta^{18}O + 10.3$ (Fritz et al., 1981). All of them are close to the GMWL (Nauditt et al., 2017).

Geothermal influences at 2950 m of elevation in the Rio Grande catchment valley are displayed in the isotopic signal in samples of the hot springs and the surrounding tributaries with more depleted values. For the 14 samples collected in November 2012, the values range from –12.2‰ to –13.6‰ for $\delta^{18}O$ and –89.1‰ to –99.9‰ for δ^2H, respectively.

The isotopic composition hence shows a strong seasonal variation in both catchments. The isotopic signal of surface waters is strongly affected by evaporation from late spring to late autumn, the seasons characterized by high temperatures and no precipitation. This leads to the enrichment in heavy isotopes—especially $\delta^{18}O$ values—in both rivers, which results in lower waterline gradients. In contrast, values are more depleted during wintertime. For the September—springtime samples show lighter values and a steeper gradient of the waterline, compared to other seasons. The reason for this is a higher effect of meteoric waters, due to more humid conditions with stronger precipitation and colder climate during wintertime. Additionally, it is affected by an increased melt water influence. Therefore,

δ^2H and $\delta^{18}O$ start to decrease at the beginning of the melting seas which is characteristic for alpine snow melt driven rivers.

Altitudinal Effect

The calculated altitudinal effect tends to be higher, above 2000 m it is higher and lower below 2000 m. However, not enough values are available to get statistically significant results for each section. Also to calculate the altitude effect after Yutsever and Gat (1981) altitude effect = lapse rate * slope of $T - \delta^{18}O$, including the $T - \delta^{18}O$ Correlation ($\delta^{18}O =$ (0.338 ± 0.028) $T_{monthly} - 11.99‰$ VSMOW) and the lapse rate of $-6°/km$ ($-0.6°C$ change per 100 m rise) is not applicable due to the lack of values and high temperature changes during a sampling day.

Fritz (1981) analyzes altitude effects for different water sources at 21° of latitude in the Andes and finds they are not well-defined and seem to change with altitude. While low altitudinal changes were observed between 2000 and 3000 m and above 4000 m, a major shift occurs at ~ 3500 m of elevation (Fritz et al., 1981). He attributes this to the moving air masses having passed the Andes as well as Mook et al. (2001) who state that the altitude effect is exceptionally high where different air masses with different source characteristics affect the precipitation at the base and mountain peaks as of the western slopes of the Andes in South America (Mook et al., 2001; Hoke et al., 2013).

Figure 8.53 plots January sampling values against elevation and shows significant altitude effects. They are, however, not well-defined and change with elevation. The altitudinal gradient gets steeper with elevation between 2500 and 3500 m. However, surface stream water samples tend to be more enriched in $\delta^{18}O$ than samples taken from the tributaries, which are mostly groundwater contributions and have been less exposed to evaporation and are hence are closer to meteoric waters (Fig. 8.54).

Hydrochemistry

Surface water chemistry in semi-arid northern-central Chile is generally controlled by evaporation, interaction of meteoric water with volcanic rocks in groundwater dominated areas and contact with evaporites in aquifer sediments (Aravena, 1995; Magaritz et al., 1989). Salinity is very low in the upper headwater source areas with conductivity rates between 30 and 150 µS7 cm. North of latitude 29°, water types are described to be dominated by $Na-SO_4$ and $Ca-SO_4$ in volcanic environments, while $Na-Cl$ values rise nearby geothermal fields, where the 'Salares' are located, the salt lakes in the desert region (Aravena et al., 1987; Fritz et al., 1990). Aravena analyzed ^{34}S values in surface waters and found out that the main source of sulphate in these regions are sulphate minerals in volcanic rocks (Aravena et al., 1999).

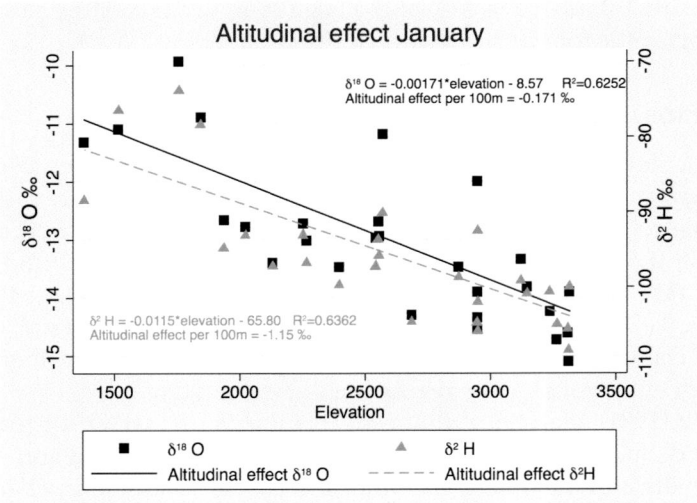

Figure 8.53 Mean altitudinal effect in δ18O and δ2H values for Rio Grande 3500–1500 masl in summer (January).

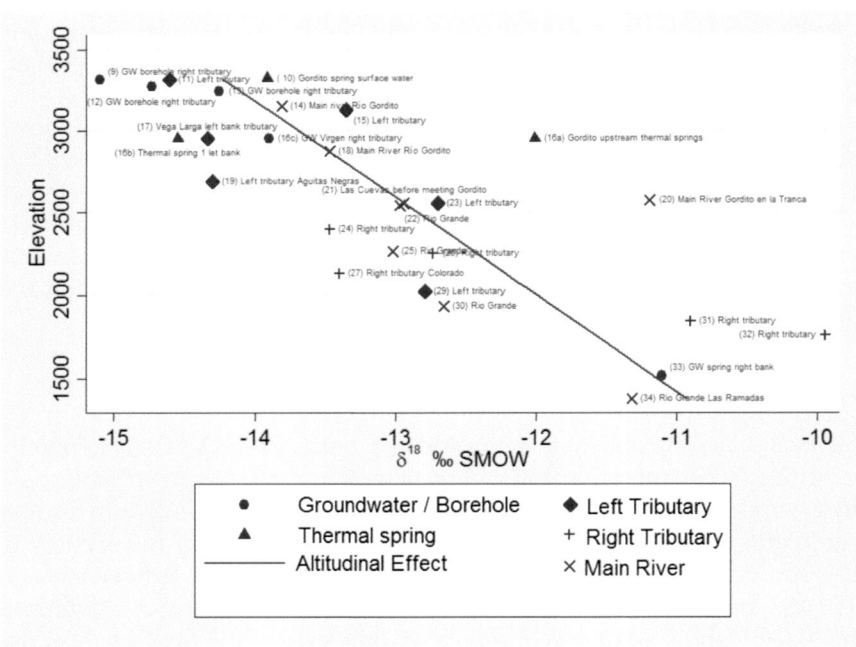

Figure 8.54 δ18O values in January 2013 plotted against elevation in the Rio Grande.

Figures 8.55 and 8.56 show piper plots for the Rio Grande and Tascadero headwaters. Values of similar chemical composition are clustered to illustrate the distribution of cations and anions (Langguth and Vogt, 2004). The samples are classified according to the method of Furtak and Langguth (1967) according to their major ionic composition.

Hydrochemistry in the Rio Grande headwaters is characterized by low mineralization where in general anions are represented by sulphate and bicarbonate and cations by calcium and sodium. This suggests that the headwaters consist of meteoric water with low residence times in groundwater in contact with host rocks and sediments. Electric conductivity values in Rio Grande stream and spring waters ranged from 38 μS/cm to 415 μS/cm. Stream water values were between 41 μS/cm to 279/cm indicating that their origin must be from meteoric rain or melt water. In thermal springs conductivity reached a value of 5.370 μS/cm also impacting nearby tributaries with increased EC of 552 μS/cm to 825 μS/cm (Nauditt et al., 2017).

For the upper headwater part, the Gordito tributary, calcium is the dominant ion originating from the weathering of sedimentary carbonate

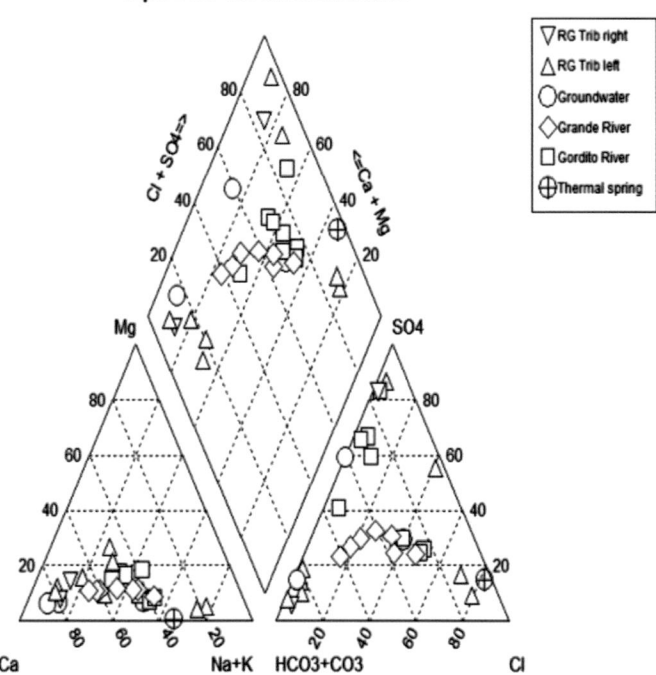

Fig. 8.55 Piper plot showing the distribution of anions and cations in stream and groundwater samples of the Grande catchment (December 2013).

rocks and calcium silicates. At an elevation between 2950 m and 2690 m water chemistry changes and is related to geothermal fields and hot springs with high contents of TDS, chloride and sodium as the dominant cation. Also, high boron contents were measured in the geothermally impacted samples. Downstream water chemistry in the Grande River is still influenced by the geothermal fields and the hot springs. Sodium levels are higher than calcium in this section and chloride dominates bicarbonate and sulphate levels.

The values located in the cation-triangle indicate the dominance of the calcium in the samples for the Grande River (Fig. 8.55). The samples in the lower middle show a nearly balanced content of calcium and sodium. Three samples are pointing out a significant dominance of sodium and potassium. Magnesium values are low in all samples. The values in the lower left corner are mainly characterized by calcium-carbonate and include the samples from tributaries. The thermal spring and the geothermally influenced tributaries are shown in the lower right corner with dominant sodium characterization. Values located in the lower middle of the right

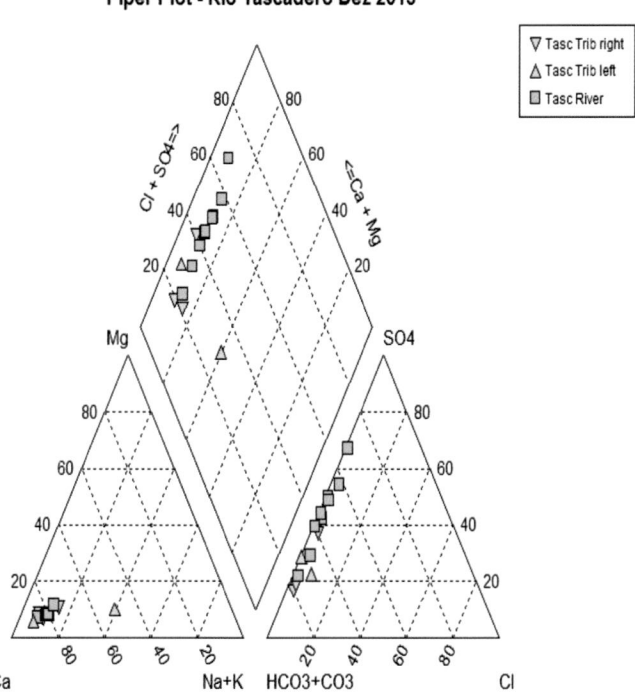

Fig. 8.56 Piper plot showing the distribution of anions and cations in the water samples of the Tascadero catchment (December 2013).

triangle belong to the main river samples showing an increase in calcium content (Fig. 8.55).

The ionic composition of the waters at 14 sampling points in the Tascadero catchment was analyzed. Calcium was found to be the dominant major cation; followed by sodium and lowest concentrations for magnesium and potassium. For anionic species bicarbonate dominates except at one location in the upper headwater where sulphate has the highest portion. Chloride concentrations are very low and the nitrate level is only slightly above zero. Clustered major ions are shown in the Piper plot in Fig. 8.56. Minor ions and trace elements were analyzed for the Tascadero catchment. Boron was detected in very low concentrations ranging between 0.07 mg/l to 0.13 mg/l as well as ammonium with approximately 0.41 mg/l in half of Tascadero samples. Dissolved iron was found in 13 samples, also with very low values between 0.06 and 0.19 mg/l.

Conclusion and Outlook

Andean headwater catchments are the main source of water supply for downstream semi-arid irrigation-dependent communities. However, only little information is available about climate, catchment characteristics, cryosphere and hydrological processes for these remote areas to support water management with discharge predictions. Therefore we used the spatial and seasonal distribution of the stable isotopic composition ($\delta^{18}O$ and δ^2H) and hydrochemistry of stream and ground waters in two perennial headwater catchments to distinguish various sources and components of stream flow. A marked spatial and seasonal variability of all measured parameters was observed. Hydrochemistry in ground water and surface water samples suggests that there is no fossil groundwater contribution to the baseflow with very low mineralization of headwaters and geothermal influences at around 3000 m of elevation. Thus the tracer survey results have provided us with a useful overview of the general flow paths in the Rio Grande. However, due to the limited number of values in space and time and the dry nature of the study year, we are unable to provide a more quantitative analysis on the exact sources and residence times of the individual sample points. We therefore suggest the following sampling strategy: (1) establish a large network of spatially distributed precipitation collectors and stream flow sampling sites for stable isotopes to determine the dynamics of dominant water sources, (2) Assess groundwater age and residence times by sampling boreholes and springs and analysing for ^3H, CFC and ^{13}C samples, (3) conduct further synoptic seasonal surveys of stream sampling sites, wells and springs for full geochemical analysis to establish the hydrogeological provenance and impact on stream flow (4) Install continuous air temperature measurements with simple sensors and data loggers at several elevation points of the Rio Grande catchment to obtain an accurate temperature lapse rate and (5) Carry out sampling campaigns in the Tascadero catchment (Nauditt et al., 2016) where grazing has been prohibited since 2014 in order to evaluate the impacts of earlier human activities on headwater catchments in tracer data from 2012 and 2013.

References

Álvarez, P., Kretschmer, N. and Oyarzun, R. 2006. Water Management for Irrigation in Chile: Causes and Consequences. Paper presented at the International Water Fair, Wasser Berlin 2006.

Aravena, R. 1996. Isotope hydrology and geochemistry of groundwater in northern Chile. Bull. Inst. Ft. etudes andines 24: 495–503.

Aravena, R. and Suzuki, O. 1990. Isotopic evolution of rivers in Northern Chile. Water Resour. Res. 26: 2887–2895.

Aravena, R., Suzuki, O., Peña, H., Pollastri, A., Fuenzalida, H. and Grilli, A. 1999. Isotopic composition and origin of the precipitation in Northern Chile. Appl. Geochem. 14: 411–422.

Arenson, L.U. and Jakob, M. 2010. The significance of rock glaciers in the dry Andes—A discussion of Azocar and Brenning. Permafrost and Periglacial Processes 21(3): 282–285.

Azócar, G.F. and Brenning, A. 2010. Hydrological and geomorphological significance of rock glaciers in the dry Andes, Chile (27°/33°). Permafrost and Periglacial Processes 21(1): 42–53.

Barnett, T.P., Adam, J.C. and Lettenmaier, D.P. 2005. Potential impacts of a warming climate on water availability in snow-dominated regions. Nature 438: 303–309, doi:10.1038/nature04141.

Barsch, D. 1996. Rockglaciers. Berlin: Springer-Verlag 331.

Bates, B.C., Kundzewicz, Z.W., Wu, S. and Palutikof, J.P. 2008. Climate Change and Water. Technical Paper of the Intergovernmental Panel on Climate Change. Geneva: IPCC Secretariat.

Blume, T., Zehe, E. and Bronstert, A. 2008. Investigation of runoff generation in a pristine, poorly gauged catchment in the Chilean Andes I: A multi-method experimental study. Hydrological Processes 22(18): 3661–3675. Doi: 10.1002/hyp.6971.

Boisier, J.P., Rondanelli, R., Garreaud, R. and Muñoz, F. 2016. Anthropogenic and natural contributions to the Southeast Pacific precipitation decline and recent megadrought in central Chile; AGU RESEARCH LETTER 10.1002/2015GL067265.

Brenning, A. 2005a. Geomorphological, hydrological and climatic significance of rock glaciers in the Andes of CentralChile (33–35-S). Permafrost and Periglacial Processes 16: 231–240. Doi: 10.1002/ppp.528.

Brenning, A. 2010. The significance of rock glaciers in the dry Andes—Reply to L. Arenson and M. Jakob. Permafrost and Periglacial Processes 21(3): 286–288.

Clark, I. and Fritz, P. 1997. Environmental Isotopes. Hydrogeology, New York: CRC Press.

Coplen, T.B. 1996. New guidelines for the reporting of stable hydrogen, carbon, and oxygen isotope ratio data: Geochimica et Cosmochimica Acta 60: 3359.

Craig, H. 1961. Isotopic variation in meteoric waters. Science 133: 1702–1703.

Dansgaard, W. 1964. Stable isotopes in precipitation. Tellus.

DGA. 2004. Diagnostico y clasificación de los cursos y cuerpos de agua según objetivo de calidad, Cuenca del Limarí, Ministerio de Obras Públicas, Gobierno de Chile 2004. http://www.sinia.cl/1292/articles-31018_Limari.pdf (accessed in October 2015).

Favier, V., Falvey, M., Rabatel, A., Praderio, E. and López, D. 2009. Interpreting discrepancies between discharge and precipitation in high-altitude area of Chile's Norte Chico region (26–32_S). Water Resources Research 45: W02424, doi:10.1029/2008WR006802.

Fritz, P., Silva, C., Suzuki, O. and Salati, E. 1981. Isotope hydrology of groundwaters in the Pampa del Tamarugal, Chile. Journal of Hydrology 53: 161–184.

Froehlich, K., Gibson, J.J. and Aggarwal, P. 2002. Deuterium excess in precipitation and its climatological significance. Proceedings of Internatiopnal Confenerence on Study of Environmental Change Using Isotope Techniques, C&S Papers Series 13/P, Vienna, Austria.

Furtak, H. and Langguth, H.R. 1967. Zur hydrochemischen Kennzeichnung von Grundwässern und Grundwassertypen mittels Kennzahlen. Intern. Assoc. Hydrogeol. 7: 89–96.

Garreaud, R.D., Vuille, M., Compagnucci, R. and Marengo, J. 2009. Present-day South American climate. Palaeogeography, Palaeoclimatology, Palaeoecology 281(3-4): 180–195.

Gat, J.R. 1987. Variability (in time) of the isotopic composition of precipitation: consequences regarding the isotopic composition of hydrologic systems. Isotope techniques in water resource development, IAEA-SM-319/39, IAEA, Vienna. 551–563.

Gat, J.R. 1996. Oxygen and hydrogen isotopes in the hydrologic cycle. Annual Review of Earth Planetary Sciences 24: 225–262.

Geyh, M., D'Amore, F., Darling, G., Paces, T., Pang, Z., Silar, J., Paces, T. and D'Amore, A. 2000. in Mook: Isotopes in Hydrology, Volume IV, Groundwater, saturated and unsaturated zone, UNESCO IHP Paris.

Genereux, D. 1998. Quantifying uncertainty in tracer-based hydrograph separations. Water Resources Research 34: 915–919.

Gharari, S., Hrachowitz, M., Fenicia, F. and Savenije, H.H.G. 2011. Hydrological landscape classification: Investigating the performance of HAND based landscape classifications in a central European meso-scale catchment. Hydrology and Earth System Sciences 15: 3275–3291.

Herczeg, A.L. and Leaney, F.W. 2010. Review: environmental tracers in arid-zone hydrology. Hydrogeology Journal.

Hoke, G.D., Aranibar, J.N., Viale, M., Araneo, D.C. and Llano, C. 2013. Seasonal moisture sources and the isotopic composition of precipitation, rivers, and carbonates across the Andes at 32.5–35.5_S. Geochemistry, Geophysics, Geosystems 14, doi:10.1002/ggge.20045.

Hrachowitz, M., Soulsby, C., Tetzlaff, D., Dawson, J.J.C., Dunn, S.M. and Malcolm, I.A. 2009. Using long-term data sets to understand transit times in contrasting headwater catchments. Journal of Hydrology 367: 237–248. Doi: 10.1016/j.jhydrol.2009.01.001.

Hrachowitz, M., Soulsby, C. and Tetzlaff, D. 2010a. Catchment transit times and landscape controls—does scale matter? Hydrological Processes 24: 117–125. Doi: 10.1002/hyp.7510.

Hrachowitz, M., Bohte, R., Mul, M.L., Bogaard, T.A., Savenije, H.H.G. and Uhlenbrook, S. 2011. On the value of combined event runoff and tracer analysis to improve understanding of catchment functioning in a data-scarce semi-arid area. Hydrology Earth System Sciences 15: 2007–2024, doi:10.5194/hess-15-2007-2011.

Hublart, P. et al. 2016. Reliability of lumped hydrological modeling in a semi-arid mountainous catchment facing water-use changes. Hydrology and Earth System Sciences 20(9): 3691–3717. Available at: http://www.hydrol-earth-syst-sci.net/20/3691/2016/ [Accessed October 22, 2016].

International Atomic Energy Agency/World Meteorological Organization. Isotope Hydrology Information System; IAEA/WMO: Vienna, Austria, 2005; accessible at http://isohis.iaea.org.

IPCC: Working group I contribution to the IPCC Fifth Assessment Report Climate Change. 2013: The Physical Science Basis, Final Draft Underlying Scientific-Technical Assessment, 2013.

Jeelani, G.H., Saravana, K.U. and Kumar, B. 2013. Variation of d^{18}O and dD in precipitation and stream waters across the Kashmir Himalaya (India) to distinguish and estimate the seasonal sources of stream flow. Journal of Hydrology 481: 157–165.

Jones, J.P., Sudicky, E.A., Brookfield, A.E. and Park, Y.-J. 2006. An assessment of the tracer-based approach to quantifying groundwater contributions to streamflow. Water Resour. Res. 42: W02407. Doi: 10.1029/2005WR004130.

Kalthoff, N., Fiebig-Wittmaack, M., Meissner, C., Kohler, M., Uriarte, M., Bischoff-Gauss, K. and Gonzales, E. 2006. The energy balance, evapo-transpiration and nocturnal dew deposition of an arid valley in the Andes. Journal of Arid Environments 65(3): 420–443.

Kendall, C. and Caldwell, E.A. 1998a. Fundamentals of isotope geochemistry. pp. 51–86. *In*: Kendall, C. and McDonnell, J.J. (eds.). Isotope Tracers in Catchment Hydrology. Elsevier Science: Amsterdam.

Kendall, C. and McDonnell, J.J. 1998. Isotope Tracers in Catchment Hydrology. Elsevier: Amsterdam.

Langguth, H.-R. & Voigt, R. (2004): Hydrogeologische Methoden. 2., überarb. & erw. Auflage. Springer Verlag, Berlin

Leibundgut, C., Maloszewski, P. and Külls, C. 2009. Tracers in Hydrology. JohnWiley & Sons Ltd, The Atrium: Southern Gate, Chichester,West Sussex, PO19 8SQ, UK.

Magaritz, M., Aravena, R., Pena, H., Suzuki, O. and Grilli, A. 1989. Water chemistry and isotope study of streams and springs in northern Chile. Journal of Hydrology 108: 323–341.

McGuire, K.J., McDonnell, J.J., Weiler, M., Kendall, C., McGlynn, B.L., Welker, J.M. and Seibert, J. 2005. The role of topography on catchment-scale water residence time. Water Resources Research 41: W05002. Doi: 10.1029/2004WR003657.

Messerli, B., Viviroli, D. and Weingartner, R. 2004. Mountains of the World: Vulnerable Water Towers for the 21st Century. AMBIO Special Report 13: 29–34.

Meza, F. 2014. Recent trends and ENSO influence on droughts in Northern Chile: An application of the Standardized Precipitation Evapotranspiration Index, Weather and Climate Extremes 51–58.

Milly, P.C.D., Dunne, K.A. and Vecchia, A.V. 2005. Global pattern of trends in streamflow and water availability in a changing climate. Nature 438: 347–350.

Masiokas, M.H., Villalba, R., Luckman, B.H., Le Quesne, C. and Aravena, J.C. 2006. Snowpack variations in the central andes of argentina and chile, 1951–2005: large-scale atmospheric influences and implications for water resources in the region. Journal of Climate 19: 6334–6352. Doi: http://dx.doi.org/10.1175/JCLI3969.1.

Masiokas, M.H., Villalba, R., Luckman, B.H. and Mauget, S. 2010. Intra- to multidecadal variations of snowpack and streamflow records in the Andes of Chile and Argentina between 30 degrees and 37 degrees S. Journal of Hydrometeorology 11: 822–831.

Mook, W.G., Gat, J.R. and Meijer, H.A.J. 2001. Environmental isotopes in the hydrological cycle: Principles and applications, v. IV: Groundwater—Saturated and unsaturated zone Technical documents. Hydrology 39, SC.2001/WS/37.

Mul, M.L., Mutiibwa, R.K., Uhlenbrook, S. and Savenije, H.H.G. 2008. Hydrograph separation using hydrochemical tracers in the Makanya catchment, Tanzania. Physics and Chemistry of the Earth 33(1-2): 151–156.

Nauditt, A., Birkel, C., Soulsby, C. and Ribbe, L. 2016. Conceptual modelling to assess the influence of hydroclimatic variability on runoff processes in data scarce semi-arid Andean catchments. Hydrological Sciences Journal Doi: 10.1080/02626667.2016.1240870.

Nauditt, A., Soulsby, C., Rusman, A., Schüth, C., Ribbe, L., Álvarez, P. and Kretschmer, N. 2017. Synoptic tracer surveys of streamwater isotopes and geochemical tracers to understand runoff processes in a semiarid Andean headwater catchment, Central Chile. Environmental Monitoring and Assessment (under review).

Nunez, J., Rivera, D., Oyarzún, R. and Arumí, J.L. 2013. Influence of Pacific Ocean multidecadal variability on the distributional properties of hydrological variables in north-central Chile. Journal of Hydrology 501(2013): 227–240.

Ohlanders, N., Rodriguez, M. and McPhee, J. 2013. Stable water isotope variation in a Central Andean watershed dominated by glacier and snowmelt. Hydrology Earth System Sciences 17: 1035–1050. Doi: 10.5194/hess-17-1035-2013.

Oyarzún, J. et al. 2003. Heavy metals in stream sediments from the Coquimbo region (Chile): Effects of sustained mining and natural processes in a semi-arid Andean Basin. Mine Water and the Environment 22(3): 155–161. Available at: http://link.springer.com/10.1007/s10230-003-0016-9 [Accessed February 14, 2014].

Oyarzun, R. 2010. Estudio de caso: Cuenca del Limarí, Región de Coquimbo, Chile, Compilación Resumida de Antecedentes. Centro de Estudios Avanzados en Zonas Aridas-Universidad de la Serena (CEAZA-ULS).

Oyarzún, R. et al. 2014. Multi-method assessment of connectivity between surface water and shallow groundwater: the case of Limarí River basin, north-central Chile. Hydrogeology Journal 22: 1857–1873. Doi: 10.1007/s10040-014-1170-9.

Oyarzún, Ricardo, Sandro Zambra, Hugo Maturana, Jorge Oyarzún, Evelyn Aguirre and Nicole Kretschmer. 2016. Chemical and isotopic assessment of surfase water-shallow groundwater interaction in the arid Grande river basin, North-Central Chile. Hydrological Sciences Journal 61(12): 2193–2204. Doi: 10.1080/02626667.2015.1093635.

Quintana, J.M. and Aceituno, P. 2012. Changes in the rainfall regime along the extratropical west coast of South America (Chile): 30–43°S. Atmosfera 25: 1–22.

Rozanski, K., Araguás-Araguás, L. and Gonfiantini, R. 1993. Isotopic patterns in modern global precipitation. In: Climate Change in Continental Isotopic Records. Geophysical Monograph 78: 1–36, Americal Geophysical Union.

Rozanski, K. and Araguás, L. 1995. Spatial and temporal variability of stable isotope composition over the South American continent. Bull. Inst. Fr. Etud. Andin. 24: 379–390.

Rusman, A. 2014. Hydrogeological assessment in the high Andean cordillera, North Central Chile. MSc thesis University of Darmstadt.

Schittek, K., Forbriger, M., Schäbitz, F. and Eitel, B. 2012. Cushion Peatlands—Fragile Water Resources in the High Andes of Southern Peru, Book chapter: Landscape and Sustainable Development, 4.

Siegenthaler, U. and Oeschger, H. 1980 Correlation of ^{18}O in precipitation with temperature and altitudes. Nature 285: 314–318.

Soulsby, C., Rodgers, P., Smart, R., Dawson, J. and Dunn, S. 2003. A tracer-based assessment of hydrological pathways at different spatial scales in a mesoscale Scottish catchment. Hydrological Processes 17: 759–777.

Soulsby, C., Tetzlaff, D. and Hrachowitz, M. 2009. Tracers and transit times: Windows for viewing catchment scale storage? Hydrological Processes 23: 3503–3507.

Soulsby, C., Birkel, C., Tetzlaff, D. and Dunn, S.M. 2011. Inferring groundwater influences on surface water in montane catchments from hydrochemical surveys of springs and streamwaters. Journal of Hydrology 333: 199–213.

Souvignet, M. 2010. Climate Change Impacts on Water Resources in Mountainous Arid Zones: A Case Study in the Central Andes, Chile. PhD University of Leipzig.

Souvignet, M., Oyarzún, R., Koen, M., Verbist, J., Gaese, H. and Heinrich, J. 2011. Hydro-meteorological trends in semi-arid North-Central Chile (29–32° S): Water resources implications for a fragile Andean region. Hydrological Sciences Journal. HSJ-2010-0077.R1.

Taylor, S., Kirchner, J.W., Osterhuber, R., Klaue, B. and Renshaw, C.E. 2001. Isotopic evolution of a seasonal snowpack and its melt. Water Resources Research 37: 759–769.

Taylor, S., Feng, X., Willimas, M. and McNamara, J. 2002. How isotopic fractionation of snowmelt effects hydrograph separation. Hydrological Processes 16: 3683–3690.

Technologies, http://www.madd.ch/index.php?option=com_deeppockets&task=contShow &id=56&Itemid=318, assessed 2014.

Uhlenbrook, S., Frey, M., Leibundgut, C. and Maloszewiski, P. 2002. Residence time based hydrograph separations in a meso-scale mountainous basin at event and seasonal time scales. Water Resources Research 38(6): 1–14.

Verbist, K., Robertson, A.W., Cornelis, W. and Gabriels, D. 2010. Seasonal predictability of daily rainfall characteristics in central-northern Chile for dry-land management. Journal for Applied Meteorology and Climate 49(9): 1938–1955.

Vicuña, S., Garreaud, R.D. and McPhee, J. 2010. Climate change impacts on the hydrology of a snowmelt driven basin in semiarid Chile. Climatic Change 105(3-4): 469–488.

Vogel, J.C. 1972. Natural isotopes in the groundwater of the Tulúm Valles, San Juan, Argentina. Hydrological Sciences Bulletin 17(1): 85–96. Doi: 10.1080/02626667209493805.

Vogel, J.C., Lerman, J.C. and Mook, W.G. 2010. Natural isotopes in surface and groundwater from Argentina. Hydrological Sciences Bulletin 20: 203–221.

Appendix

Appendix 1: Various streams sampled in December 2013.

1 Gordito headwaters (RG011);
2 Vega Larga (RG017);
3 Aguitas Negras (RG019/-1);
4 Grande and Colorado River confluence (RG027);
5 Tascadero headwaters at confluence with Calderoncito River (TA01).

Pictures during fieldwork in Chile. 1. Las Ramadas DGA gauging station; 2. + 3. Tascadero DGA gauging station; 4. Tascadero high elevation climate monitoring station; 5. Borehole close to Gordito River at 3450 m; 6. Hot spring at Los Banos (Photographs by André Rusman).

Pictures: Various streams sampled in December 2013. 1. Gordito headwaters (RG011); 2. Vega Larga (RG017); 3. Aguitas Negras (RG019/-1); 4. Grande and Colorado River confluence (RG027); 5. Tascadero headwaters at confluence with Calderoncito River (TA01) (Photographs by André Rusman).

CHAPTER 9

Isotopic Characterization of Waters Across Chile

R. Sánchez-Murillo,[1,]* *E. Aguirre-Dueñas,*[2] *M. Gallardo-Amestica,*[2]
P. Moya-Vega,[2] *C. Birkel,*[3] *G. Esquivel-Hernández*[1] *and J. Boll*[4]

Introduction

Chile is a long, narrow strip of land between the southeastern Pacific Ocean (west) and southern Andes Cordillera (18°–67° S, east). This large geographic and orographic (up to ~ 6,900 m a.s.l.) spectrum comprises a wide range of climatic scenarios from warm and cold desert in the north, to temperate and cold oceanic climate in the east and southeast, and temperate Mediterranean climate in the central region, offering a unique setting to study latitudinal water stable isotope variations governed by geographic features and potentially reflecting shifts in modern climatic patterns. The recognition that modern precipitation dynamics may help to understand past climate conditions preserved in paleo-archives has increased the number of studies across the entire Andes Cordillera (Wolfe et al., 2001; Hoffmann et al., 2003; Vuille et al., 2003; Vimeux et al., 2009; Perry et al., 2014; Fiorella et al., 2015a, b). In Chile, isotopic studies have been highly concentrated in the northern region due to the intriguing extreme dryness

[1] Stable Isotope Research Group, National University of Costa Rica, Heredia, Costa Rica.
[2] Chilean Nuclear Energy Commission, Laboratory of Environmental Isotopes, Santiago, Chile.
[3] Department of Geography, University of Costa Rica, San José, Costa Rica, Northern Rivers Institute, University of Aberdeen, Aberdeen, Scotland.
[4] Department of Civil and Environmental Engineering, Washington State University, Pullman, USA.
* Corresponding author: ricardo.sanchez.murillo@una.cr

of the Atacama Desert, the proximity of the Altiplano (i.e., Andean Plateau) within the Central Andes Cordillera, and the need for better understanding of groundwater recharge processes to ultimately improve water resources management in the arid northern region. In the south of Chile, however, abundant surface water and groundwater resources coupled with remote locations have resulted in less frequent studies compared to the central and northern regions (Arumí and Oyarzún, 2006). Nevertheless, water-related stable isotope studies in the latter region are receiving increased attention (Hervé-Fernández et al., 2016; Lavergne et al., 2017).

A pioneering three-year data set (1984–1986) of precipitation isotopes from northern Chile by Aravena et al. (1999) showed $\delta^{18}O$ values ranging between −18‰ and −15‰ at high altitude stations, compared to −10‰ and −6‰ at lower elevations. The $\delta^{18}O$-depleted values observed in the high altitude area, the Altiplano, were related to processes that affect the air masses that (i) originated over the Atlantic Ocean, (ii) cross the Amazon Basin (continental effect), (iii) ascended the Andes (altitude effect) and (iv) precipitated (convective effect) in the Altiplano. They also identified a second source of moisture, associated with air masses from the southeastern Pacific Ocean, which may have contributed to $\delta^{18}O$-enriched values observed in lower altitude areas. Similar isotopic patterns were documented in springs and groundwater showing the representation of the long-term isotopic composition of rain in northern Chile. The relationship between the isotopic and chemical composition of rain, spring, and stream water in the high Andes of northern Chile was first studied by Fritz et al. (1981) and Magaritz et al. (1989). Fritz et al. (1981) reported that groundwater at Pampa del Tamarugal (within the Atacama Desert) originates from infiltrated surface water rather than directly infiltrated precipitation. Based on their low ^{14}C activities, the authors suggested that most of the water pumped was fossil. Magaritz et al. (1989) showed that the isotopic pattern of the springs in the high Andes of northern Chile is mainly a reflection of the altitude of their recharge areas, whereas processes that occur during snow melt seem to play a major role in the high-altitude springs. Similarly, the streams show analogous patterns to the springs in the higher part of the basins, but their isotopic composition is modified along the river course, mainly due to secondary evaporation processes along the river network. Hydrochemical studies in the Limarí River Basin in northern Chile (Oyarzún et al., 2014) showed an active interaction between surface water and shallow groundwater, and a minor effect of local precipitation events on the hydrological behavior in the study area. A recent study by Uribe et al. (2015) in closed basins within Salar del Huasco (northern Chile) estimated a long-term average recharge of 22 mm/yr and demonstrated no hydrogeological connectivity between the aquifer of the Salar del Huasco

Basin and the aquifer that feeds the springs of the nearby town of Pica. In central Chile, Ohlanders et al. (2013) determined glacier and snowmelt contributions to streamflow using stable isotopes in precipitation and surface water, whereby glacier and snowmelt inputs ranged from 50 up to 90% during dry La Niña years.

From a larger scale perspective, Bershaw et al. (2016) conducted an extensive study in modern surface water samples (including the northern Andean Plateau and surrounding regions) to elucidate patterns and causes of isotope fractionation in this continental environment. The authors reported a progressive increase in $\delta^{18}O$ of stream water west of the eastern Cordillera (~ 1‰/70 km), which they attributed to a larger fraction of moisture recycling and a potential evaporative enrichment downwind, concluding that elevation is a primary control on the isotopic composition of surface water across the entire Andean Plateau and its surrounding areas. Consistent with the early findings by Aravena et al. (1999), Bershaw et al. (2016) and Fiorella et al. (2015a, b) suggested that precipitation patterns in the central Andes Cordillera are mainly governed by the easterly winds, which provide a large supply of moisture. The southeastern Pacific-derived moisture only contributes a minor amount at low elevations near the coast of, for example, La Serena. Similarly, Hoke et al. (2013) conducted a study on the eastern flank of the Andes in the Mendoza Province of Argentina, including a sampling transect in the western flank of the border with Chile (Las Cuevas, 3,200 m a.s.l.). Their results indicated that precipitation on the eastern slopes of the Andes at ~ 33° S, at elevations above 2 km, is largely derived from a westerly Pacific-source component and a mixture of easterly and westerly sources below 2 km.

The main goal of this chapter is to present a long-term analysis of water stable isotope ($\delta^{18}O$, δ^2H, *d*-excess, and lc-excess) variations in precipitation across the extreme latitudinal and altitudinal gradients of Chile coupled with representative surface water, groundwater, geothermal and ice coring isotopic data. The core of temporal and spatial analysis is based on a 24-year (1991–2015) continuous record of monthly precipitation samples (N = 684) across four stations (from north to south): La Serena, Santiago, Puerto Montt and Punta Arenas. The isotopic values were obtained from the Isotopes Monitoring in Precipitation database of the Environmental Isotopes Laboratory of the Chilean Nuclear Energy Commission (GNIP-CCHEN) (http://www.cchen.cl/). Five discontinued short-term GNIP-CCHEN stations (1988–1991) were included for a better spatial coverage: Valparaíso, Temuco, Concepción, Chillán and Coyhaique. Additionally, representative precipitation, surface water, groundwater, geothermal and ice core isotopic data were obtained from the existing literature (Fritz et al., 1981; Aravena and Suzuki, 1990; Alpers and Whittemore, 1990; Aravena, 1995; Aravena et al., 1999; Leybourne and Cameron, 2006; Ohlanders et al., 2013; Hoke et al., 2013; Oyarzún et al., 2014; Uribe et al., 2015; Fernández-Hervé et al., 2016) and from the isotopic archives of the International Atomic Energy Agency (IAEA, 2016). Long-term seasonal and temporal diagnostics are coupled with 10-day representative Lagrangian air mass back trajectories to highlight prevailing moisture sources and distinguish transport mechanisms as well as the influence of latitudinal isotopic effects during the wettest months. Analysis of these long-term water stable isotope data provides a fundamental baseline and revision for future isotope-informed modeling efforts and paleoclimate interpretations across the Pacific and Atlantic slopes within the southern Andes Cordillera biomes.

Climate Generalities

Chile is characterized by strong climatic gradients due to its unique geographical setting that extends over 4,000 km from around 18° S to almost 67° S (Fig. 9.57; Table 9.14). Such a longitudinal extension is coupled to an extreme topographical gradient from sea level up to ~ 6,900 m a.s.l. (i.e., Ojos del Salado volcano) with the Andes Cordillera traversing the continent and all of Chile. The Andes Cordillera act as an orographic barrier and separate Chile from air mass movements from the Atlantic Ocean (Aravena et al., 1999). Furthermore, Chile's climate is strongly influenced by the subtropical southeastern Pacific anticyclone (high pressure area) with cold sea currents (i.e., Humboldt Current) and low pressure systems forming off the Antarctic Sea (i.e., circumpolar low pressure area). Four main morphological units condition the existence of 11 different types of climate (from warm desert to polar/tundra) and associated vegetation: Coastal Plains, Coastal Mountains, Intermediate Depression and the Andes Cordillera (Smith and Evans, 2007).

As a consequence, northern Chile is characterized by the hyper-arid Atacama Desert with very low precipitation and high temperatures extending close to the city of La Serena (Verbist et al., 2010). According to the modified Köppen-Geiger climate classification by Peel et al. (2007), the area of La Serena is a cold desert climate (*BWk*) with annual rainfall of around 100 mm and a mean air temperature of 14°C (Fig. 9.57; Table 9.14). Peak rainfall occurs in winter (July) with virtually no rain during the summer months from November through April. Further to the south at an elevation of around 500 m a.s.l., the city of Santiago is characterized by a semi-arid cold steppe (*BSk*) climate with slightly higher annual precipitation of 329 mm on average and a mean annual air temperature of 13.9°C (Fig. 9.57; Table 9.14). Towards the central southern region of Chile rainfall is abundant with a mean annual amount of 1,952 mm at Puerto Montt and an average annual air temperature of 10.3°C (Fig. 9.57; Table 9.14).

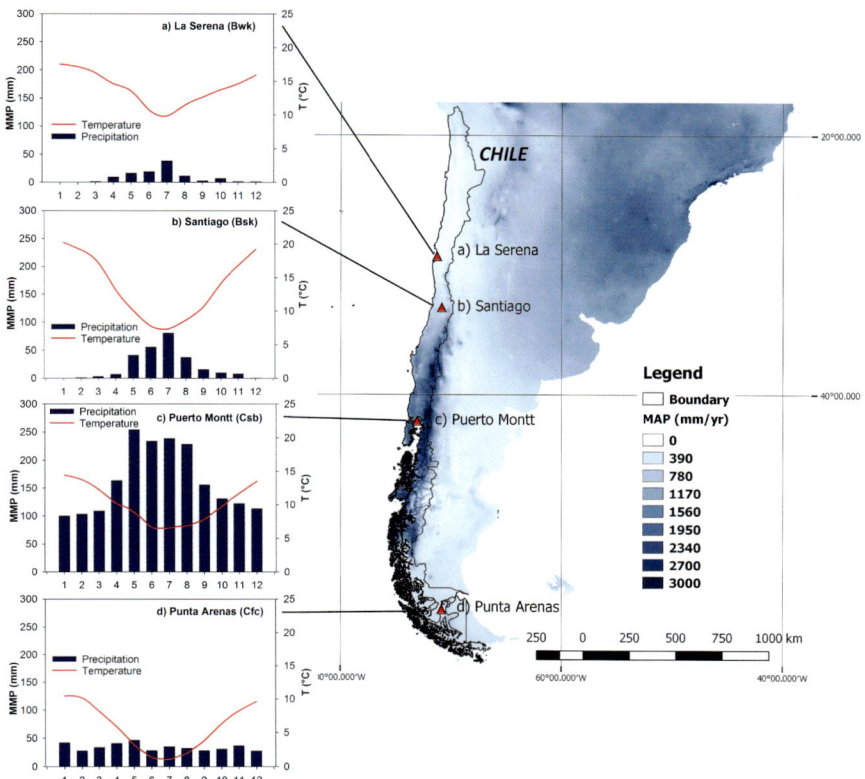

Fig. 9.57 Map of Chile including: long-term (24-years) stable isotope monitoring stations from GNIP-CCHEN (La Serena, Santiago, Puerto Montt and Punta Arenas) (red triangles), gridded mean annual precipitation (MAP) (mm yr[-1]), and four climographs showing monthly mean precipitation (MMP) (blue bars) and monthly mean temperature (red line) for each station (both variables were derived from the most recent monthly gridded Global Historical Climatology Network, GHCN, version 3.0 product; Peterson and Vose, 1997). Based on the updated Köppen-Geiger world climate classification (Peel et al., 2007), the study site climates are classified as: mid-latitude cold desert (*BWk*, La Serena), semi-arid cold steppe (*BSk*, Santiago), dry-summer or Mediterranean (*Csb*, Puerto Montt), and maritime temperate/sub-polar or oceanic (*Cfc*, Punta Arenas).

The climate of Puerto Montt is classified as dry-summer Mediterranean (*Csb*) with rainfall peaking in winter from June to September (Fig. 9.57; Table 9.14). In the Patagonia region, the climate of Punta Arenas can be described as maritime temperate (*Cfc*) with average annual rainfall of 410 mm and an average temperature of 5.8°C (Fig. 9.57; Table 9.14). Generally, precipitation increases with altitude and falls mostly as snow in the southern hemisphere winter, despite few high altitude measurements (Favier et al., 2009).

Table 9.14 Long-term (1991–2015) stable isotope characteristics in precipitation across Chile at four continuous monitoring stations.

Unit (‰)	La Serena (N = 40)				Santiago (N = 145)				Puerto Montt (N = 241)				Punta Arenas (N = 258)			
	$\delta^{18}O$	$\delta^{2}H$	d-ex	lc-ex	$\delta^{18}O$	$\delta^{2}H$	d-ex	lc-ex	$\delta^{18}O$	$\delta^{2}H$	d-ex	lc-ex	$\delta^{18}O$	$\delta^{2}H$	d-ex	lc-ex
Mean	-4.9	-29.6	9.8	1.7	-7.4	-50.6	8.3	-0.4	-5.5	-36.6	7.7	-0.5	-9.1	-70.5	2.3	-6.9
SD	2.2	15.9	5.5	11.6	2.7	21.3	5.1	17.5	2.7	18.7	5.8	18.8	3.3	23.8	6.9	11.7
Max	1.0	4.4	19.6	-11.7	-2.1	-3.0	25.1	-14.7	3.5	18.4	25.9	-19.8	3.5	18.4	19.2	-24.7
Min	-10.0	-76.0	-5.3	5.2	-13.7	-99.0	-6.3	5.0	-14.5	-101.1	-13.4	5.4	-21.4	-162.7	-17.7	6.5
w-mean	-5.6	-34.5	10.4	-0.2	-8.2	-55.0	10.9	1.7	-6.6	-43.1	9.7	0.6	-9.1	-72.1	0.6	-1.6

Methods

The 24-years (1991–2015) continuous record of monthly precipitation samples (N = 684) was compiled from data of four stations: La Serena (northern region, ~ 30° S, 142 m a.s.l., N = 40), Santiago (inter-mountainous central region, ~ 33° S, 520 m a.s.l., N = 145), Puerto Montt (southern coastal region, ~ 41° S, 81 m a.s.l., N = 241) and Punta Arenas (Patagonia region, ~ 53° S, 37 m a.s.l., N = 258) (Fig. 9.57). Monthly stable isotope archives ($\delta^{18}O$ and δ^2H) of precipitation (1991–2015) were obtained from the GNIP-CCHEN (http://www.cchen.cl/) in cooperation with the Meteorological Directorate of Chile under the General Directorate of Civil Aviation of Chile (DGAC). The Chilean network is part of the GNIP initiative (GNIP-IAEA-WMO, 2016). This database also includes observed mean monthly air temperature (°C) and monthly precipitation amount (mm). Mean monthly precipitation P (mm) and temperature T (°C) characteristics were used to construct individual climographs (Fig. 9.57) were derived from the most recent monthly gridded Global Historical Climatology Network (GHCN) version 3.0 product (Peterson and Vose, 1997). In addition, gridded mean annual $\delta^{18}O$ (‰) in precipitation below ~ 20° S for South America was derived from Bowen and Revenaugh (2003) for comparison purposes.

Precipitation samples collected before 2009 were analyzed in a FINNIGAN Mat 252 Isotope Ratio Mass Spectrometer (IRMS) with an automatic equilibrium method (CO_2–H_2O; H_2–H_2O), the analytical precision was ±0.2‰ for $^{18}O/^{16}O$ and ±1‰ for $^2H/^1H$. After 2009, isotope compositions were measured by laser spectrometry using a LWIA-LGR DLT-100 (Los Gatos Research, USA) with an analytical precision of ±0.08‰ for $^{18}O/^{16}O$ and ±1‰ for $^2H/^1H$. Isotopic compositions were normalized to the VSMOW-SLAP scales, through the use of calibrated secondary laboratory standards and are defined as:

$$\delta^2H_{sample} = \frac{(^2H/^1H)_{sample} - (^2H/^1H)_{VSMOW}}{(^2H/^1H)_{VSMOW}} \tag{1}$$

$$\delta^{18}O_{sample} = \frac{(^{18}O/^{16}O)_{sample} - (^{18}O/^{16}O)_{VSMOW}}{(^{18}O/^{16}O)_{VSMOW}} \tag{2}$$

Deuterium excess (hereafter *d*-excess; Dansgaard, 1964) was calculated for each monthly sample (Equation 3). In addition, to determine the degree of deviation of monthly precipitation samples from regional/Local Meteoric Water Lines (LMWL), the line-conditioned excess (lc-excess) was calculated according to Landwehr and Coplen (2006) (Equation 4) (La Serena: a = 6.97, b = 4.68; Santiago: a = 7.72, b = 6.64; Puerto Montt: a = 6.72, b = 0.63; Punta Arenas: a = 7.02, b = –6.62). This calculation uses the LMWL as a reference rather than simply using the deviation from the GMWL (Sprenger et al.,

2017). The coefficients *a* and *b* are the slope and the y-intercept of the LMWL, respectively. Based on the analytical precision reported, the estimated average uncertainties are ±1.1‰ (*d*-excess) and ±1.5‰ (lc-excess).

$$d - excess = \delta^2H - 8 \cdot \delta^{18}O \tag{3}$$

$$lc - excess = \delta^2H - a \cdot \delta^{18}O - b \tag{4}$$

Five discontinued short-term GNIP-CCHEN stations (1988–1991) were included to achieve a better spatial coverage: Valparaíso (central coastal region, ~ 33° S, 41 m a.s.l., N = 16), Temuco (southern-central region, ~ 39° S, 114 m a.s.l., N = 27), Concepción (southern coastal region, ~ 37° S, 11 m a.s.l., N = 19), Chillán (southern-central region, ~ 36° S, 147 m a.s.l., N = 27) and Coyhaique (Patagonia region, ~ 45° S, 310 m a.s.l., N = 117) (Fig. 9.58). Additionally, representative precipitation, surface water, groundwater, geothermal and ice core isotopic data were obtained from the existing literature (Fritz et al., 1981; Aravena and Suzuki, 1990; Alpers and Whittemore, 1990; Aravena, 1995; Aravena et al., 1999; Leybourne and Cameron, 2006; Ohlanders et al., 2013; Hoke et al., 2013; Oyarzún et al., 2014; Uribe et al., 2015; Fernández-Hervé et al., 2016) and from the isotopic archives of IAEA (2016) (Fig. 9.58).

The influence of atmospheric trajectory and source meteorological conditions on the subsequent stable isotope composition of precipitation was analyzed using the HYSPLIT Lagrangian model (Stein et al., 2015) developed by the Air Resources Laboratory of NOAA (USA). The HYSPLIT model uses a three-dimensional Lagrangian air mass vertical velocity algorithm to determine the position of the air mass and reports these values at an hourly time-resolution over the trajectory (Soderberg et al., 2013). Representative 10-day air mass back trajectories were calculated for the three wettest months in 2015 at each monitoring station due to the nature of the monthly sampling. To compute a trajectory, the HYSPLIT model requires a starting time, location and altitude as well as NOAA meteorological data files (e.g., GDAS, global data assimilation system, 0.5° resolution: 2006-present; Su et al., 2015).

Results and Discussion

Regional isotopic characteristics

The best-fit continental meteoric water line to data from Chile (Chile-LMWL) is described as: $\delta^2H = 7.66 \cdot \delta^{18}O + 3.42$ ($r^2 = 0.94$, N = 684, p < 0.001) (Fig. 9.59A, top panel). Overall in the 24-years continuous record and across the four monitoring stations, $\delta^{18}O$ and δ^2H ranged from −21.4‰ to +3.5‰ and from −162.7‰ to +18.4‰, respectively (Figs. 9.59B and 9.59C, top panel), while *d*-excess ranged from −17.7‰ to +25.9‰

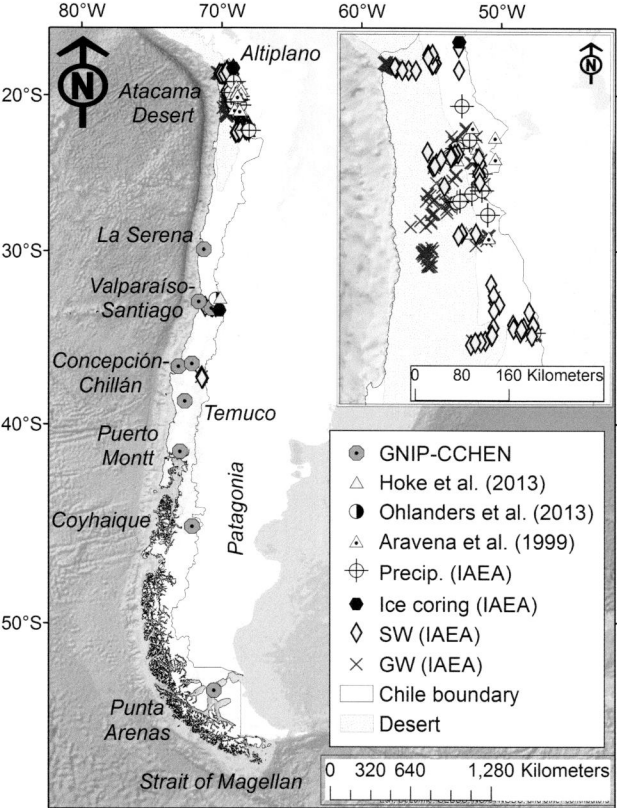

Fig. 9.58 Overview map of Chile including (a) the core GNIP-CCHEN monitoring stations and 5 discontinued short-term GNIP-CCHEN stations (pink octagons), (b) other precipitation samples from IAEA archives (green crossed-circles), (c) published precipitation data with available locations by Aravena et al. (1999) (yellow triangles), Hoke et al. (2013) (pink triangles), and Ohlanders et al. (2013) (yellow-black circles), (d) surface water (SW, cyan rhombi), groundwater (GW, blue crosses), and ice coring (red hexagons) isotope samples from IAEA archives. The inset shows the high concentration of isotopic sampling in northern Chile.

(mean = +7.0‰) and lc-excess varied from –6.9‰ to +1.7‰ (mean = –1.5‰) (Table 9.14). For the La Serena station, located near the southeastern Pacific coast (Fig. 9.57), the LMWL is described as: $\delta^2H = 6.97 \cdot \delta^{18}O + 4.68$ ($r^2 = 0.90$, N = 40, p < 0.001) (Fig. 9.59A, top panel). Since the northern region of Chile is semi-permanently under the influence of a high pressure system known as the sub-tropical southeastern Pacific anticyclone in combination with the cold Humboldt Current (Montecinos and Aceituno, 2003) and the isolation of the Atlantic Ocean moisture by the Andes Cordillera (Aravena et al., 1999), the precipitation events are less intense and occur mostly from May through August. Although the long-term *d*-excess at La

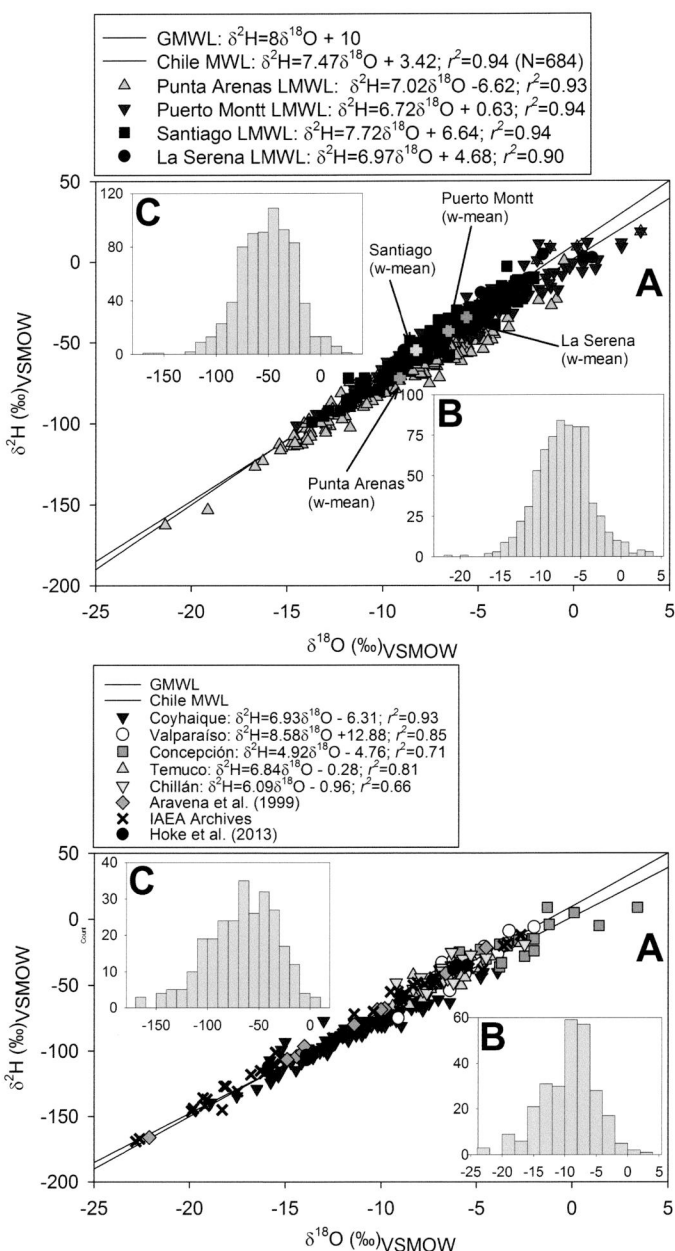

Fig. 9.59 Top panel: (A) Local Meteoric Water Lines (LMWLs) for each 24-years monitoring station. Bottom panel: (A) Local Meteoric Water Lines (LMWLs) for each discontinued short-term GNIP-CCHEN monitoring station and available data from Aravena et al. (1999), Hoke et al. (2013), and IAEA archives for Chile. The GMWL (black line) and Chile MWL (blue line) are plotted as references. Insets (B) and (C) in both panels show histograms for $\delta^{18}O$ (‰) and δ^2H (‰), respectively.

Serena (+9.8‰) is close to the global mean (+10‰; Craig, 1961), potential secondary evaporation processes below the cloud base during small rainfall events (MAP = 106 mm) coupled with a moderate mean annual air temperature range (10–20°C; Fig. 9.57) may introduce an artifact effect (Sánchez-Murillo et al., 2016a) which is responsible for the relatively lower slope and intercept values. For the Santiago station, located at a higher elevation (520 m a.s.l.) and approximately ~ 100 km from the Pacific coast, the LMWL is described as: $\delta^2H = 7.72 \cdot \delta^{18}O + 6.64$ ($r^2 = 0.94$, N = 145, p < 0.001) (Fig. 9.59A, top panel). Enhanced orographic distillation and greater precipitation amounts (MAP = 329; Fig. 9.57; Table 9.14) resulted in more depleted values with a mean $\delta^{18}O$ of –7.4‰. In the southern coastal region of Puerto Montt, the LMWL can be described as $\delta^2H = 6.72 \cdot \delta^{18}O + 0.63$ ($r^2 = 0.94$, N = 241, p < 0.001) (Fig. 9.59A, top panel). This location experienced a typical temperate oceanic climate with large precipitation amounts year round (MAP = 1,952 mm; Fig. 9.57) and moderate mean annual air temperatures, ranging from 15°C down to 5°C. In the coastal Patagonia region of Punta Arenas, the LMWL is described as: $\delta^2H = 7.02 \cdot \delta^{18}O - 6.62$ ($r^2 = 0.93$, N = 258, p < 0.001) (Fig. 9.59A, top panel). Mean annual precipitation at Punta Arenas (410 mm) as well as mean annual temperature (ranging from 10°C down to 0°C) are considerably lower than at Puerto Montt (Fig. 9.57). In the last two locations, the lower intercepts may represent enhanced non-equilibrium processes due the influence of the circumpolar low pressure and temperature conditions from the Antarctic Sea. Overall, a clear latitudinal effect was observed in the isotopic composition along Chile. For instance, long-term *d*-excess and lc-excess exhibited a consistent decreasing latitudinal trend: +9.8‰ (+1.5‰) (La Serena), +8.3‰ (–0.4‰) (Santiago), +7.7‰ (–0.5‰) (Puerto Montt), and +2.3‰ (–6.9‰) (Punta Arenas) (Table 9.14), reflecting the influence of non-equilibrium processes as the mean annual temperature decreases towards the southern region, which favors snow formation and greater kinetic fractionation (Dansgaard, 1964). A similar decreasing trend was observed for $\delta^{18}O$ and δ^2H (Table 9.14).

For the five discontinued short-term (1988–1991) GNIP-CCHEN monitoring stations (Fig. 9.58), a similar isotopic pattern was observed when analyzing the LMWLs. For instance, the LMWL at Coyhaique (Patagonia region, Fig. 9.58) is described as: $\delta^2H = 6.93 \cdot \delta^{18}O - 6.31$ ($r^2 = 0.93$, N = 117, p < 0.001) (Fig. 9.59A, bottom panel) in concordance with the Punta Arenas-LMWL. The LMWLs at Temuco, Chillán, and Concepción (Fig. 9.58) also exhibited relatively low slopes (4.92 up to 6.84) and intercept (–4.76 up to –0.26) values (Fig. 9.59A, bottom panel). Valparaíso-LMWL is the only one across Chile from the GNIP-CCHEN database with a relatively high slope and intercept ($\delta^2H = 8.58 \cdot \delta^{18}O + 12.88$ ($r^2 = 0.85$, N = 16, p < 0.001). However, the lower sample size may not be sufficient to establish further inferences. In general, the spectrum of isotopic composition in precipitation across Chile ranged from –22.8‰ up to +3.5‰ for $\delta^{18}O$ and from –169‰ up to +18.4‰

for δ^2H (Fig. 9.59B and 9.59C, in both panels). By combining all available isotopic precipitation records, the continental MWL can be described as: $\delta^2H = 7.59 \cdot \delta^{18}O + 3.25$ ($r^2 = 0.95$, N = 957, p < 0.001), which is quite similar to the one described by the 24-years continuous monitoring stations.

Long-term Spatial Variability

Dual relationships between δ^2H, *d*-excess, and lc-excess (Figs. 9.66S; 9.67S) revealed a strong gradient from north to south and suggested a large influence of kinetic processes likely related to snow formation, particularly, in the southern region of Chile. In the arid region of La Serena, $\delta^{18}O$ and δ^2H ranged from −10.0‰ to +1.0‰ (mean = −4.9 ± 2.2‰) and −76.0‰ to +4.4‰ (mean = −29.6 ± 15.9‰), respectively. The values of *d*-excess and lc-excess ranged from +19.6‰ up to −5.3‰ (mean = +9.8 ± 5.5‰) and from +5.2‰ up to −11.7‰ (mean = +1.7 ± 11.6‰), respectively (Fig. 9.60a; Table 9.14). In the central region of Santiago, $\delta^{18}O$ and δ^2H ranged from −13.7‰ to −2.1‰ (mean = −7.4 ± 2.7‰) and −99.0‰ to −3.0‰ (mean = −50.6 ± 21.3‰), respectively, whereas *d*-excess and lc-excess fluctuated from +25.1‰ up to −6.3‰ (mean = +8.3 ± 5.1‰) and from +5.0‰ up to −14.7‰ (mean = −0.4 ± 17.5‰) (Fig. 9.60b; Table 9.14). In the southern coastal region of Puerto Montt, $\delta^{18}O$ and δ^2H ranged from −14.5‰ to +3.5‰ (mean = −5.5 ± 2.7‰) and −101.1‰ to +18.4‰ (mean = −36.6 ± 18.4‰), respectively. Likewise, *d*-excess and lc-excess varied from −13.4‰ up to +25.9‰ (mean = +7.7 ± 5.8‰) and from +5.4‰ up to −19.8‰ (mean = −0.5 ± 18.8‰) (Fig. 9.60c; Table 9.14). In Punta Arenas, $\delta^{18}O$ and δ^2H ranged from −21.4‰ to +3.5‰ (mean = −9.1 ± 3.3‰) and −162.7‰ to +18.4‰ (mean = −70.5 ± 23.8‰), respectively, whereas *d*-excess and lc-excess ranged from −17.7‰ up to +19.2‰ (mean = +2.3 ± 6.9‰) and from +6.5‰ up to −24.7‰ (mean = −6.9 ± 11.7‰) (Fig. 9.60d; Table 9.14).

Long-term Seasonal Variability

Long-term seasonality is well constrained across the four continuous monitoring stations (Fig. 9.57). Isotopic composition decreased from summer (DJF) to a minimum in winter (JJA) and increased again towards the spring season (SON) (Fig. 9.61; Table 9.15). In general, enriched and less variable isotopic values (−4.5‰ to −5.20‰ for $\delta^{18}O$ and −23.7‰ to −32.6‰ δ^2H) were observed at La Serena (Fig. 9.61). This location is isolated from the influence of the Atlantic Ocean moisture by the central Andes Cordillera and received no rainfall in DJF. The most depleted values were observed at Punta Arenas monitoring station, where $\delta^{18}O$ composition decreased from −7.2‰ in summer (DFJ) to −9.2‰ in autumn (MAM), reached a minimum of −10.7‰ in winter (JJA) and increased again to −9.1‰ in spring (SON).

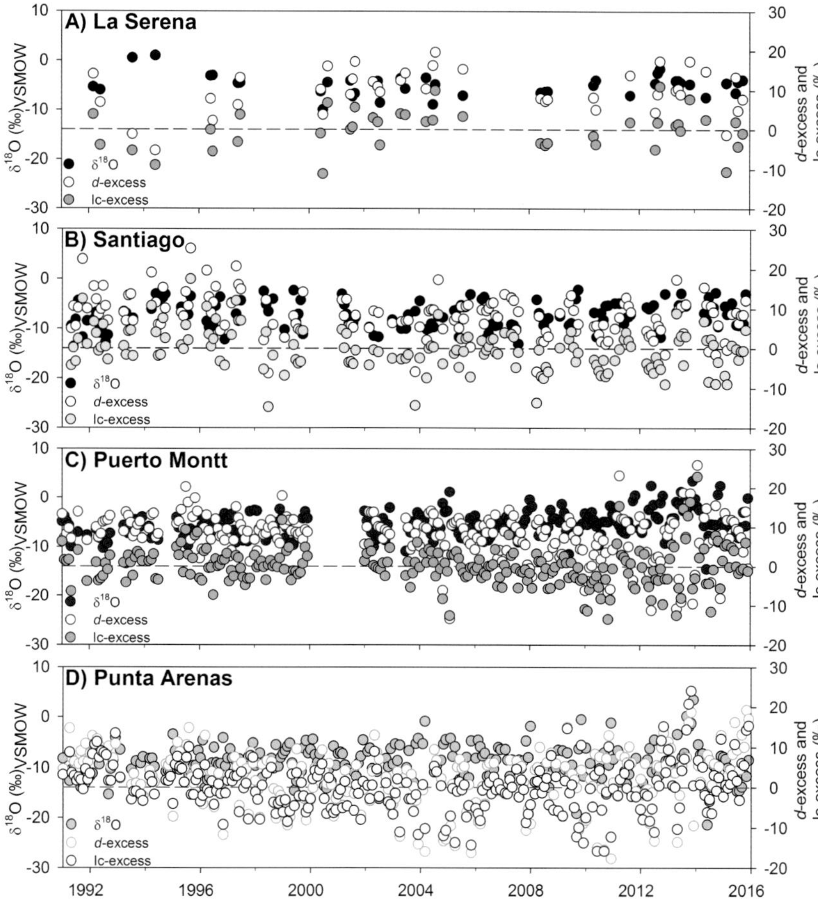

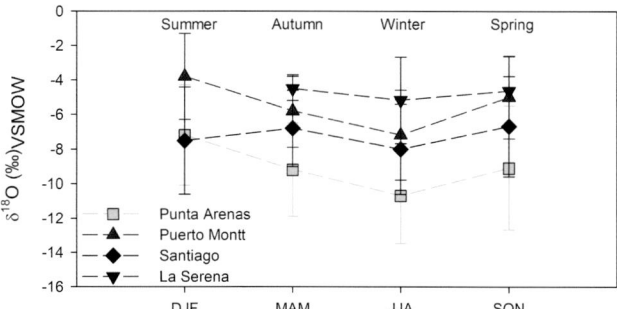

Fig. 9.60 Monthly $\delta^{18}O$ (‰), *d*-excess (‰), and lc-excess (‰) time series (1991–2015) of (A) La Serena, (B) Santiago, (C) Puerto Montt, and (D) Punta Arenas. Dashed-lines represent the lc-excess value of 0‰.

Fig. 9.61 Long-term seasonal $\delta^{18}O$ (‰) variability for each station. Seasons are grouped as: DJF (summer), MAM (autumn), JJA (winter), and SON (spring).

Table 9.15 Seasonal stable isotope characteristics in precipitation across Chile during 1991–2015.

Unit (‰)	La Serena			Santiago			Puerto Montt			Punta Arenas		
	$\delta^{18}O$	δ^2H	d-ex	$\delta^{18}O$	δ^2H	d-ex	$\delta^{18}O$	δ^2H	d-ex	$\delta^{18}O$	δ^2H	d-ex
Set-Oct-Nov (spring)												
Mean	-4.6	-23.7	13.3	-6.7	-45.4	8.0	-5.0	-33.2	6.8	-9.1	-70.6	2.3
SD	2.0	16.2	4.4	2.9	23.2	6.1	2.4	17.2	5.5	3.6	28.3	7.5
Max	-1.6	4.4	17.4	-2.1	-3.0	25.1	0.7	12.0	18.6	0.2	9.0	19.2
Min	-7.0	-42.5	7.7	-11.8	-88.9	-5.9	-9.3	-63.0	-9.6	-19.1	-153.2	-16.3
N	8	8	8	35	35	35	62	62	62	70	70	70
Dec-Jan-Feb (summer)												
Mean	---	---	---	-7.5	-54.3	5.8	-3.8	-25.4	5.2	-7.2	-58.0	-0.4
SD	---	---	---	3.1	22.3	4.4	2.5	17.2	6.9	2.9	20.9	6.9
Max	---	---	---	-3.0	-25.1	13.7	3.5	18.4	25.9	3.5	18.4	17.0
Min	---	---	---	-12.2	-92.0	-1.5	-9.2	-64.4	-13.4	-14.5	-113.5	-17.7
N	---	---	---	10	10	10	52	52	52	60	60	60
Mar-Apr-May (autumn)												
Mean	-4.5	-26.7	8.9	-6.8	-45.0	9.1	-5.8	-36.9	9.5	-9.2	-71.2	2.8
SD	0.7	7.1	5.4	2.2	17.3	5.7	2.1	15.1	4.8	2.7	17.7	6.9
Max	-3.5	-17.8	14.0	-2.5	-12.2	20.6	-0.8	-7.2	23.2	-0.9	-23.2	14.9
Min	-5.4	-37.2	-1.3	-11.5	-82.5	-5.9	-13.5	-94.0	-10.8	-14.7	-111.9	-16.0
N	6	6	6	35	35	35	62	62	62	64	64	64

	Jun-Jul-Aug (winter)											
Mean	-5.2	-32.6	8.8	-8.0	-55.4	8.4	-7.2	-48.6	8.9	-10.7	-81.6	4.3
SD	2.5	17.1	5.6	2.6	21.1	4.2	2.6	17.6	5.4	2.8	21.0	5.3
Max	1.0	2.8	19.6	-2.3	-10.4	16.0	2.5	8.8	20.2	-4.5	-46.0	16.1
Min	-10.0	-76.0	-5.3	-13.7	-99.0	-6.3	-14.5	-101.1	-11.2	-21.4	-162.7	-15.0
N	26	26	26	65	65	65	65	65	65	64	64	64

Although, the seasonal pattern at Puerto Montt and Santiago followed the same trend, the isotopic composition at Santiago was more depleted due to its higher elevation and larger orographic distillation from the Pacific coast (~ 100 km), whereas at Puerto Montt, the isotopic composition is controlled mainly by the temperate maritime climate conditions (Fig. 9.61; Table 9.16).

Table 9.16 Mean monthly precipitation P (mm) and temperature T (°C) characteristics were derived from the most recent monthly gridded Global Historical Climatology Network (GHCN) version 3.0 product (Peterson and Vose, 1997). Total annual precipitation and annual average temperature are also given.

Month	La Serena		Santiago		Puerto Montt		Punta Arenas	
	P (mm)	T (°C)	P (mm)	T (°C)	P (mm)	T (°C)	P (mm)	T (°C)
January	0.0	17.5	10.0	21.0	238.7	6.6	35.2	1.1
February	0.0	17.5	10.0	20.0	228.5	6.8	32.2	2.0
March	1.0	16.5	12.0	18.0	155.9	7.8	28.2	3.8
April	9.4	15.0	16.0	15.0	130.9	9.6	30.7	6.3
May	16.2	13.5	52.0	12.0	121.9	11.6	36.7	8.2
June	18.9	12.0	42.0	9.0	113.1	13.4	27.6	9.6
July	38.2	12.0	86.0	9.0	100.1	14.3	41.9	10.4
August	11.4	12.0	45.0	10.0	103.3	13.6	27.8	10.1
September	3.0	13.0	24.0	12.0	108.9	12.1	33.6	8.1
October	7.1	14.0	12.0	15.0	163.3	10.1	41.0	5.8
November	1.3	15.0	10.0	17.0	254.1	8.9	46.6	3.2
December	0.8	16.5	10.0	19.0	233.8	6.7	28.2	1.4
Annual	106.0	14.0	329.0	13.9	1952.0	10.3	410.0	5.8

Summary of Isotopic Effects and Wider Implications

Despite the latitudinal difference, the Santiago station exhibited more depleted isotope values than Puerto Montt, most likely due to the orographic distillation in a ~ 100 km transect from the coast to approximately 520 m a.s.l. Nevertheless, the location of the four stations along a unique latitudinal transect (30° S–53° S) resulted in clear isotopic trends from north to south (Fig. 9.62). Overall, long-term $\delta^{18}O$ (δ^2H) values ranged as follows: –4.9‰ (–29.6‰) (La Serena), –7.4‰ (–50.6‰) (Santiago), –5.5‰ (–36.6‰) (Puerto Montt) and –9.1‰ (–70.5‰) (Punta Arenas). Likewise, long-term *d*-excess and lc-excess exhibited a latitudinal decreasing trend: +9.8‰

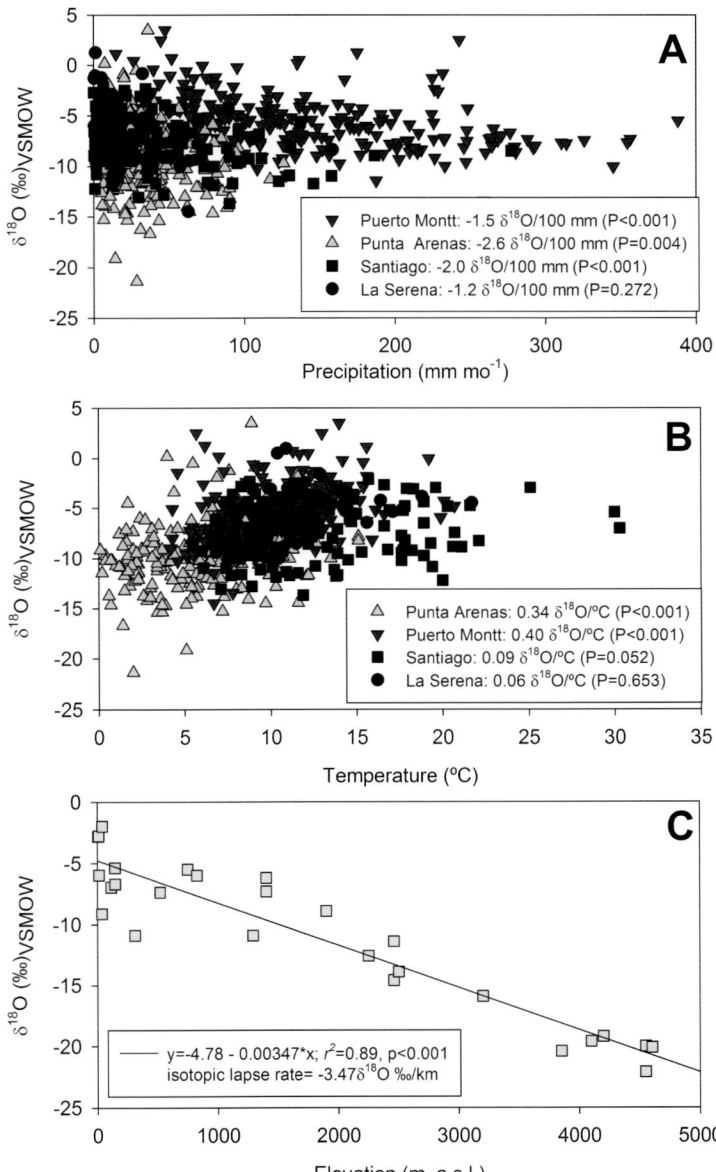

Fig. 9.62 (A) Monthly precipitation (mm) and $\delta^{18}O$ (‰) relationship for each station. (B) Air temperature (°C) and $\delta^{18}O$ (‰) relationship for each station. (C) Elevation (m a.s.l.) and $\delta^{18}O$ (‰) relationship across Chile.

(+1.7‰) (La Serena) +8.3‰ (–0.4‰) (Santiago), +7.7‰ (–0.5‰) (Puerto Montt) and +2.3‰ (–6.9‰) (Punta Arenas), reflecting the influence of non-equilibrium processes as mean annual temperature decreases and snow formation increases towards the southern regions (Table 9.16). Normally, the latitudinal effect is in the order of –0.6‰/degree and up to –2‰/degree in the colder Antartic continent (Mook, 2006). The latitudinal effect from La Serena (30° S, $\delta^{18}O$ mean = –4.9‰) and Punta Arenas (53° S, $\delta^{18}O$ mean = –9.1‰) was ~ 0.2‰/degree. The distinct monthly precipitation amounts converged in significant amount relationships at three locations: –2.6‰/100 mm (Punta Arenas; P = 0.004), –1.5‰/100 mm (Puerto Mont; P < 0.001) and –2.0‰/100 mm (Santiago) for $\delta^{18}O$. Similar amount effects have been reported in high elevation cordilleras (up to 3,820 m a.s.l.; Sánchez-Murillo et al., 2016b). No significant precipitation amount relationship was found at La Serena station (Fig. 9.62A). A significant temperature effect was observed only at two stations: Punta Arenas (+0.34‰/°C) and Puerto Montt (+0.40‰/°C) $\delta^{18}O$, respectively (Fig. 9.62B). These values are in agreement with global temperature effects of +0.4‰/°C for $\delta^{18}O$ (Mook, 2006). However, Fiorella et al. (2015) conducted a precipitation collection in the central Andes Cordillera during 2008–2013 (elevation range: 395 to 4,340 m a.s.l.) and reported an isotopic lapse rate of –1.9 ± 0.5‰/km. In addition, Poage and Chamberlain (2001) compiled 68 studies throughout many of the world's mountain belts and found an empirically consistent, linear relationship between change in elevation and change in the isotopic composition of precipitation along altitudinal transects. They concluded that there were no significant differences in isotopic lapse rates from most regions of the world (~ –2.8‰/km). Figure 9.62C shows the relationship of $\delta^{18}O$ versus elevation (only for sites where elevations were properly reported). The elevation range covers ~ 5,000 m a.s.l. Based on this significant linear regression, the orographic effect across the western slope of the southern Andes Cordillera can be described as –3.47‰/km for $\delta^{18}O$ in contrast to an average lapse rate of –1.45‰/km derived from Bowen and Revenaugh (2003).

The western slope of the Andes Cordillera is an exceptional case, because precipitation at higher elevation is controlled by continental air mass trajectories from the Atlantic Ocean, which travel through a wide range of biomes from the Amazon Basin to the Patagonia region, resulting in depleted precipitation. In contrast, in the coastal and central lowland regions, precipitation is mainly governed by the southeastern Pacific Ocean dynamics, resulting in enriched precipitation (Aravena et al., 1999). The combination of both processes is reflected in large apparent orographic effects, but also in remarkable spatial isotopic differences. Figure 9.63 shows a gridded mean annual $\delta^{18}O$ (‰) below ~ 20° S for South America according to Bowen and Revenaugh (2003). Although, this precipitation isoscape captures the isotopic difference within the western and eastern slopes of the Andes Cordillera, the relative magnitude of the isotopic composition appears to be strongly biased by the temperature effect, which is based solely on latitude and altitude estimations (e.g., $\delta^{18}O_p = a/Lat_x/^2 + b/Lat_x/ + cAlt_x$; Bowen and Revenaugh, 2003), without taking into consideration the strong influence of the southeastern Pacific anticyclone (high pressure area) and circumpolar low pressure area from

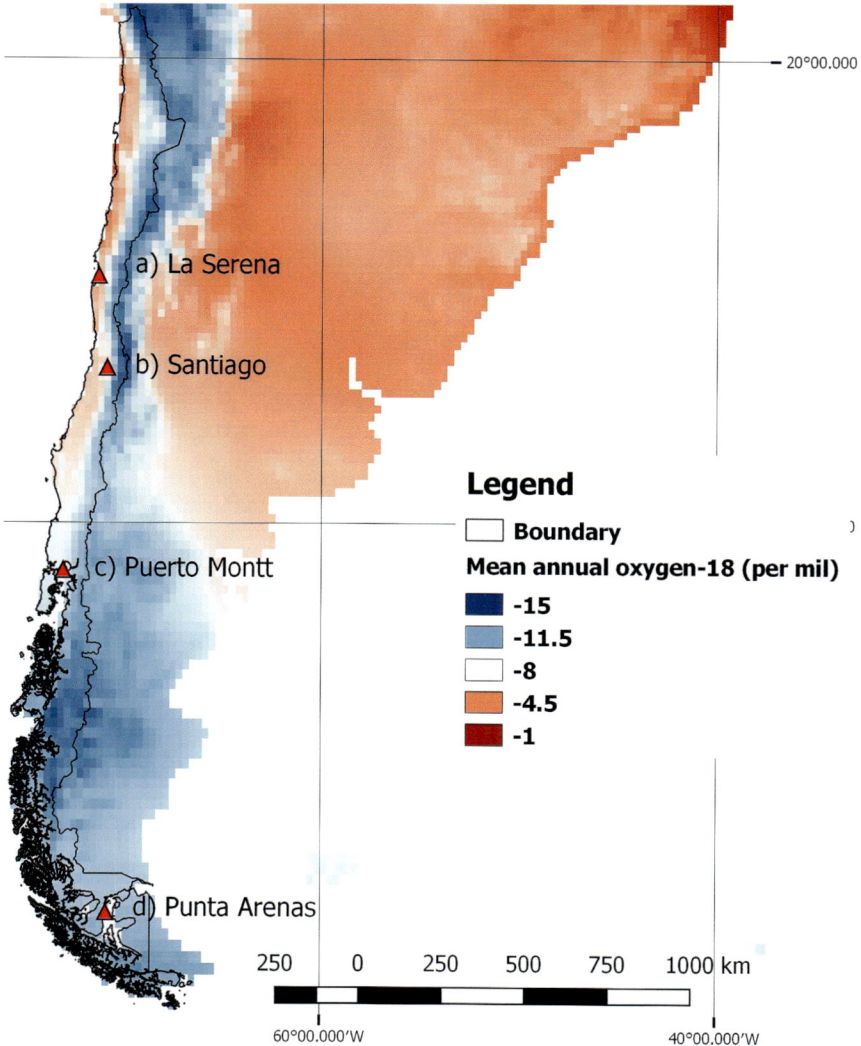

Fig. 9.63 Gridded mean annual δ18O (‰) below ~ 20°S for South America according to Bowen and Revenaugh (2003). Long-term stable isotope monitoring stations are defined by red triangles.

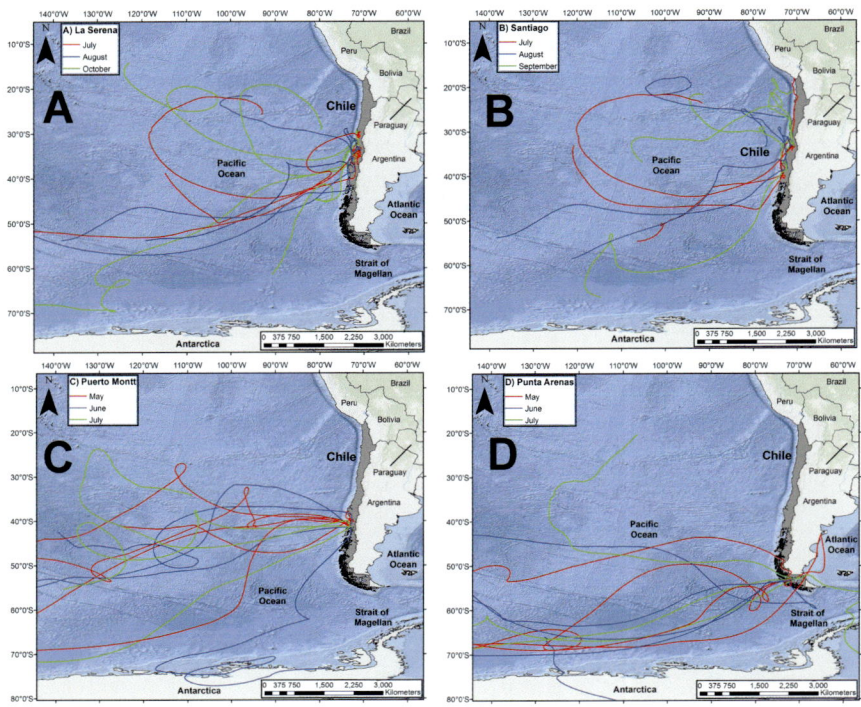

Fig. 9.64 Representative 10-day air mass back trajectories calculated using the HYSPLIT model (Stein et al., 2015) for the three wettest months in 2015 at La Serena (A), Santiago (B), Puerto Montt (C), and Punta Arenas (D) monitoring station. Months are color coded.

the Antarctic Sea. The latter is represented in the HYSPLIT air mass back trajectories (Fig. 9.64). During the wettest months, precipitation at La Serena and Santiago mainly originated from within the southeastern Pacific Ocean, whereas a strong Antarctic Sea influence was observed at Puerto Montt and Punta Arenas. Figure 9.65 shows representative isotopic data in different hydrological components. Although, historical isotopic studies in Chile have been highly concentrated in the northern region, the available data serve as a fundamental reference. As expected, ice coring data at high elevation sites presented the most depleted compositions but in a similar range of modern precipitation (Fig. 9.59). In the northern region, groundwater and surface water exhibited strong secondary evaporation processes; however, the isotopic spectrum (Fig. 9.65) also highlighted the relevance of spring recharge at high elevations in several locations in the western slope of the Andes Cordillera.

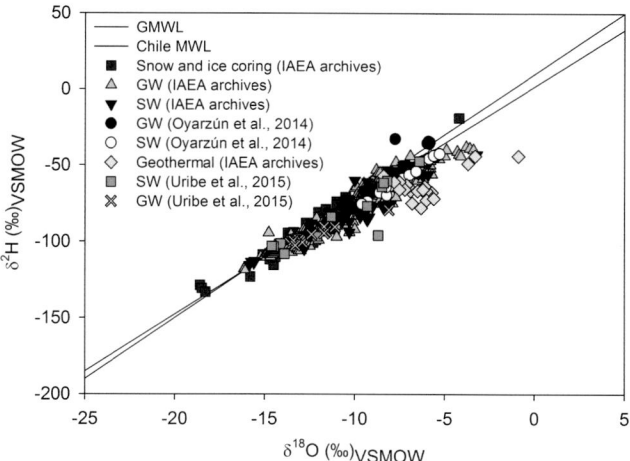

Fig. 9.65 Dual δ^2H–δ^{18}O (‰) diagram including: available isotopic data (i.e., surface water, groundwater, ice coring, and geothermal) from IAEA archives for Chile, Uribe et al. (2015), and Oyarzún et al. (2014). Samples are symbol and color coded.

Conclusions

Continuous (> 10-years) stable isotope records in precipitation along the Andes Cordillera are scarce. However, the recognition that modern precipitation dynamics may help to understand past climate conditions in paleo-archives has increased the number of monitoring efforts and related-hydrological studies. This study presents a long-term analysis (24-years) of monthly stable isotopes in precipitation across four stations in Chile (La Serena, Santiago, Puerto Montt and Punta Arenas). Overall, the 24-year continental meteoric water line of Chile is described as: δ^2H = 7.66 δ^{18}O + 3.42 (r^2 = 0.94) with a mean *d*-excess of +7.0‰. The second-order variables, *d*-excess (lc-excess), exhibited a strong latitudinal decreasing trend: +9.8‰ (+1.5‰) (La Serena), +8.3‰ (–0.4‰) (Santiago), +7.7‰ (–0.5‰) (Puerto Montt) and +2.3‰ (–6.9‰) (Punta Arenas), likely reflecting the influence of non-equilibrium processes as mean annual temperature decreases and snow formation is favored towards the southern regions. Long-term δ^{18}O ranged from enriched values within the northern semi-arid region of La Serena to a more depleted composition in the coastal Patagonia region of Punta Arenas, whereas intermediate δ^{18}O compositions were observed within the central and higher elevation region of Santiago and the southern coastal region of Puerto Montt. This isotopic pattern is well depicted in the surface water and groundwater domains.

Representative 10-day HYSPLIT air mass back trajectories revealed the strong influence of two major atmospheric transport mechanisms: the sub-tropical southeastern Pacific anticyclone (high pressure area) and the circumpolar low pressure area from the Antarctic Sea. The latitudinal spectrum among the monitoring network resulted in (a) a significant precipitation amount effect ($P < 0.01$) in three stations: $-2.6‰$ $\delta^{18}O/100$ mm (Punta Arenas), $-1.5‰$ $\delta^{18}O/100$ mm (Puerto Montt), and $-2.0‰$ $\delta^{18}O/100$ mm (Santiago), and (b) a temperature effect only significant ($P < 0.01$) at Punta Arenas ($+0.34‰$ $\delta^{18}O)/°C$) and Puerto Montt ($+0.40‰$ $\delta^{18}O/°C$) stations. The relationship of $\delta^{18}O$ versus elevation (only for sites where elevations were properly reported) resulted in an orographic effect across the western slope of the southern Andes Cordillera of $-3.47‰/km$ for $\delta^{18}O$. Finally, the absence of a continuous sampling transect from the coast to the high elevations within the Andes Cordillera of Chile invokes the need of further investigation of altitude and/or orographic effects under a changing climate. This investigation should be complemented with the implementation of surface water and groundwater isoscapes, which can significantly improve the spatial understanding of hydrological connectivity between high elevation recharge and lowland discharge within the Pacific slope of the Andes Cordillera.

Acknowledgements

The authors would like to thank the support from the Environmental Isotopes Laboratory of the Chilean Nuclear Energy Commission (GNIP-CCHEN) (http://www.cchen.cl/) and the cooperation of the Meteorological Directorate of Chile belonging to General Directorate of Civil Aviation of Chile (DGAC). RS would like to recognize the support from the IAEA grant CRP-19747 under the initiative "Stable isotopes in precipitation and paleoclimatic archives in tropical areas to improve regional hydrological and climatic impact models" and the Research Office of the National University of Costa Rica through grants SIA-0482-13, SIA-0378-14, and SIA-0101-14. Similarly, the authors would like to thank the Water Resources Programme at the IAEA and the Nuclear Energy Commission of Chile for providing access to isotopic archives of Chile. CB was supported by the University of Costa Rica Research Council under project 237-B4-39. Support from the Isotope Network for Tropical Ecosystem Studies (ISONet) funded by the University of Costa Rica Research Council is also acknowledged.

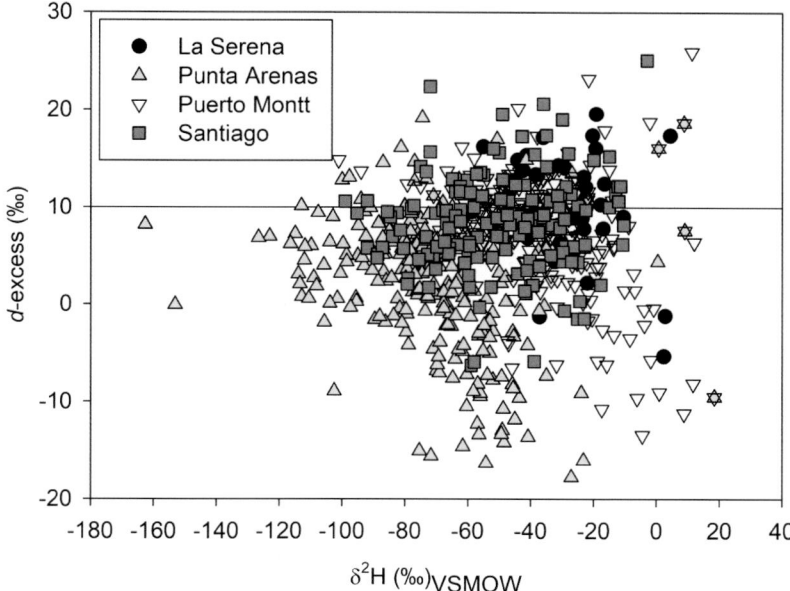

Fig. 9.66S δ²H (‰) and *d*-excess (‰) relationship for each long-term station. The horizontal black line denotes the *d*-excess value of +10(‰).

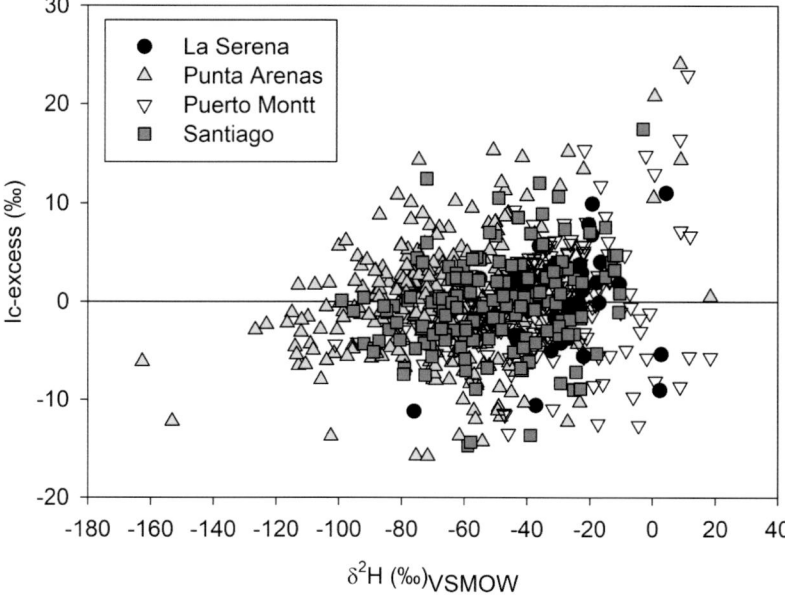

Fig. 9.67S δ²H (‰) and Ic-excess (‰) relationship for each long-term station. The horizontal black line denotes the Ic-excess value of 0 (‰).

References

Alpers, C.N. and Whittemore, D.O. 1990. Hydrogeochemistry and stable isotopes of ground and surface waters from two adjacent closed basins, Atacama Desert, northern Chile. Applied Geochemistry 5(5-6): 719–734. http://dx.doi.org/10.1016/0883-2927(90)90067-F.

Aravena, R. and Suzuki, O. 1990. Isotopic evolution of river water in the northern Chile region. Water Resources Research 26(12): 2887–2895. http://dx.doi.org/10.1016/0168-9622(89)90008-0.

Aravena, R. 1995. Isotope hydrology and geochemistry of northern Chile groundwaters. Bull. IFEA 24(3): 495–503.

Aravena, R., Suzuki, O., Pena, H., Pollastri, A., Fuenzalida, H. and Grilli, A. 1999. Isotopic composition and origin of the precipitation in Northern Chile. Applied Geochemistry 14(4): 411–422. http://dx.doi.org/10.1016/S0883-2927(98)00067-5.

Arumí Ribera, J.L. and Oyarzún Lucero, R.A. 2006. Groundwaters of Chile. Boletín Geológico y Minero 117(1): 37–45. Last accessed 2/1/2017. Available at: http://aprchile.cl/pdfs/LAs_aguas_subterraneas_de_chile.PDF.

Bershaw, J., Saylor, J.E., Garzione, C.N., Leier, A. and Sundell, K.E. 2016. Stable isotope variations ($\delta 18O$ and δD) in modern waters across the Andean Plateau. Geochimica et Cosmochimica Acta 194: 310–324. http://dx.doi.org/10.1016/j.gca.2016.08.011.

Bowen, G. and Revenaugh, J. 2003. Interpolating the isotopic composition of modern meteoric precipitation. Water Resour. Res. 39: 1299. doi:10.1029/2003WR002086.

Craig, H. 1961. Isotopic variations in meteoric waters. Science 133(3465): 1702–1703. Doi: 10.1126/science.133.3465.1702. http://www.jstor.org/stable/1708089.

Dansgaard, W. 1964. Stable isotopes in precipitation. Tellus 16: 436–468. Doi: 10.1111/j.2153-3490.1964.tb00181.x.

Favier, V., Falvey, M., Rabatel, A., Praderio, E. and López, D. 2009. Interpreting discrepancies between discharge and precipitation in high-altitude area of Chile's Norte Chico region (26–32° S). Water Resources Research 45: W02 424. Doi: 10.1029/2008WR006802.

Fernández-Hervé, P., Oyarzún, C., Brumbt, C., Huygens, D., Bodé, S., Verhoest, N.E.C. and Boeckx, P. 2016. Assessing the 'two water worlds' hypothesis and water sources for native and exotic evergreen species in south-central Chile. Hydrological Processes 30(23): 4227–4241. Doi: 10.1002/hyp.10984.

Fiorella, R.P., Poulsen, C.J., Zolá, R.S.P., Jeffery, M.L. and Ehlers, T.A. 2015a. Modern and long-term evaporation of central Andes surface waters suggests paleo archives underestimate Neogene elevations. Earth and Planetary Science Letters 432: 59–72. http://dx.doi.org/10.1016/j.epsl.2015.09.045.

Fiorella, R.P., Poulsen, C.J., Pillco Zolá, R.S., Barnes, J.B., Tabor, C.R. and Ehlers, T.A. 2015b. Spatiotemporal variability of modern precipitation $\delta 18O$ in the central Andes and implications for paleoclimate and paleoaltimetry estimates. J. Geophys. Res. Atmos. 120: 4630–4656. Doi: 10.1002/2014JD022893.

Fritz, P., Suzuki, O., Silva, C. and Salati, E. 1981. Isotope hydrology of groundwaters in the Pampa del Tamarugal, Chile. Journal of Hydrology 53(1-2): 161–184. http://dx.doi.org/10.1016/0022-1694(81)90043-3.

GNIP/IAEA/WMO. 2016. Global Network of Isotopes in Precipitation and Global Network of Isotopes in River. The GNIP and GNIR Databases. Accessible at: http://www.iaea.org/water. Last accessed: 21, December, 2016.

Hoffmann, G., Ramirez, E., Taupin, J.D., Francou, B., Ribstein, P., Delmas, R., Dürr, H., Gallaire, R., Simôes, J., Schotterer, U. and Stievenard, M. 2003. Coherent isotope history of Andean ice cores over the last century. Geophysical Research Letters 30(4). Doi: 10.1029/2002GL01487.

Hoke, G.D., Aranibar, J.N., Viale, M., Araneo, D.C. and Llano, C. 2013. Seasonal moisture sources and the isotopic composition of precipitation, rivers, and carbonates across the

Andes at 32.5–35.5° S. Geochemistry, Geophysics, Geosystems 14: 962–978. Doi: 10.1002/ggge.20045.

Lavergne, A., Daux, V., Villalba, R., Pierre, M., Stievenard, M. and Srur, A.M. 2017. Improvement of isotope-based climate reconstructions in Patagonia through a better understanding of climate influences on isotopic fractionation in tree rings. Earth and Planetary Science Letters 459: 372–380. http://dx.doi.org/10.1016/j.epsl.2016.11.045.

Leybourne, M.I. and Cameron, E.M. 2006. Composition of groundwaters associated with porphyry-Cu deposits, Atacama Desert, Chile: elemental and isotopic constraints on water sources and water–rock reactions. Geochimica et Cosmochimica Acta 70(7): 1616–1635. http://dx.doi.org/10.1016/j.gca.2005.12.003.

Magaritz, M., Aravena, R., Peña, H., Suzuki, O. and Grilli, A. 1989. Water chemistry and isotope study of streams and springs in northern Chile. Journal of Hydrology 108: 323–341. http://dx.doi.org/10.1016/0022-1694(89)90292-8.

Montecinos, A. and Aceituno, P. 2003. Seasonality of the ENSO-related rainfall variability in central Chile and associated circulation anomalies. Journal of Climate 16(2): 281–96. http://dx.doi.org/10.1175/1520-0442(2003)016<0281:SOTERR>2.0.CO;2.

Mook, W.G. 2006. Introduction to Isotope Hydrology. Taylor & Francis. London, UK. 226 pp.

Ohlanders, N., Rodriguez, M. and McPhee, J. 2013. Stable water isotope variation in a Central Andean watershed dominated by glacier and snowmelt. Hydrology and Earth System Sciences 17: 1035–1050. Doi: 10.5194/hess-17-1035-2013.

Oyarzún, R., Barrera, F., Salazar, P., Maturana, H., Oyarzún, J., Aguirre, E., Alvarez, P., Jourde, H. and Kretschmer, N. 2014. Multi-method assessment of connectivity between surface water and shallow groundwater: the case of Limarí River basin, north-central Chile. Hydrogeology Journal 22: 1857–1873. Doi: 10.1007/s10040-014-1170-9.

Oyarzún, R., Jofré, E., Morales, P., Maturana, H., Oyarzún, J., Kretschmer, N., Aguirre, E., Gallardo, P., Toro, L.E., Muñoz, J.F. and Aravena, R. 2015. A hydrogeochemistry and isotopic approach for the assessment of surface water–groundwater dynamics in an arid basin: the Limarí watershed, North-Central Chile. Environmental Earth Sciences 73(1): 39–55. Doi: 10.1007/s12665-014-3393-4.

Oyarzún, R., Zambra, S., Maturana, H., Oyarzún, J., Aguirre, E. and Kretschmer, N. 2016. Chemical and isotopic assessment of surface water–shallow groundwater interaction in the arid Grande river basin, North-Central Chile. Hydrological Sciences Journal 61(12): 2193–2204. Doi: 10.1080/02626667.2015.1093635.

Peel, M.C., Finlayson, B.L. and McMahon, T.A. 2007. Updated world map of the Köppen-Geiger climate classification. Hydrol. Earth Syst. Sci. 11: 1633–1644. Doi: 10.5194/hess-11-1633-2007, 2007.

Perry, L.B., Seimon, A. and Kelly, G.M. 2014. Precipitation delivery in the tropical high Andes of southern Peru: new findings and paleoclimatic implications. International Journal of Climatology 34(1): 197–215. Doi: 10.1002/joc.3679.

Peterson, T.C. and Vose, R.S. 1997. An overview of the global historical climatology network temperature database. Bull. Am. Meteorol. Soc. 78(12): 2837–2849.

Poage, M.A. and Chamberlain, C.P. 2001. Empirical relationships between elevation and the stable isotope composition of precipitation and surface waters: considerations for studies of paleoelevation change. American Journal of Science 301(1): 1–15. Doi: 10.2475/ajs.301.1.1.

Rozanski, K. and Araguás, L. 1995. Spatial and temporal variability of stable isotope composition over the South American continent. Bulletin de l'Institut français d'études andines 24(3): 379–390.

Sánchez-Murillo, R., Birkel, C., Welsh, K., Esquivel-Hernández, G., Corrales-Salazar, J., Boll, J., Brooks, E.S., Roupsard, O., Sáenz-Rosales, O., Katchan, I. and Arce-Mesén, R. 2016a. Key drivers controlling stable isotope variations in daily precipitation of Costa Rica:

Caribbean Sea versus Eastern Pacific Ocean moisture sources. Quaternary Sci. Rev. 131(Part B): 250–261. Doi: 10.1016/j.quascirev.2015.08.028.

Sánchez-Murillo, R., Esquivel-Hernández, G., Sáenz-Rosales, O., Piedra-Marín, G., Fonseca-Sánchez, A., Madrigal-Solís, H., Ulloa-Chaverri, F., Rojas-Jiménez, L.D. and Vargas-Víquez, J.A. 2016b. Isotopic composition in precipitation and groundwater in the northern mountainous region of the Central Valley of Costa Rica. Isotopes in Environmental and Health Studies. Doi: 10.1080/10256016.2016.1193503.

Schauwecker, S. 2011. Near-surface temperature lapse rates in a mountainous catchment in the Chilean Andes (Doctoral dissertation, Swiss Federal Institute of Technology Zurich).

Smith, R.B. and Evans, J.P. 2007. Orographic precipitation and water vapor fractionation over the Southern Andes. Journal of Hydrometeorology 8: 3–19. Doi: 10.1175/JHM555.1.

Sprenger, M., Tetzlaff, D., Tunaley, C., Dick, J. and Soulsby, C. 2017. Evaporation fractionation in a peatland drainage network affects stream water isotope composition. Water Resources Research 53. Doi: 10.1002/2016WR019258.

Stein, A.F., Draxler, R.R., Rolph, G.D., Stunder, B.J.B., Cohen, M.D. and Ngan, F. 2015. NOAA's HYSPLIT atmospheric transport and dispersion modeling system. Bulletin of the American Meteorological Society 96: 2059–2077. Doi: 10.1175/BAMS-D-14-00110.1.

Su, L., Yuan, Z., Fung, J.C.H. and Lau, A.K.H. 2015. A comparison of HYSPLIT backward trajectories generated from two GDAS datasets. Science of the Total Environment 506-507: 527–537. Doi: 10.1016/j.scitotenv.2014.11.072.

Suzuki, O. and Aravena, R. 1990. Isotopic evolution of river water in the northern Chile region. Water Resources Research 26(12): 2887–2895.

Uribe, J., Muñoz, J.F., Gironás, J., Oyarzún, R., Aguirre, E. and Aravena, R. 2015. Assessing groundwater recharge in an Andean closed basin using isotopic characterization and a rainfall-runoff model: Salar del Huasco basin, Chile. Hydrogeology Journal 23(7): 1535–1551. Doi: 10.1007/s10040-015-1300-z.

Verbist, K., Robertson, A.W., Cornelis, W.M. and Gabriels, D. 2010. Seasonal predictability of daily rainfall characteristics in Central Northern Chile for dry-land management. Journal of Applied Meteorology and Climatology 49: 1938–1955. Doi: 10.1175/2010JAMC2372.1, 35 2010.

Vimeux, F., Ginot, P., Schwikowski, M., Vuille, M., Hoffmann, G., Thompson, L.G. and Schotterer, U. 2009. Climate variability during the last 1000 years inferred from Andean ice cores: A review of methodology and recent results. Palaeogeography, Palaeoclimatology, Palaeoecology 281(3): 229–241. http://dx.doi.org/10.1016/j.palaeo.2008.03.054.

Vuille, M., Bradley, R.S., Healy, R., Werner, M., Hardy, D.R., Thompson, L.G. and Keimig, F. 2003. Modeling δ18O in precipitation over the tropical Americas: 2. Simulation of the stable isotope signal in Andean ice cores. J. Geophys. Res. 108: 4175. Doi: 10.1029/2001JD002039, D6.

Wolfe, B.B., Aravena, R., Abbott, M.B., Seltzer, G.O. and Gibson, J.J. 2001. Reconstruction of paleohydrology and paleohumidity from oxygen isotope records in the Bolivian Andes. Palaeogeography, Palaeoclimatology, Palaeoecology 176(1): 177–192. http://dx.doi.org/10.1016/S0031-0182(01)00337-6.

CHAPTER 10

Hydrological Modeling to Assess Runoff in a Semi-arid Andean Headwater Catchment for Water Management in Central Chile

Penedo-Julien, S.,[1,*] *Nauditt, A.,*[1] *Künne, A.,*[2]
Souvignet, M.,[3] and *Krause, P.*[4]

Introduction

Population growth and economic development have led to a degradation of water resources worldwide. Water stress is likely to be exacerbated by climate change by increasing the number and intensity of climate-driven hazards, such as droughts and floods. In this context, arid and semi-arid mountainous areas are especially vulnerable to changes in climate due to their high fragile dependency on precipitation and temperature patterns. Furthermore, these regions often host basins' headwaters playing a key role as water providers for downstream communities and ecosystems (Barnett et al., 2005; Price and Egan, 2014).

Hence adaptation strategies to climate change are particularly relevant for these areas. However, policies are usually designed at the national level

---

[1] Institute for Technology and Resources Management in the Tropics and Subtropics, Technical University Cologne.
[2] Department of Geoinformatic, Hydrology and Modelling, Friedrich Schiller University of Jena.
[3] United Nations University (UNU), Institute for Environment and Human Security.
[4] Thüringer Landesanstalt für Umwelt und Geologie (TLUG).
* Corresponding author: santiago.penedo@th-koeln.de

and there is a lack of implementation at the watershed level. Therefore, climate change impact assessments at the river basin scale are needed to support planning and decision-making.

Hydrological models have been widely used to assess the impacts of climate change on water availability at the watershed level (Jiang et al., 2007; Vicuña et al., 2011; Zhu and Ringler, 2012; Vargas et al., 2013; Alam et al., 2015; Mourato et al., 2015; Versini et al., 2016). For instance, Bai et al. (2009) have described models as useful tools for testing hypotheses about watershed behavior and Ji (2008) for scenario analysis to support water management in the context of climate change. Hydrological modeling can provide a better understanding of the water cycle and give insights to hydrologic processes in remote mountainous catchments.

However, arid mountainous and poorly gauged regions still pose a major challenge due to their high spatial and temporal climatic variability as well as the lack of good quality and statistically representative *in situ* monitored data. Furthermore, models have to deal with large uncertainties related to unknown storage conditions and hydraulic properties of the associated geological layers. For this reason, hydrological model research in such areas is particularly relevant. Once models have proved their predictive ability they can be used to properly design and assess ad hoc sustainable management plans, considering climate change impacts.

Rainfall-Runoff Modeling in Semi-Arid and Snow Melt Driven Mountainous Catchments

Robust rainfall-runoff simulations provide valuable information for long term and seasonal water availability projections as well as for the management of hydrological extremes such as floods and droughts. In arid and semi-arid environment conditions for hydrological modeling are unfavorable. This can be mainly attributed to the lack of representative rainfall values to force rainfall-runoff modeling as well as general data scarcity with mostly unknown contributions from near-surface, sub-surface and groundwater flow paths (Pilgrim et al., 1988; Lidén and Harlin, 2000; Van Loon and Van Lanen, 2012). Due to their limited accessibility, monitored data are even sparser in high elevation cryospheric headwater catchments increasing uncertainty (Zambrano-Bigiarini et al., 2016).

This might explain why most of the available modeling work based on data from snowmelt driven mountainous catchments has been carried out at colder high latitude regions, where precipitation rates are higher than 500 mm per year and considerable snowfall accumulation occurs during the winter months as for, e.g., the Northern Colorado River Basin, the Rocky Mountains and the Alps (Pomeroy et al., 2004; Lehning et al.,

2006; Adam et al., 2009). Fewer modeling exercises have been carried out in mountainous areas draining to subtropical semi-arid or arid regions at low and middle latitudes with the purpose to assess potential climate change impacts on discharge—mostly incorporating snow cover data satellite imagery such as MODIS (Immerzeel, 2010; Chauvin et al., 2011; Bocchiola et al., 2011; Duethmann et al., 2014). In Chile, hydrological modeling has been applied to few glaciered catchments at latitudes further south from our study region for which more comprehensive data sets and observations are available. For instance, modeling case studies were conducted using the Water Evaluation and Planning Tool (WEAP), the physically-based and semi-distributed TOPKAPI-ETH as well as the SWAT model (Pellicciotti et al., 2008; Stehr et al., 2009; Ragettli et al., 2014; Omani et al., 2016).

Rainfall-Runoff Modeling in the Andean Headwaters of the Limarí River Basin

For the headwater catchments of the Limarí Basin the WEAP model has already been applied on a monthly scale. The model was used to assess the impacts of climate projections—obtained with statistical downscaling—on streamflow. The results showed that increasing temperatures and changing precipitation patterns are likely to significantly reduce long-term runoff due to accelerated snow and glacier melting and hence lead to lower water availability, especially in the summer time (Vicuña et al., 2010).

Empirical observations show that some physical aspects of high elevation environments, such as limited vegetation cover and sparse soil formation have a rather limited influence on the hydrological system, and hence they do not require complex model representation. Therefore, recently the conceptual rainfall-runoff model HBV light was used to test the hydrological processes dominated by snow melt and groundwater movements in the Rio Grande and Tascadero headwaters. By calibrating the model across different periods of extreme climatic conditions, impacts on model performance, parameter sensitivity and identifiability were investigated. The results suggest that, independently of a dry or wet period of calibration, the streamflow response is mostly consistent with flux from groundwater storage, while only a small fraction comes from direct routing of snowmelt. The variation of model parameters, such as the groundwater rate coefficient, was found to be consistent with differing recharge in wet and dry years. The resulting snowmelt–groundwater model represents a useful tool for predictions of seasonal water availability and a basis for further field studies data scarce and semi-arid Andean catchments (Nauditt et al., 2016).

However, to represent rainfall-runoff relationships for discharge predictions on a daily scale more reliably, a better representation of precipitation inputs and catchment behavior are needed. The small number of precipitation events recorded at the only station in the catchment outlet cannot represent areal precipitation of Andean catchments with a steep elevation gradient. Additional inputs at higher elevations might therefore be provided by satellite based precipitation products (Zambrano-Bigiarini et al., 2016). Also the missing information about *in situ* catchment specific characteristics such as snow, land use and soils might be provided by satellite products. Although there are many applications of lumped hydrological models with the incorporation of landscape classification (Savenjie, 2010) or tracer analyses results (Soulsby et al., 2003; Hrachowitz et al., 2010) for higher moisture rich latitudes, there is a demand to test if these approaches are appropriate for data sparse semi-arid snowmelt driven catchments.

In this study we tested the performance of two models incorporating spatially distributed catchment information in the same headwater catchment of the Limarí: the Upper Hurtado Basin (UHB).

Rainfall-Runoff Modeling in the Upper Hurtado Basin (UHB)

We used two hydrological models (SWAT and J2000) with different levels of complexity and spatial distribution approaches to simulate discharge for the Hurtado headwater catchment to compare their performance considering several criteria. As aforementioned, SWAT has only been applied to more moisture rich Andean catchments (Stehr et al., 2009; Omani et al., 2016) while the J2000 model has not yet been used in this region. The UHB in semi-arid central Chile was selected for this study since it combines four interesting challenges for hydrological modeling in particular and for water resources management in general: (i) snow-driven mountainous river basin, (ii) semi-arid climate, (iii) high spatial heterogeneity, and (iv) data-scarcity.

Study Area

The UHB is located within the Coquimbo region and has a drainage area of 670 km² (see Fig. 10.66). The elevation varies from 2,000 to 5,500 m a.s.l. with a mean elevation of 3,724. According to the Köppen-Geiger classification the UHB belongs to the Bsk subgroup which is characterized by a long dry season (7–10 months) and that precipitation is less than potential evapotranspiration (Souvignet, 2010). The regional climate is driven by three main factors: (i) the southeast Pacific anticyclone, (ii) the cold Humboldt Current along the Pacific coast, (ii) the mountain range of the Andes (Oyarzún et al., 2003). Precipitation and temperature patterns follow an E-W axis driven by changes in altitude. Mean annual precipitation, including snowfall, for the period 1979–2006 measured at Hurtado was 147 mm concentrating in the winter months of June and July. The dry season

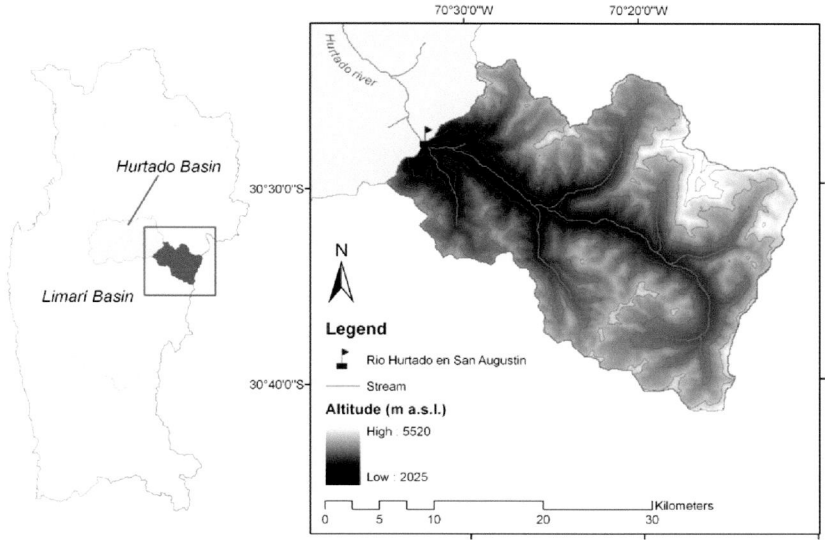

Fig. 10.66 Location of the Upper Hurtado Basin (UHB).

goes from September until April. Precipitation is strongly influenced by orography, El Niño Southern Oscillation (ENSO) and the Pacific Decadal Oscillation (PDO) contributing to a high inter- and intra-annual variability (Souvignet, 2010; Núñez et al., 2013). For instance, ENSO events can multiply annual precipitation by a factor of 2–3 (Oyarzún et al., 2003). Moreover, local basin features such as valley orientation, hill slope exposure, abrupt elevation changes result in relevant, small-scale variability in precipitation and snow accumulation spatial patterns (Favier et al., 2009). Mean annual temperature, also measured at Hurtado for the same period, was 17°C reaching its minimum in July (13°C), coinciding with the precipitation maximum, while the maximum temperature was reached in February (21°C).

The Hurtado River has an average flow (1979–2006) of 3.28 $m^3 \cdot s^{-1}$. It is a snow-driven basin concentrating precipitation in winter but showing its highest discharge values in spring and early summer due to snow melt. A recent study conducted in north-central Chile found a streamflow regime shift related to a change in the PDO (Núñez et al., 2013) which needs to be considered for future hydrological analysis. The UHB hosts the headwaters of the Limarí Basin even though its glacier coverage is relatively low (7%, Souvignet, 2010). The basin contributes to the downstream agricultural (e.g., the Recoleta irrigation system) and municipal water demand making it relevant for the socio-economic development of the region.

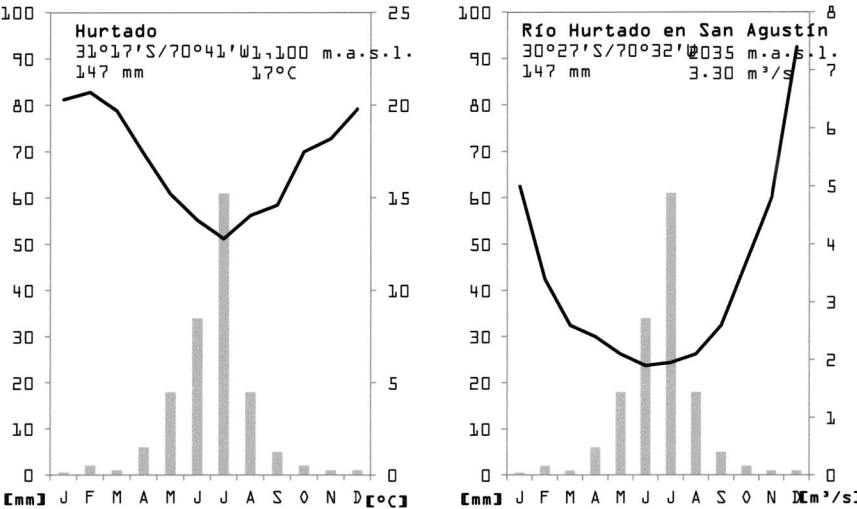

Fig. 10.67 Temperature diagram (1979-2006) at Hurtado (left) and hydrological station at San Agustín (right) with monthly precipitation as reference. Data: Dirección General de Aguas (DGA, Chilean Water Authority)

Data and Methods

Selecting the appropriate model is a key step in order to accurately represent the most important features of the watershed and be able to predict variables such as streamflow, evapotranspiration, snowmelt, among others (Bai et al., 2009). Cunderlik (2003) suggests the complexity as a key factor to be taken into account since simpler models require less physical parameters and input data as well as setup efforts and computational resources. By increasing complexity more parameters are needed. Hence, the potential for equifinality—i.e., the similar model predictions can be obtained with different sets of parameters (Beven and Freer, 2001).

One of the main challenges while modeling a cold mountainous watershed is a good estimation of snow accumulation and snowmelt. For instance, in the UHB, snowmelt is the main driver of streamflow and its good representation is determinant to obtain satisfactory modeling results. In order to accurately represent hydrological-related snow processes and reduce uncertainty, good quality monitored data is required. This is especially difficult in basins with heterogeneous orography and lack of financial resources to install long-term monitoring stations, especially at high altitudes with limited accessibility.

The best approach to model snow accumulation and snowmelt should include an energy balance but the variables needed for this (e.g., solar radiation, surface albedo, snow density, wind speed, atmospheric vapor pressure) are usually not available (Zeinivand and De Smedt, 2009). Therefore, models generally approximate this problem by using air temperature-based calculations to simulate snow accumulation and snowmelt (Zeinivand and De Smedt, 2009).

For this study, two hydrological models with different levels of complexity, snowmelt computation and distribution approach were implemented in the UHB in order to compare their performance in such a challenging environment. Streamflow was selected as hydrological variable to compare both models due to its utmost relevance for the agricultural production downstream as well as for the environment.

The Soil and Water Assessment Tool (SWAT) is a basin-scale, continuous time model, which operates on a daily time step and is designed to predict the impact of management on water, sediment, and agriculture (Gassman et al., 2007). SWAT is a physically based, spatially semi-distributed and computationally efficient model which can simulate long periods of time (Arnold et al., 1998). SWAT components can be categorized into two main classes: (i) the land phase of the hydrological cycle and (ii) the routing of runoff through the stream network (Stehr et al., 2008). The land phase considers processes such as surface runoff, snow accumulation and melt, evapotranspiration, percolation, groundwater while the routing phase includes methods to determine how water is channelized through the river network (see Table 10.17). An extended description of the model and its components can be found in Neitsch et al. (2005).

Table 10.17 Hydrological processes and the calculation methods used for this study (SWAT and J2000).

Process	SWAT	J2000
Evapotranspiration	Hargreaves and Samani (1985)	
Snow melt	Temperature index (Fontaine et al., 2002)	Accumulation, metamorphosis, and snow melt (Krause, 2001)
Infiltration	Based on the Curve number (CN) after SCS (1972)	Maximum infiltration capacity (three different cases: summer, winter, snow)
Surface runoff	Curve number (CN) after SCS (Soil Conservation Service, 1972)	Linear storage
Interception	Based on LAI	Based on LAI (Dickinson, 1984)
Interflow and percolation	Kinetic storage model	Maximum percolation rate Interflow: Linear storage
Routing	Muskingum method (Cunge, 1969)	Kinematic wave normal mode
Regionalization of point measurements (e.g., temp, precip)	Elevation bands	Elevation correction

J2000 is a process-based fully distributed hydrological model developed by Krause (2001) to understand and describe the hydrological processes taking place at the basin level. J2000 was designed for dynamic simulations of water transport in meso- to macro-scale river basins (Krause, 2001). It was implemented using the Jena Adaptable Modeling System (JAMS) framework (Kralisch et al., 2007). J2000 has different modules representing hydrological processes such as evapotranspiration, snow accumulation and melt, infiltration, interception, groundwater and reach routing (see Table 10.17). J2000 generates four different runoff components: (i) RD1 which is the fast direct runoff flowing over the surface, (ii) RD2 which is the 'fast' subsurface runoff (also known as interflow (1) which can be regarded as a lateral subsurface flow (Nepal, 2012), (iii) RG1 which is the 'slow' subsurface runoff component (also known as interflow (2) originating in the weathering layer of the lithological zone, (iv) RG2 which is the 'slow' baseflow component originating in the consolidated bedrock. The complete description of the model and its components is available in Krause (2001).

Both models apply the Hydrological Response Units (HRUs) concept which was first proposed by Leavesley et al. (1983) trying to decompose the basin's heterogeneity into modeling entities with relatively homogenous topographic characteristics. However, each hydrologic model uses a specific approach to delimit and delineate the HRUs. SWAT treats the HRUs as lumped land areas within the sub-basins and are comprised of unique land cover and soil (Neitsch et al., 2005). J2000 defines the HRUs based on Flügel (1996) as land use and pedogeological homogenous units controlling water dynamics within the basin. The HRU delineation for the J2000 was achieved by using topographic values such as slope, aspect and the topographic wetness index (TWI) which were derived from the DEM. In addition, geology, soil and land use classes were overlaid to derive the final HRUs via cluster analysis using a tool developed by Pfennig and Wolf (2007).

Daily records for temperature and precipitation for the period 1974–2006 were used from surrounding stations for different locations and altitudes (Table 10.18). Furthermore, a discharge station with daily values for the period 1963–2006 was used to calibrate and validate both models.

The topography was derived from the Digital Elevation Model (DEM) which was provided by the Centre for Advanced Studies in Arid Zones (CEAZA) with a cell size of 90 × 90 m and a vertical accuracy varying from 6 to 10 m. CEAZA processed the DEM from the Shuttle Radar Topography Mission (SRTM).

The soil map was obtained from the Soil and Terrain Database for South America which is a subset of the FAO/UNESCO Soil Map of the World (FAO, 1995). The derived properties include parameters such as pH, organic carbon content, C/N ratio, clay mineralogy, depth, moisture capacity, drainage class. Some of these properties are important to parameterize the

Table 10.18 Location of meteorological and one discharge stations (T = temperature, P = precipitation).

Station	Lat (S)	Long (W)	Elevation (m.a.s.l.)	Parameter	Period	Resolution
Caren	30°51	70°46	740	T	1967–2006	Daily
Hurtado	30°17	70°41	1,100	T, P	1974–2006	Daily
La Laguna Emb.	30°12	70°02	3,160	T, P	1974–2006	Daily
La Paloma Emb.	30°41	71°02	320	Sunshine, Wind	1974–2003	Monthly
Las Ramadas	31°01	70°34	1,380	T, P	1974–2006	Daily
Pabellón	30°24	70°34	1,920	P	1968–2006	Daily
Pichasca	30°23	70°52	725	P	1946–2006	Daily
Rapel	30°43	70°46	870	P	1969–2006	Daily
San Agustín	30°27	70°32	2,035	Discharge	1963–2006	Daily

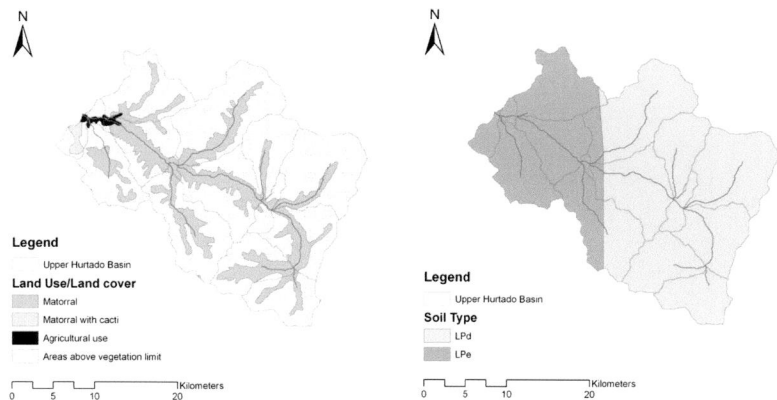

Fig. 10.68 Land use and soil maps for the UHB.

two models concerning soil properties governing hydrological processes (e.g., infiltration). Due to the low resolution of the soil map only two soil classes could be delimited for the UHB (see Fig. 10.68). The two major soil classes in the region area are eutric and dystric Leptosols (LPe and LPd, respectively).

The land cover map was composed after data originally sampled by the Institute for Agricultural Research (INIA, 2008). The dominant land cover in the basin is classified as areas above vegetation limit (i.e., no or almost no vegetation) with 73% of the total surface. Matorral or shrubland (25%) follows as land cover mainly located in the vicinity of the main rivers. Matorral with cacti and agriculture, practised in the lower parts of

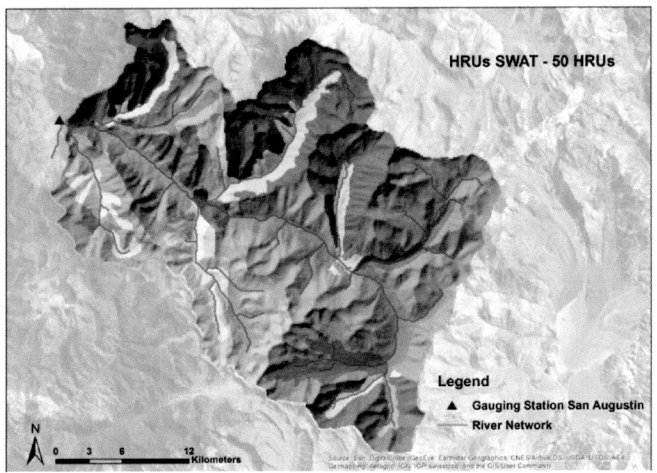

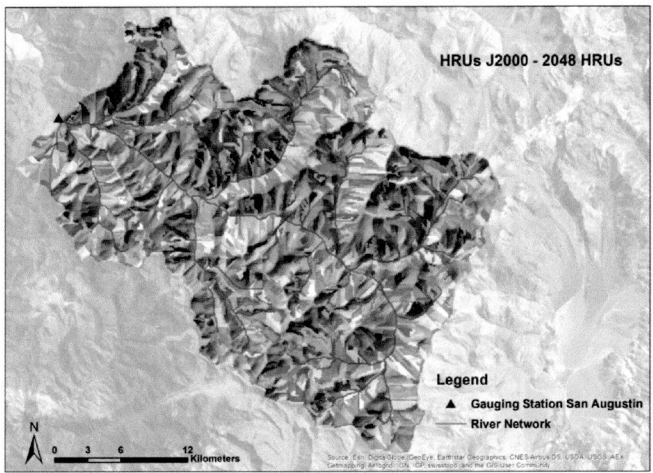

Fig. 10.69 HRUs used to model with SWAT (top) and J2000 (bottom).

the basin (mainly orchards and vineyards), cover around 6% of the basin, respectively. The different GIS layers (topography, soil type, land cover) were used to delineate the HRUs for modelling purposes: 50 HRUs for SWAT, 2,048 HRUs for J2000 (see Fig. 10.69).

Calibration and validation of hydrological models is necessary to adjust the large number of parameters (23 for SWAT and 48 total parameters for J2000) and then to evaluate the predicted values with a selected parameter set. For this study, the split-sample calibration-validation approach suggested by Klemeš (1986) was applied which is widely used in hydrological simulations. The period 1984–1990 was considered to adjust the different parameters (calibration) while 1991–2003 to test and assess the response of the models with the previously defined set of parameters (validation). According to Gupta et al. (2006) there are three calibration methods available: (i) manual trial-and-error, (ii) automatic or numerical parameter optimization, and (iii) combination of both. The third option was used for both models doing first a manual trial-and-error followed by a numerical optimization. The automatic calibration was done minimizing the sum of squares of the Nash-Sutcliffe (NS) goodness of fit (which is explained below). The model simulations for both models were performed daily.

The comparison of observed and simulated values helps to evaluate the performance of the model during the calibration and validation periods. There are several indicators available to evaluate model performance. In order to compare the results of both models—SWAT and J2000—four efficiency criteria were selected based on published hydrological studies (Krause et al., 2005; Legates and McCabe, 1999; Yapo et al., 1996): r^2—to measure dispersion, Nash-Sutcliffe (NS)—to assess peaks, logarithmic Nash-Sutcliffe (LNS)—to assess low flows—and the percentage bias (PBIAS)—to assess over- or underestimation.

Results and Discussion

A sensitivity analysis was conducted for both models to reduce the number of parameters to be calibrated. The most sensitive parameters (p value < 0.05) were adjusted to fit the simulated to the observed streamflow. For SWAT, the sensitive parameters included baseflow alfa factor, initial SCS runoff curve number, melt factors of snow, slope length for lateral subflow, soil evaporation compensation factor, surface runoff lag coefficient, groundwater delay and snowmelt base temperature. As it can be observed, these parameters cover a broad scope of processes such as snowmelt, surface runoff and groundwater movement as well as evapotranspiration. Furthermore, the most sensitive parameters for J2000 included base temperature, temperature, rain and soil heat factors for snowmelt calculation as well as reduction coefficient for evapotranspiration calculation and RG1 and RG2 outflow adaptation factors. These parameters also cover a wide scope of hydrological processes such as snowmelt, evapotranspiration (important for the water balance) and groundwater flows.

The graphical results of both simulations can be seen in Fig. 10.70 for the calibration period. Both models can fairly reproduce the monthly hydrograph. Van Liew et al. (2005) suggest a scale to interpret NS values: (i) higher than 0.75 as 'good', (ii) between 0.75 and 0.36 as 'satisfactory', (iii) and below 0.36 as 'not satisfactory'. Based on this scale SWAT (NS = 0.75) has 'satisfactory' performance while J2000 (NS = 0.89) performs as 'good' (Table 10.19).

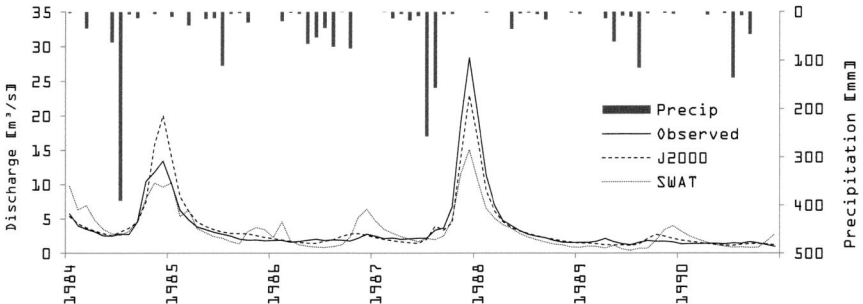

Fig. 10.70 Results of the calibration period (1984–1990) for both models.

Table 10.19 Efficiency criteria for the calibration period (1984–1990).

Model	r²	NS	LNS	PBIAS
J2000	0.89	0.89	0.90	0.53
SWAT	0.75	0.70	0.45	−9.85

However, none of the models was able to properly simulate extreme peak flows, especially the one from 1988 which is largely underpredicted (SWAT only predicts half of the peak while J2000 only 0.8). This is very likely a consequence of an underprediction of total precipitation in the basin due to the low number of precipitation stations, especially in high altitudes. Furthermore, SWAT can better reach moderate peak flows (e.g., 1985) than J2000 which overpredicts it. The most significant difference in performance can be observed in the simulation of the baseflow during low flow periods. SWAT systematically under- and overpredicts the baseflow (LNS = 0.45) while J2000 is able to represent the dynamic and magnitude (LNS = 0.85). This might be a consequence of a better representation of groundwater and storage processes. Finally, PBIAS shows how the predicted values were larger or smaller than their observed counterparts (Yapo et al., 1996). J2000 overpredicts only 0.53% of the values while SWAT's underpredicts them by almost 10% (Table 10.19).

Figure 10.71 shows the performance of both models during the validation period (1991–2003). Based on the efficiency criteria, excluding LNS (Table 10.20), both models present a comparable performance. SWAT has a slightly better r² and NS. A graphical analysis, however, allows detecting that SWAT overpredicts moderate peak flows while underpredicting extreme ones while J2000 simulated fairly the moderate peak flows but it also underpredicts extreme ones.

There was a long dry period from 1994 until 1998 where discharge was mainly driven by baseflow. Again, as for the calibration period, SWAT was not able to reproduce the hydrological dynamics during dry years (LNS =

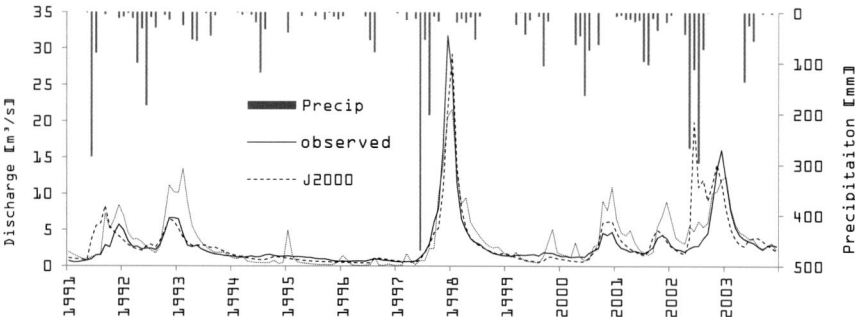

Fig. 10.71 Results of the validation period (1991–2003) for both models.

Table 10.20 Efficiency criteria for the validation period (1991–2003).

Model	r^2	NS	LNS	MAE	PBIAS
J2000	0.70	0.67	0.68	1.03	6.20
SWAT	0.74	0.70	0.01	1.37	18.73

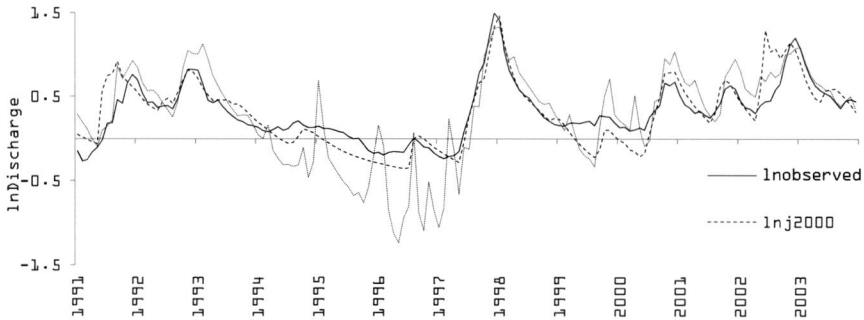

Fig. 10.72 Results of the logarithmic values for the validation period (1991–2003).

0.01, 'not satisfactory') while J2000 did it 'satisfactory' (LNS = 0.68). During the validation period the difference in low flow simulation is even clearer. Furthermore, J2000 overpredicts in 6.19% the observed values while SWAT does it in more than double (18.73%).

Figure 10.72 shows the logarithmic values for the measured and predicted values to reduce the sensitivity to extreme peak flows and better visualize the performance of both models during low flow periods (Krause et al., 2005). It can clearly be seen that J2000 reproduces fairly the hydrological pattern while SWAT does it poorly with a systematic under- and overprediction. This fundamental difference might be a consequence of the two sub-surface runoff components (RG1 and RG2) modeled by J2000 with different velocities. This approach appears to better simulate the timing and amount of water contributing to the baseflow during the dry period in the basin. For the SWAT, on the other hand, the automated baseflow separation and recession analysis technique developed by Arnold and Allen (1999) was considered. The results, however, present great difficulties of baseflow simulation during dry periods.

In general, if only the NS was considered, both models were able to simulate the monthly streamflow in the UHB. However, there are significant differences if the hydrograph is looked at closer. For instance, as mentioned before none of the models was able to accurately represent streamflow peaks. This suggests that both models are melting snow too slowly during extreme events and surface runoff is being underpredicted. Furthermore, this might be an indication of poor aerial representation of precipitation distribution and, therefore, not enough snow is accumulated during precipitation events. Also underprediction of regionalized precipitation due to lack of stations at high altitudes generates high uncertainty of the input data. Finally, a source of uncertainty is introduced by deriving streamflow data from water level observations through a rating curve. Moreover, discharge measurements are often not possible during high flows, introducing extrapolation errors for peak flows in the hydrograph (Domeneghetti et al., 2012). On the other hand, J2000 performs much better than SWAT during low flow periods suggesting a better representation of groundwater processes. This, however, requires further research on groundwater pathways and their validation with hydrological models.

Another important difference between the two models is the distribution approach. SWAT used only 195 HRUs being unable to properly represent the orographic heterogeneity of the basin. J2000 increased by a factor of 40 the number of HRUs and allowed a better representation of water distribution and precipitation distribution.

The Role of Areal Precipitation in Modeling Data Scarce Snowmelt Driven Catchments

Rainfall-runoff modeling in arid to semi-arid catchments is a major challenge mainly attributable to a lack of observed precipitation as well as longer periods of no rainfall and high climatic variability (Pilgrim et al., 1988; Lidén et al., 2011; Van Loon and Van Lanen, 2012). For semi-arid data scarce Andean catchments, model calibration is extremely dependent on how precipitation inputs at higher elevations are calculated. In the semi distributed conceptual HBV-light model (Seibert and Vis, 2012) for example, attempts to fix the precipitation calibration parameter Pcalt value to 10 or 20% rainfall increase per 100 m of elevation provided insufficient moisture to sustain flows over dry periods. Thus, Nauditt et al. (2016) used the precipitation parameter Pcalt for Monte Carlo calibration with a range of up to 50% increase per 100 meters for the Rio Grande and Tascadero headwater catchments. Whilst this provided a pragmatic solution to close the water balance, as it calculates percentages of observed rainfall at Las Ramadas station, precipitation at higher elevations during the days when no rain was recorded at this station could not be estimated. In the absence of more direct measurements of precipitation (or inferences from remote sensing) at high altitudes, this limits the ability of HBV to provide good simulations, especially when precipitation inputs are low with many 0 values in the input time series. With additional precipitation data from satellite, radar or weather models and more advanced spatial interpolation methods its performance might be improved (Zambrano et al., 2016).

Despite this, the study suggests that HBV can provide reasonable simulations of both the melt and dry periods. The structure of HBV is well suited to conceptualize the catchment as a snowmelt store essentially recharging a groundwater body which sustains stream flows. The model struggled with the timing of peak flows, but this likely reflects the simplifying assumptions of the day degree method in HBV light and probably less well performing snow melt simulations (Nauditt et al., 2016).

For modeling the Hurtado with J2000, stations outside the basin were used, including two stations at higher elevations (Table 10.18) to establish regressions for altitudinal differences. The regionalization of the precipitation was then achieved by using mainly one station (Pabellón) and doing elevation corrections for each HRU, based on the regressions established by the model.

For the SWAT modeling, regionalization was approached by applying the concept of elevation bands with fixed lapse rates in mm/km. This approach has the advantage that precipitation inputs can be added to the 0 values.

Apparently, the method of J2000 using other stations in the region was more successful to spatially estimate the areal precipitation. Nonetheless, the extrapolation of monitored precipitation data to higher altitudes poses the strongest challenge to outweigh uncertainties related to the input data. This might be also the main reason why both models underestimated ENSO-related extreme streamflow peaks (e.g., 1998), although this might be also a consequence of a poor estimation of high flows, as mentioned before.

The application of models such as J2000, SWAT and HBV light proved that, in this type of environment, the proper representation of precipitation and groundwater processes are of utmost importance. Satellite products could support these regionalization procedures while the use of tracers might give an insight on the pathways of the different runoff components. Given the complex topography, high short wave radiation inputs and the effects of aspect, spatially-distributed energy balance-based snowmelt models might be also useful for the high Andes (Pellicciotti et al., 2008; Ragletti et al., 2014).

How Can the Results of Hydrological Modeling Support Water Management in the Limarí Basin?

A key objective of the presented study was to contribute to the development of models that can give a reasonable prediction of flows and water resource availability across periods of hydro-climatic variability in semi-arid Andean catchments.

The calibrated models were capable of providing streamflow forecasts for the melting (October to December) and dry season (September to May) based on precipitation during the wet season (May–August). In the larger Limarí River Basin where downstream water use has significantly increased during the last decades and where is a high degree of dependence on dry season flows, such models are clearly of value. Also, the ability of the models to produce reasonable predictions across a range of climatic extremes encompassing El Niño and La Niña years implies that they provide a basis for tentative projections of flows under climate change scenarios. These would enable evaluation of anticipated climatic changes due to temperature increase on inter-annual discharge variability. Such analyses are of increasing importance as according to the IPCC (2013), in the long term (i.e., period 2081–2100), the mean annual temperatures in Central Chile are expected to increase by +1 to +4°C, with increases of up to 5°C for daily maximum temperatures. This is consistent with the strong trends of increasing temperatures during the past 60 years at high elevation areas in the extratropical Andes (Vuille et al., 2015). Global circulation models forced with the high-emission scenario RCP8.5 for 2050 project an annual precipitation decrease of 15–20% in central Chile (Collins et al., 2013). However, climate model projections for precipitation and related variables remain uncertain, though in some places in the central Andes they are being used to make projections of likely impacts on monthly and annual stream flow regimes (Vicuña et al., 2010; Bonelli et al., 2014; Bozkurt et al., 2017).

An alternative, or complementary approach, could be to use seasonal discharge scenarios based on winter precipitation and exploratory long term scenarios which are based on hydro-climatic variability. These could use estimated input data from historical records for periods that reflect extreme climatic conditions with higher mean daily temperatures, and precipitation extremes from the wet and dry phases. The likely flow response to such situations which could be used as proxies for more common situations expected from climate change projections and associated temperature increases and precipitation changes. Such scenarios could give guidance to inform stakeholders on local and national level for long term water management and allocation strategies.

The models are therefore most useful to develop discharge scenarios based on estimations of hydro-climatic variability such as anticipated rising temperatures and lower precipitation rates for central Chile (Cortés et al., 2011; Boisier et al., 2016). They can be used to test the long term sensitivity of the local water resource systems against climate change. Such scenarios would serve as decision support for disaster risk management to deal with climatic extremes.

Conclusions

Andean headwater catchments are the main source of water supply for downstream semi-arid irrigation dependent communities. Only little information is available about catchment characteristics, cryosphere and hydrological processes for these remote areas to support water management with discharge predictions.

Arid mountainous river basins represent a challenge to hydrological models. First, the lack of representative meteorological data poses an uncertainty problem especially at high altitudes where no information is available (there is not even a precipitation station at least at the mean elevation). Furthermore, the spatial and temporal variability of precipitation events and other meteorological variables driven by ENSO is difficult to account using hydrological models.

Based on hydro-meteorological time series of almost 50 years at the catchment outlets and *in situ* measurements, runoff processes in three perennial Andean headwater catchments—the Rio Grande, Tascadero and Hurtado—in central Chile were assessed using statistical analyses, different modeling approaches, field observations and synoptic tracer surveys as tools to understand dominant hydrological processes in two headwater catchments of the 11.696 km² Limarí Basin.

Arid regions are characterized by long dry periods where baseflow plays a preponderant role. Therefore, modeling low flows (which may go on for years) are of utmost relevant to assess water availability, especially during droughts. The results from this study show that models can fairly simulate the hydrograph for the calibration and validation periods. However, they systemically underpredict extreme peak flows related to ENSO events. This may be induced by the lack of representative precipitation data in high altitudes in the basin and also by a misrepresentation of snow-related processes. Moreover, SWAT shows slightly better statistics based on r² and NS for both periods (calibration and validation). However, SWAT could not properly simulate low flows during the long dry periods—an important limitation for this basin, while J2000 showed a good baseflow simulation performance also during dry periods. These models are hence not recommended for flood management in the region since the peak flows were systematically underpredicted.

We also concluded that the selection and evaluation of efficiency criteria is paramount to assess the accuracy of a model and its usefulness for a specific basin. A wrong selection can lead to wrong assumptions and wrong decisions. Hence, a combination of different efficiency criteria is the best approach for a sound scientific model calibration and validation (Krause et al., 2005).

Groundwater processes are still a major research gap in the area and its representation has to be improved (Hublart et al., 2016). Moreover, due to the difficulty of identifying and estimating precipitation in this region isotope-based studies could help quantifying contribution from snowmelt, rainfall, groundwater and glaciers to streamflow (Ohlanders et al., 2013). This information could also help to physically determine the value of model parameters and reduce their predictive uncertainty.

Although the lack of information can pose a problem for the implementation of hydrological models in poorly gauged basins, newer technologies such as remote sensing can contribute to improve the quality of the data available. Moreover, physically-based models can help in gaining understanding about hydrological processes and can be used as a management tool in fragile and vulnerable basins in arid regions facing new challenges such as climate change, over use and land use changes.

In this context, besides the recommendations in the previous chapter, further research would add to the urgently needed evidence base for sustainable water resource management. Immediate priorities based on the above described findings, would include to use the calibrated models for water management and improve the representation of rainfall inputs in the water balance of high elevation headwater catchments by: (1) incorporating satellite based precipitation data to recalibrate the HBV light model for Rio Grande and Tascadero and SWAT and J2000 for Hurtado and evaluate the results against the simulation performance with station data. Zambrano-Bigiarini et al. (2016) identified CHIRPSv2 as best performing product in that region; Beck et al. (2016) will soon release a promising improved MSWEP vs product. (2) A real-time precipitation product as PERSIAN CSS should be applied to test its ability to close the water balance and be used for discharge predictions based on real-time data. Satellite based precipitation estimates can provide input data for hydrological models in regions where rain or snow gauges are not installed—and will likely not be available in the future. (3) Simulating scenarios with synthetic data of rising temperatures changing the snowmelt onset.

It can be concluded that the methods and models evaluated and used in this study can serve as tools for decision support in water availability and allocation. Further scenarios and thresholds for water management will be developed after discussing and updating research demand with the stakeholders in charge in both regions. However, the modeling results, although useful, need to be recognized as tentative and uncertain. Future work therefore also needs to be based on stronger uncertainty analysis on the entire modeling process (Refsgaard et al., 2004).

References

Adam, J.C., Hamlet, A.F. and Lettenmaier, D.P. 2009. Implications of global climate change for snowmelt hydrology in the twenty-first century. Hydrolog. Process 23(7): 962–972.

Alam, S., Ali, M.M. and Islam, Z. 2015. Potential impact of climate change on water availability of Brahmaputra River Basin. Proceedings of Canadian Society for Civil Engineering Annual General Conference, Regina, Canada, pp. 937–946.

Arnold, J.G., Srinivasan, S., Muttiah, R.S. and Williams, J.R. 1998. Large area hydrologic modeling and assessment Part I: Model development. J. Am. Water Resour. Assoc. 34(1): 73–89.

Arnold, J.G. and Allen, P.M. 1999. Automated methods for estimating baseflow and ground water recharge from streamflow records. J. Am. Water Resour. Assoc. 35(2): 411–424.

Bai, Y., Wagener, T. and Reed, P. 2009. A top-down framework for watershed model evaluation and selection under uncertainty. Environ. Model Software 24(8): 901–916.

Barnett, T.P., Adam, J.C. and Lettenmaier, D.P. 2005. Potential impacts of a warming climate on water availability in snow-dominated regions. Nature 438: 303–309.

Beck, H.E., van Dijk, A.I.J.M., Levizzani, V., Schellekens, J., Miralles, D.G., Martens, B. and de Roo, A. 2016. MSWEP: 3-hourly 0.25 global gridded precipitation (1979–2015) by merging gauge, satellite, and reanalysis data. Hydrol. Earth Syst. Sci. Discuss, p. 1.

Beven, K. and Freer, J. 2001. Equifinality, data assimilation, and uncertainty estimation in mechanistic modelling of complex environmental systems using the GLUE methodology. J. Hydrol. 249(1-4): 11–29.

Bocchiola, D., Diolaiuti, G., Soncini, A., Mihalcea, C., D'Agata, C., Mayer, C., Lambrecht, A., Rosso, R. and Smiraglia, C. 2011. Prediction of future hydrological regimes in poorly gauged high altitude basins: the case study of the upper Indus, Pakistan. Hydrol. Earth Syst. Sci. 15: 2059–2075.

Boisier, J.P., Rondanelli, R., Garreaud, R. and Muñoz, F. 2016. Anthropogenic and natural contributions to the Southeast Pacific precipitation decline and recent megadrought in central Chile. Geophys. Res. Lett. 43(1): 413–421.

Bonelli, S., Vicuña, S., Meza, F., Gironás, J. and Barton, J. 2014. Incorporating climate change adaptation strategies in urban water supply planning: the case of central Chile. Journal of Water and Climate Change 5(3): 357–376.

Bozkurt, D., Rojas, M., Boisier, J.P. and Valdivieso, J. 2017. Climate change impacts on hydroclimatic regimes and extremes over Andean basins in central Chile. Hydrol. Earth Syst. Sci. Discuss, 29 pp.

Chauvin, G.M., Flerchinger, G.N., Link, T.E., Marks, D., Winstral, A.H. and Seyfried, M.S. 2011. Long-term water balance and conceptual model of a semi-arid mountainous catchment. J. Hydrol. 400(1-2): 133–143.

Collins, M., Knutti, R., Arblaster, J., Dufresne, J.-L., Fichefet, T., Friedlingstein, P., Gao, X., Gutowski, W.J., Johns, T., Krinner, G., Shongwe, M., Tebaldi, C., Weaver, A.J. and Wehner, M. 2013. Long-term climate change: Projections, commitments and irreversibility. pp. 1029–1136. *In*: Stocker, T.F., Qin, D., Plattner, G.-K., Tignor, M., Allen, S.K., Doschung, J., Nauels, A., Xia, Y., Bex, V. and Midgley, P.M. (eds.). Climate Change 2013: The Physical Science Basis. Contribution of Working Group I to the Fifth Assessment Report of the Intergovernmental Panel on Climate Change, Cambridge University Press.

Cortés, G., Vargas, X. and McPhee, J. 2011. Climatic sensitivity of streamflow timing in the extratropical western Andes Cordillera. J. Hydrol. 405: 93–109.

Cunderlik, J.M. 2003. Hydrologic Model Selection for the CFCAS Project: Assessment of Water Resources Risk and Vulnerability to Changing Climatic Conditions, Engineering, University of Western Ontario. Dept. of Civil and Environmental.

Cunge, J.A. 1969. On the subject of a flood propogation method (Muskingum method). Journal of Hydraulics Research, International Association of Hydraulics Research 7(2): 205–230.

Dickinson, R.E. 1984. Climate processes and climate sensitivity. *In*: Hansen, J.E. and Takahashi, T. (eds.). American Geophysical Union, Geophysical Monograph Series, Washington, D.C. 29: 368 pp.

Domeneghetti, A., Castellarin, A. and Brath, A. 2012. Assessing rating-curve uncertainty and its effect on hydraulic model calibration. Hydrol. Earth Syst. Sci. 16: 1191–1202.

Duethmann, D., Peters, J., Blume, T., Vorogushyn, S. and Güntner, A. 2014. The value of satellite derived snow cover images for calibrating a hydrological model in snow-dominated catchments in Central Asia. Water Resour. Res. 50(3): 2002–2021.

FAO. 1995. Digital Soil Map of the World, Version 3.5, Rome, Italy.

Favier, V., Falvey, M., Rabatel, A., Praderio, E. and López, D. 2009. Interpreting discrepancies between discharge and precipitation in high-altitude area of Chile's Norte Chico region (26–32_S). Water Resour. Res. 45(2): W02424, 20 pp.

Fontaine, T.A., Cruickshank, T.S., Arnold, J.G. and Hotchkiss, R.H. 2002. Development of a snowfall-snowmelt routine for mountainous Terrain for the soil water assessment tool (SWAT). J. Hydrol. 262(1-4): 209–223.

Flügel, W.-A. 1996. Hydrological Response Units (HRU) as modelling entities for 28 hydrological river basin simulations and their methodological potential for modelling 29 complex environmental process systems. Results from the Sieg catchment, In: DIE ERDE 127: 42–62.

Gassman, P., Reyes, M.R., Green, C.H. and Arnold, J. 2007. The soil and water assessment tool: Historical development, applications, and future research directions. American Society of Agricultural and Biological Engineers 50(4): 1211–1250.

Gupta, H.V., Beven, K.J. and Wagener, T. 2006. Model Calibration and Uncertainty Estimation. In Encyclopedia of Hydrological Sciences. John Wiley & Sons, Ltd.

Hargreaves, G.H. and Samani, Z.A. 1985. Reference crop evapotranspiration from temperature. Appl. Eng. Agr. 1(2): 96–99.

Hrachowitz, M., Bohte, R., Mul, M.L., Bogaard, T.A., Savenije, H.H.G. and Uhlenbrook, S. 2011. On the value of combined event runoff and tracer analysis to improve understanding of catchment functioning in a data-scarce semi-arid area. Hydrol. Earth Syst. Sci. 15: 2007–2024.

Hublart, P., Ruelland, D., García de Cortázar-Atauri, I., Gascoin, S., Lhermitte, S. and Ibacache, A. 2016. Reliability of lumped hydrological modeling in a semi-arid mountainous catchment facing water-use changes. Hydrol. Earth Syst. Sci. 20(9): 3691–3717.

IPCC. 2013. Climate Change 2013. The Physical Science Basis. Contribution of Working Group I to the Fifth Assessment Report of the Intergovernmental Panel on Climate Change. *In*: Stocker, T.F., Qin, D., Plattner, G.-K., Tignor, M., Allen, S.K., Boschung, J., Nauels, A., Xia, Y., Bex, V. and Midgley, P.M. (eds.). Cambridge University Press, Cambridge, United Kingdom and New York, NY, USA, 1535 pp.

Immerzeel, W., Van Beek, W. and Bierkens, M. 2010. Climate change will affect the Asian water towers. Science 328(5984): 1382–1385.

Jasechko, S., Kirchner, J.W., Welker, J.M. and McDonnell, J. 2016. Substantial proportion of global streamflow less than three months old. Nature Geoscience 9: 126–129.

Jiang, T., Chen, Y.D., Xu, C., Chen, X., Chen, X. and Singh, V.P. 2007. Comparison of hydrological impacts of climate change simulated by six hydrological models in the Dongjiang Basin, South China. J. Hydrol. 336(3-4): 316–333.

Klemeš, V. 1986. Operational testing of hydrological simulation models. Hydrolog. Sci. J. 31(1): 13–24.

Kralisch, S., Krause, P., Fink, M., Fischer, C. and Flügel, W.-A. 2007. Component based environmental modelling using the JAMS framework. *In*: Kulasiri, D. and Oxley, L. (eds.). Proceedings of the MODSIM 2007 International Congress on Modelling and Simulation. Christchurch, New Zealand.

Krause, P. 2001. Das hydrologische Modellsystem J2000: Ein Modellsystem zur physikalisch basierten Nachbildung der hydrologischen Prozesse in großen Flusseinzugsgebieten Doctoral Dissertation. Albert-Ludwigs-Universität Freiburg.

Krause, P. 2002. Quantifying the impact of land use changes on the water balance of large catchments using the J2000 model. Phys. Chem. Earth 27: 663–673.

Krause, P., Boyle, D.P. and Bäse, F. 2005. Comparison of different efficiency criteria for hydrological model assessment. ADGEO 5: 89–97.

Leavesley, G.H., Lichty, R.W., Troutman, B.M. and Saindon, L.G. 1983. Precipitation-runoff modeling system—User's manual, Denver, Colorado. Available at: http://pubs.usgs. gov/wri/1983/4238/report.pdf.

Legates, D.R. and McCabe, G.J. 1999. Evaluating the use of "goodness-of-fit" Measures in hydrologic and hydroclimatic model validation. Water Resour. Res. 35(1): 233–241.

Lehning, M., Völksch, I., Gustafsson, D., Nguyen, T.A., Stähli, M. and Zappa, M. 2006. ALPINE3D: a detailed model of mountain surface processes and its application to snow hydrology. Hydrolog. Process 20: 2111–2128.

Lidén, R. and Harlin, J. 2000. Analysis of conceptual rainfall–runoff modelling performance in different climates. J. Hydrol. 238: 231–247.

Mourato, S., Moreira, M. and Corte-Real, J. 2015. Water resources impact assessment under climate change Scenarios. Mediterranean Watersheds 29(7): 2377–2391.

Nash, J.E. and Sutcliffe, J.V. 1970. River flow forecasting through conceptual models part I—A discussion of principles. J. Hydrol. 10(3): 282–290.

Nauditt, A., Birkel, C., Soulsby, C. and Ribbe, L. 2016. Conceptual modelling to assess the influence of hydroclimatic variability on runoff processes in data scarce semi-arid Andean catchments. Hydrolog. Sci. J. 62: 515–532.

Neitsch, S.L. et al. 2005. Soil and Water Assessment Tool Theoretical Documentation—Version 2005, Grassland/Temple, TX: Blackland Research Center/Soil and Water Research Laboratory, Agricultural Research Service.

Nepal, S. 2012. Evaluating Upstream-Downstream Linkages of Hydrological Dynamics in the Himalayan Region. Doctoral Dissertation. Friedrich-Schiller-University Jena. 201 pp.

Núñez, J., Rivera, D., Oyarzún, R. and Arumí, J.L. 2013. Influence of Pacific Ocean multidecadal variability on the distributional properties of hydrological variables in north-central Chile. J. Hydrol. 501: 227–240.

Ohlanders, N., Rodriguez, M. and McPhee, J. 2013. Stable water isotope variation in a Central Andean watershed dominated by glacier and snowmelt. Hydrol. Earth Syst. Sci. 17(3): 1035–1050.

Omani, N., Srinivasan, R., Karthikeyan, R., Venkata Reddy, K. and Smith, P.K. 2016. Impacts of climate change on the glacier melt runoff from five river basins. Transactions of the American Society of Agricultural and Biological Engineers 59(4): 829–848.

Oyarzún, J., Maturana, H., Paulo, A. and Pasieczna, A. 2003. Heavy metals in stream sediments from the Coquimbo region (Chile): Effects of sustained mining and natural processes in a semi-arid Andean basin. Mine Water and the Environment 22(3): 155–161.

Peel, M.C., Finlayson, B.L. and McMahon, T.A. 2007. Updated world map of the Köppen-Geiger climate classification. Hydrol. Earth Syst. Sci. 11: 1633–1644.

Pellicciotti, F., Helbing, J., Rivera, A., Favier, V., Corripio, J., Araos, J. and Sicart, J.E. 2008. A study of the energy-balance and melt regime of Juncal Norte glacier, semi-arid Andes of central Chile, using models of different complexity. Hydrolog. Process 22(19): 3980–3997.

Pfennig, B. and Wolf, M. 2007. Extraction of process-based topographic model units using SRTM elevation data for Prediction in Ungauged Basins (PUB) in different landscapes.

pp. 685–691. *In*: Kulasiri, D. and Oxley, L. (eds.). MODSIM 2007 International Congress on Modelling and Simulation. Christchurch, New Zealand.

Pilgrim, D.H., Chapman, T.G. and Doran, D.G. 1988. Problems of rainfall runoff modelling in arid and semiarid regions. Hydrolog. Sci. J. 33(4): 379–400.

Pomeroy, J.W., Gray, D.M., Brown, T., Hedstrom, N.R., Quinton, W.L., Granger, R.J. and Carey, S.K. 2007. The cold regions hydrological model: a platform for basing process representation and model structure on physical evidence. Hydrolog. Process 21: 2650–2667.

Price, M.F. and Egan, P.A. 2014. Policy brief: Our global water towers: ensuring ecosystem services from mountains under climate change. Policy brief. *In*: Mishra, A., Demuth, S., Ávila, B. and Cárdenas, M.R. (eds.). International Hydrological Programme (IHP), Man and the Biosphere Programme (MAB) of UNESCO.

Ragettli, S., Cortés, G., McPhee, J. and Pellicciotti, F. 2014. An evaluation of approaches for modelling hydrological processes in high-elevation, glacierized Andean watersheds. Hydrolog. Process 28: 5674–5695.

Refsgaard, J.C. and Henriksen, H.J. 2004. Modelling guidelines: Terminology and guiding principles. Advanced Water Resources 27(1): 71–82.

Savenije, H.H.G. 2010. HESS opinions topography driven conceptual modelling (FLEX-Topo). Hydrol. Earth Syst. Sci. 14: 2681–2692.

Seibert, J. and Vis, M.J.P. 2012. Teaching hydrological modeling with a user-friendly catchment-runoff-model software package. Hydrol. Earth Syst. Sci. 16: 3315–3325.

Soil Conservation Service. 1972. Section 4: Hydrology. In National Engineering Handbook.

Soulsby, C., Rodgers, P., Smart, R., Dawson, J. and Dunn, S. 2003. A tracer-based assessment of hydrological pathways at different spatial scales in a mesoscale Scottish catchment. Hydrolog. Process 17: 759–777.

Souvignet, M. 2010. Climate Change Impacts on Water Resources in Mountainous Arid Zones: A Case Study in the Central Andes, Chile. Doctoral Dissertation. University of Leipzig. 182 pp.

Stehr, A., Debels, P., Romer, F. and Alcayaga, H. 2008. Hydrological modelling with SWAT under conditions of limited data availability: evaluation of results from a Chilean case study. Hydrolog. Sci. J. 53(3): 588–601.

Stehr, A., Debels, P., Arumi, J.L., Romero, F. and Alcayaga, H. 2009. Combining the Soil and Water Assessment Tool (SWAT) and MODIS imagery to estimate monthly flows in a data-scarce Chilean Andean basin. Hydrolog. Sci. J. 54(6): 1053–1067.

Uhlenbrook, S., Seibert, J., Leibundgut, C. and Rodhe, A. 1999. Prediction uncertainty of conceptual rainfall–runoff models caused by problems in identifying model parameters and structure. Hydrolog. Sci. J. 44: 779–797.

Uhlenbrook, S., Mohamed, Y. and Gragne, A. 2010. Analyzing catchment behavior through catchment modelling in the Gilgel Abay, Upper Blue Nile River Basin, Ethiopia. Hydrol. Earth Syst. Sci. 14: 2153–2165.

Van Liew, M.W., Arnold, J.G. and Bosch, D.D. 2005. Problems and potential of autocalibrating a hydrologic model. In Transactions of the ASAE 48(3): 1025–1040.

Van Loon, A.F. and Van Lanen, H.A.J. 2012. A process-based typology of hydrological drought, Hydrol. Earth Syst. Sci. 16: 1915–1946.

Vargas, X., Gómez, T., Ahumada, F., Rubio, E., Cartes, M. and Gibbs, M. 2013. Water availability in a mountainous Andean watershed under CMIP5 climate change scenarios. Proceedings of H02, IAHS-IAPSO-IASPEI Assembly, Gothenburg, Sweden, July 2013 (IAHS Publ. 360, 2013): 33–38.

Versini, P.A., Pouget, L., McEnnis, S., Custodio, E. and Escaler, I. 2016. Climate change impact on water resources availability: case study of the Llobregat River basin (Spain). Hydrolog. Sci. J. 61(14): 2496–2508.

Vicuña, S., Garreaud, R.D. and McPhee, J. 2010. Climate change impacts on the hydrology of a snowmelt driven basin in semiarid Chile. Climatic Change 105(3): 469–488.

Vicuña, S., Alvarez, P., Melo, O., Dale, L. and Meza, F. 2014. Irrigation infrastructure development in the Limarí Basin in Central Chile: Implications for adaptation to climate variability and climate change. Water International 39(5): 620–634.

Vuille, M., Franquist, E., Garreaud, R., Lavado, W. and Cáceres, B. 2015. Impact of the global warming hiatus on Andean temperature. J. Geophys. Res. 120(9): 3745–3757.

Yapo, P.O., Gupta, H.V. and Sorooshian, S. 1996. Automatic calibration of conceptual rainfall-runoff models: sensitivity to calibration data. J. Hydrol. 181(1-4): 23–48.

Zambrano-Bigiarini, M., Nauditt, A., Birkel, C., Verbist, K. and Ribbe, L. 2016. Temporal and spatial evaluation of satellite-based rainfall estimates across the complex topographical and climatic gradients of Chile. Hydrol. Earth Syst. Sci. 21: 1295–1320.

Zeinivand, H. and De Smedt, F. 2009. Hydrological modeling of snow accumulation and melting on river basin scale. Water Resour. Manag. 23(11): 2271–2287.

Zhu, T. and Ringler, C. 2012. Climate change impacts on water availability and use in the Limpopo River Basin. Water 4(1): 63–84.

Index